Sustainable and Green Technologies for Industrial Chemical Engineering

Sustainable and Green Technologies for Industrial Chemical Engineering

Guest Editor
Antonio Gil Bravo

Basel • Beijing • Wuhan • Barcelona • Belgrade • Novi Sad • Cluj • Manchester

Guest Editor
Antonio Gil Bravo
Department of Science
Public University of Navarra
Pamplona
Spain

Editorial Office
MDPI AG
Grosspeteranlage 5
4052 Basel, Switzerland

This is a reprint of the Special Issue, published open access by the journal *Eng* (ISSN 2673-4117), freely accessible at: www.mdpi.com/journal/eng/special_issues/UNQJFL1U61.

For citation purposes, cite each article independently as indicated on the article page online and using the guide below:

Lastname, A.A.; Lastname, B.B. Article Title. *Journal Name* **Year**, *Volume Number*, Page Range.

ISBN 978-3-7258-3178-4 (Hbk)
ISBN 978-3-7258-3177-7 (PDF)
https://doi.org/10.3390/books978-3-7258-3177-7

© 2025 by the authors. Articles in this book are Open Access and distributed under the Creative Commons Attribution (CC BY) license. The book as a whole is distributed by MDPI under the terms and conditions of the Creative Commons Attribution-NonCommercial-NoDerivs (CC BY-NC-ND) license (https://creativecommons.org/licenses/by-nc-nd/4.0/).

Contents

Antonio Gil Bravo
Sustainable and Green Technologies for Industrial Chemical Engineering
Reprinted from: *Eng* **2025**, 6, 16, https://doi.org/10.3390/eng6010016 1

Martha Mantiniotou, Bogdan-Cristian Bujor, Vassilis Athanasiadis, Theodoros Chatzimitakos, Dimitrios Kalompatsios and Konstantina Kotsou et al.
Response Surface Methodology-Aided Optimization of Bioactive Compound Extraction from Apple Peels Through Pulsed Electric Field Pretreatment and Ultrasonication
Reprinted from: *Eng* **2024**, 5, 2886-2901, https://doi.org/10.3390/eng5040150 4

Julles Mitoura dos Santos Junior and Adriano Pinto Mariano
Gasification of Lignocellulosic Waste in Supercritical Water: Study of Thermodynamic Equilibrium as a Nonlinear Programming Problem
Reprinted from: *Eng* **2024**, 5, 1096-1111, https://doi.org/10.3390/eng5020060 20

Grazielly Maria Didier de Vasconcelos, Isabela Karina Della-Flora, Maikon Kelbert, Lidiane Maria de Andrade, Débora de Oliveira and Selene Maria de Arruda Guelli Ulson de Souza et al.
Screening of Azo-Dye-Degrading Bacteria from Textile Industry Wastewater-Activated Sludge
Reprinted from: *Eng* **2024**, 5, 116-132, https://doi.org/10.3390/eng5010008 36

Stamatia Gavela and Georgios Papadakos
Activity Concentration Index Values for Concrete Multistory Residences in Greece Due to Fly Ash Addition in Cement
Reprinted from: *Eng* **2023**, 4, 2926-2940, https://doi.org/10.3390/eng4040164 53

Iryna Kovalchuk, Oleg Zakutevskyy, Volodymyr Sydorchuk, Olena Diyuk and Andrey Lakhnik
The Effect of High-Energy Ball Milling of Montmorillonite for Adsorptive Removal of Cesium, Strontium, and Uranium Ions from Aqueous Solution
Reprinted from: *Eng* **2023**, 4, 2812-2825, https://doi.org/10.3390/eng4040158 68

Valentin Y. Basevich, Sergey M. Frolov, Vladislav S. Ivanov, Fedor S. Frolov and Ilya V. Semenov
The Effects of Multistage Fuel-Oxidation Chemistry, Soot Radiation, and Real Gas Properties on the Operation Process of Compression Ignition Engines
Reprinted from: *Eng* **2023**, 4, 2682-2710, https://doi.org/10.3390/eng4040153 82

Simón Yunes, Jeffrey Kenvin and Antonio Gil
On the Genesis of a Catalyst: A Brief Review with an Experimental Case Study
Reprinted from: *Eng* **2023**, 4, 2375-2406, https://doi.org/10.3390/eng4030136 111

Iryna Kovalchuk
Performance of Thermal-, Acid-, and Mechanochemical- Activated Montmorillonite for Environmental Protection from Radionuclides U(VI) and Sr(II)
Reprinted from: *Eng* **2023**, 4, 2141-2152, https://doi.org/10.3390/eng4030122 143

Ioannis Giovanoudis, Vassilis Athanasiadis, Theodoros Chatzimitakos, Dimitrios Kalompatsios, Eleni Bozinou and Olga Gortzi et al.
Implementation of Cloud Point Extraction Using Surfactants in the Recovery of Polyphenols from Apricot Cannery Waste
Reprinted from: *Eng* **2023**, 4, 1225-1235, https://doi.org/10.3390/eng4020072 155

Saad E. Kaskah, Gitta Ehrenhaft, Jörg Gollnick and Christian B. Fischer
Prediction Model for Optimal Efficiency of the Green Corrosion Inhibitor Oleoylsarcosine: Optimization by Statistical Testing of the Relevant Influencing Factors
Reprinted from: *Eng* **2023**, *4*, 635-649, https://doi.org/10.3390/eng4010038 . . . . **166**

Editorial

Sustainable and Green Technologies for Industrial Chemical Engineering

Antonio Gil Bravo

INAMAT2, Science Department, Campus of Arrosadia, Public University of Navarra, Building Los Acebos, E-31006 Pamplona, Spain; andoni@unavarra.es

1. Introduction

The aim of this *Eng* Special Issue is to collect experimental and theoretical research relating engineering science and technology to the general topics of *Eng*. In this Special Issue, many of these general topics have been selected with the idea of contributing to the discourse on Sustainable and Green Technologies. Therefore, these aspects are being addressed from various points of view and have the support of this fascinating field of engineering and its applications.

In recent years, there has been a paradigm shift towards greener processes/products, with a particular emphasis on sustainability. Manufacturing processes require reworking and adapting new methods to ensure they are more efficient, cleaner, less polluting and ideally cost effective. There are multiple sectors of the bio-based and chemical market that can be adapted towards greener initiatives, for example, bio-based chemicals, biofuels and biomaterials and the concept of waste-to-energy [1]. This Special Issue focuses on greener and more sustainable processes for chemical and feedstock substitutes, with a focus on where bio-based products can be introduced into the petroleum-driven market as a drop-in replacement or functional replacement and/or through novel products.

For many years, environmental problems were considered to be caused by the economic system and the rapid use of natural resources [2]. It has taken many years to establish the use of materials (raw materials), the initial design of chemical processes, hazardous properties of products, energy consumption and other parameters involved in the manufacture of products, such as life cycle, recycling, etc.

The rapid development of new chemical technologies and the huge amounts of innovative chemical products available have forced attention to focus on corrective actions for harmful impacts. However, the most efficient way to reduce negative impacts is to design and innovate manufacturing processes, taking into account, for example, energy, materials, the use and generation of secondary materials that are hazardous and, finally, the life cycles of products and their recycling into new materials [1].

Green chemistry involves other synthetic routes that are more environmentally friendly. Green technologies investigate alternative reaction conditions, alternative media (solvent-free) and even alternative energy sources [3]. Designing chemical products and processes that reduce or eliminate the use and generation of hazardous substances is the most fundamental approach to pollution prevention. Green chemistry addresses the need to produce the goods and services that society depends on in a more environmentally friendly manner. Pharmaceuticals can be produced while minimizing the amount of waste generated, biodegradable plastics can be synthesized from plants and reactions can be performed in water instead of traditional organic solvents by applying the principles of green chemistry to chemical products and processes.

The principles of green engineering are based on the principles of green chemistry, established by Paul Anastas and Julie Zimmerman [2,3]. These principles recommend a basis for researchers and technologists to apply in the design of new materials, products, processes and systems. Products, processes and systems can be made more intrinsically benign by changing the inherent nature of the system or the circumstances/conditions of the system to reduce the problem of toxins and associated exposure to harmful effects (or both).

These are some of the aspects that have been covered in this Special Issue and that may allow for greater discussion among potential readers of it. For more information, please see the List of Contributions.

2. Overview of the Published Articles

This Special Issue contains up to 10 papers, including 1 *Review*, published by several authors interested in new cutting-edge developments in the field of engineering. The submitted papers are from authors from 10 countries, namely Brazil, Germany, Greece, Iraq, Morocco, Romania, Russia, Spain, Ukraine and USA.

3. Conclusions

The articles published in this Special Issue present important advancements in the topics of our journal. I would like to express my sincere gratitude to all the authors, who have professionally and enthusiastically wanted to contribute to this Special Issue, and I would also like to thank the Managing Editors and reviewers who helped to improve the papers and made important contributions to this Special Issue. I hope that the articles showcased in this Special Issue are interesting and inspiring for its readers, especially young scholars who are eager to learn about recent advances and contribute future research in the field.

Funding: The author is grateful for financial support from the Spanish Ministry of Science and Innovation (MCIN/AEI/10.13039/501100011033) through project PID2023-146935OB-C21.

Conflicts of Interest: The author declares no conflicts of interest.

List of Contributions:

1. Kaskah, S.E; Ehrenhaft, G.; Gollnick, J.; Fischer, Ch.B. Corrosion Inhibitor Oleoylsarcosine: Optimization by Statistical Testing of the Relevant. *Eng* **2023**, *4*, 635–649. https://doi.org/10.3390/eng4010038.
2. Giovanoudis, I.; Athanasiadis, V.; Chatzimitakos, Th.; Kalompatsios, D.; Bozinou, E.; Gortzi, O.; Nanos. G.D.; Lalas, S.I. Implementation of Cloud Point Extraction Using Surfactants in the Recovery of Polyphenols from Apricot Cannery Waste. *Eng* **2023**, *4*, 1225–1235. https://doi.org/10.3390/eng4020072.
3. Kovalchuk, I. Performance of Thermal-, Acid-, and Mechanochemical-Activated Montmorillonite for Environmental Protection from Radionuclides U(VI) and Sr(II). *Eng* **2023**, *4*, 2141–2152. https://doi.org/10.3390/eng4030122.
4. Yunes, S.; Kenvin, J.; Gil, A. On the Genesis of a Catalyst: A Brief Review with an Experimental Case Study. *Eng* **2023**, *4*, 2375–2406. https://doi.org/10.3390/eng4030136.
5. Basevich, V.Y.; Frolov, S.M.; Ivanov, V.S.; Frolov, F.S.; Semenov, I.V. The Effects of Multistage Fuel-Oxidation Chemistry, Soot Radiation, and Real Gas Properties on the Operation Process of Compression Ignition Engines. *Eng* **2023**, *4*, 2682–2710. https://doi.org/10.3390/eng4040153.
6. Kovalchuk, I.; Zakutevskyy, O.; Sydorchuk, V.; Diyuk, O.; Lakhnik, A. The Effect of High-Energy Ball Milling of Montmorillonite for Adsorptive Removal of Cesium, Strontium, and Uranium Ions from Aqueous Solution. *Eng* **2023**, *4*, 2812–2825. https://doi.org/10.3390/eng4040158.

7. Gavela, S.; Papadakos, G. Activity Concentration Index Values for Concrete Multistory Residences in Greece Due to Fly Ash Addition in Cement. *Eng* **2023**, *4*, 2926–2940. https://doi.org/10.3390/eng4040164.
8. Didier de Vasconcelos, G.M.; Della-Flora, I.K.; Kelbert, M.; de Andrade, L.M.; de Oliveira, D.; de Arruda Guelli Ulson de Souza, S.M.; Ulson de Souza, A.A.; de Andrade, C.J. Screening of Azo-Dye-Degrading Bacteria from Textile Industry Wastewater-Activated Sludge. *Eng* **2024**, *5*, 116–132. https://doi.org/10.3390/eng5010008.
9. dos Santos Junior, J.M.; Pinto Mariano, A. Gasification of Lignocellulosic Waste in Supercritical-Water: Study of Thermodynamic Equilibrium as a Nonlinear Programming Problem. *Eng* **2024**, *5*, 1236–1264. https://doi.org/10.3390/eng5020060.
10. Mantiniotou, M.; Bujor, B.-C.; Athanasiadis, V.; Chatzimitakos, Th.; Kalompatsios, D.; Kotsou, K.; Bozinou, E.; Lalas, S.I., Response Surface Methodology-Aided Optimization of Bioactive Compound Extraction from Apple Peels Through Pulsed Electric Field Pretreatment and Ultrasonication. *Eng* **2024**, *5*, 2886–2901. https://doi.org/10.3390/eng5040150.

References

1. Gil, A. Challenges on Waste-to-Energy for the Valorization of Industrial Wastes: Electricity, Heat and Cold, Bioliquids and Biofuels. *Environ. Nanotechnol. Monit. Manag.* **2022**, *17*, 100615. [CrossRef]
2. Anastas, P.T.; Williamson, T.C. *Green Chemistry: Designing Chemistry for the Environment*; ACS Publications: Washington, DC, USA, 1996.
3. Anastas, P.T.; Zimmerman, J.B. Design through the Twelve Principles of Green Engineering. *Environ. Sci. Technol.* **2003**, *37*, 94A–101A. [CrossRef] [PubMed]

Disclaimer/Publisher's Note: The statements, opinions and data contained in all publications are solely those of the individual author(s) and contributor(s) and not of MDPI and/or the editor(s). MDPI and/or the editor(s) disclaim responsibility for any injury to people or property resulting from any ideas, methods, instructions or products referred to in the content.

Article

Response Surface Methodology-Aided Optimization of Bioactive Compound Extraction from Apple Peels Through Pulsed Electric Field Pretreatment and Ultrasonication

Martha Mantiniotou [1], Bogdan-Cristian Bujor [2], Vassilis Athanasiadis [1], Theodoros Chatzimitakos [1], Dimitrios Kalompatsios [1], Konstantina Kotsou [1], Eleni Bozinou [1] and Stavros I. Lalas [1,*]

[1] Department of Food Science and Nutrition, University of Thessaly, Terma N. Temponera Street, 43100 Karditsa, Greece; mmantiniotou@uth.gr (M.M.); vaathanasiadis@uth.gr (V.A.); tchatzimitakos@uth.gr (T.C.); dkalompatsios@uth.gr (D.K.); kkotsou@agr.uth.gr (K.K.); empozinou@uth.gr (E.B.)

[2] Faculty of Animal Productions Engineering and Management, University of Agronomic Sciences and Veterinary Medicine of Bucharest, 59 Mărăşti Boulevard, District 1, 011464 Bucharest, Romania; bogdanbjrbogdan@gmail.com

* Correspondence: slalas@uth.gr; Tel.: +30-24410-64783

Citation: Mantiniotou, M.; Bujor, B.-C.; Athanasiadis, V.; Chatzimitakos, T.; Kalompatsios, D.; Kotsou, K.; Bozinou, E.; Lalas, S.I. Response Surface Methodology-Aided Optimization of Bioactive Compound Extraction from Apple Peels Through Pulsed Electric Field Pretreatment and Ultrasonication. *Eng* 2024, 5, 2886–2901. https://doi.org/10.3390/eng5040150

Academic Editor: Antonio Gil Bravo

Received: 30 September 2024
Revised: 30 October 2024
Accepted: 5 November 2024
Published: 6 November 2024

Copyright: © 2024 by the authors. Licensee MDPI, Basel, Switzerland. This article is an open access article distributed under the terms and conditions of the Creative Commons Attribution (CC BY) license (https://creativecommons.org/licenses/by/4.0/).

Abstract: Apple by-products (i.e., peels) are often thrown away, yet they are highly nutritious and provide numerous advantages as they contain a variety of nutrients such as vitamins, minerals, and antioxidants. Apple peels also comprise a high level of antioxidants, particularly polyphenols and flavonoids. This research aimed to determine the most efficacious extraction techniques and parameters to accomplish maximum bioactive compounds recovery from apple peels. Several extractions were conducted, including stirring, ultrasonication, and pulsed electric field-assisted extractions. Response surface methodology and several factors such as temperature, extraction duration, and solvent composition were considered to have a major impact on the isolation of bioactive compounds. The findings indicated that the most practical and efficient approach was to combine the pulsed electric field process with ultrasonication and stirring at 80 °C for 30 min, while 75% aqueous ethanol comprised the optimal solvent concentration, demonstrating the critical role of the solvent in optimizing extraction efficiency. The optimal conditions were obtained through response surface methodology with a statistical significance of $p < 0.05$. The extract exhibited a total polyphenolic content (TPC) of 17.23 mg gallic acid equivalents (GAE) per g of dry weight (dw), an ascorbic acid content (AAC) of 3.99 mg/g dw, and antioxidant activity of 130.87 μmol ascorbic acid equivalents (AAE)/g dw, as determined by FRAP and 95.38 μmol AAE/g dw from the DPPH assay. The measured antioxidant activity highlighted the significant potential of apple peels as a cost-effective source of exceptionally potent extracts.

Keywords: *Malus domestica*; PEF; ultrasound; polyphenols; ascorbic acid; antioxidants; response surface methodology; binary solvents; principal component analysis; Starking delicious

1. Introduction

The apple (*Malus domestica* Borkh) is a globally renowned and extensively cultivated fruit [1]. It is a member of the Rosaceae family and produces a vast array of goods, including juices, jams, compotes, tea, wine, and dried apples [2]. Currently, the cultivated apple ranks as the third most widely cultivated fruit crop globally [3]. In 2018, the Food and Agriculture Organization (FAO) reported a total apple production exceeding 86 million tons [4], of which 11 million were generated in the Mediterranean area [5]. Apple peels are commonly regarded as waste materials in the canned apple and apple sauce manufacturing sectors, despite their untapped potential and possible health advantages. Therefore, there is a need to improve the exploitation of these peels [6]. Apple peels include five primary categories of polyphenolic chemicals (Figure S1), including phenolic acids (mostly chlorogenic acid), flavan-3-ols (specifically (+)-catechin, and (−)-epicatechin), flavonols (primarily various quercetin glycosides), flavanones (specifically hesperidin), and anthocyanins (mostly pelargonin chloride) [7]. Monomeric and polymeric flavan-3-ols compounds provide a significant proportion (about 60%) of the overall polyphenol content found in apple peels. In contrast, flavonols, hydroxycinnamic acids, dihydrochalcones, and anthocyanins contribute to 18%, 9%, 8%, and 5% of the total polyphenol content, respectively [7].

Utilizing by-products as additives for the development of new goods could reduce waste within the food processing sector [8]. Moreover, wastes usually exhibit notably elevated levels of polyphenols [9], which could effectively be used as natural antioxidants for humans. Thus, antioxidant and antibacterial compounds found in apple peels, such as ferulic acid, dopamine, and caffeic acid, could make them suitable for food preservatives. The appeal of apple peel lies in its antioxidant characteristics, which have the potential to facilitate the creation of innovative goods. These goods may consist of either human-consumable foods that can improve the health of the human body or animal feed with comparable characteristics [10]. Furthermore, apple peels could be employed to enrich apple juices with bioactive and antioxidant compounds.

Over the past few years, efforts have been undertaken by industries and researchers to apply green extraction techniques on waste, due to ecological sustainability. Green technologies such as pulsed electric field (PEF) and ultrasonication (US), generally, necessitate lower energy consumption in comparison to conventional processes, such as stirring [11]. PEF is a highly efficient method that has multiple benefits. It helps inactivate microorganisms, improving mass transfer in food products and recovering valuable bioactive compounds from food waste [12]. PEF has gained significant popularity among different food industries in recent years, offering extra motivation for businesses to minimize waste and the resulting harm to the environment [13]. Ultrasonication, on the other hand, is also a green technique that has several advantages, including reduced extraction duration, less energy and power consumption, minimized bioactive thermal degradation, and the production of extracts characterized by their high quality [14]. Tian et al. [15] reported a total polyphenol content of ~42 mg/g extracted from apple peels utilizing 75% methanol as a solvent and 20 min ultrasonication in triplicate. Furthermore, Sethi et al. [16] determined polyphenol content at a range of 708–1232 mg GAE/100 g dw and ascorbic acid content ranging from 6.81 to 27.68 mg/100 g dw from the peels of twelve different apple cultivars. Fotirić Akšić et al. [17] also reported a total polyphenol content ranging from ~3 to 21 mg GAE/kg dw in apple peels. Therefore, studying the optimal extraction method is important because apple peels contain sufficient bioactive compounds.

Green extraction techniques may be employed either as a pretreatment or as an exclusive extraction technique in order to recover various bioactive compounds from plant sources in a sustainable manner [18]. Moreover, green techniques might enhance the efficiency of the extraction, when compared to conventional ones. Recently, an increasing number of environmentally conscious extraction techniques have been applied to eliminate toxic organic solvents and reduce extraction duration and energy use [19]. PEF has been utilized in various food waste and by-products, such as orange peels, lemon peels, olive leaves and kernels, and many others [12], resulting in increased polyphenolic and antioxidant compound yields. Although apple peels have numerous advantages for humans, there is insufficient research concerning the environmentally sustainable production of extracts rich in antioxidant polyphenols.

The main objective of this investigation was to determine the most efficient combination of green pretreatment techniques for the production of extracts that are abundant in antioxidant compounds, including ascorbic acid and polyphenols. These extracts are a highly promising alternative for the food and pharmaceutical sectors. Although the extraction of bioactive and antioxidant compounds from apple peels has been extensively studied, to the authors' best knowledge, not enough attention has been devoted to combining green pretreatment techniques to maximize yield in specific byproducts. Even though

PEF [20] and US [21] have been employed in apple peels, the combination of the two has not been studied yet. Moreover, it is important to emphasize that environmentally friendly solvents are used to maintain an environmentally conscious profile, free of toxic organic solvents. The optimization of the extraction was accomplished through response surface methodology (RSM). The study focused on investigating the effects of green solvent combinations comprising water and ethanol, as well as the temperature impact and extraction duration on the extraction process. In addition to conventional extraction through stirring (ST), green sample pretreatment techniques, including PEF and US, were employed to further optimize this procedure. Moreover, a partial least squares (PLS) model was utilized to determine the optimal conditions.

2. Materials and Methods

2.1. Chemicals and Reagents

Iron (III) chloride was obtained from Merck (Darmstadt, Germany). Trichloroacetic acid, hydrochloric acid, methanol, aluminum chloride, 2,2-diphenyl-1-picrylhydrazyl (DPPH), L-ascorbic acid, 2,4,6-tris(2-pyridyl)-s-triazine (TPTZ), and all polyphenolic standards for the HPLC determination were bought from Sigma-Aldrich (Darmstadt, Germany). Gallic acid, Folin–Ciocalteu reagent, and ethanol were procured from Panreac Co. (Barcelona, Spain). Finally, anhydrous sodium carbonate was bought from Penta (Prague, Czech Republic). A deionizing column was used to generate deionized water, which was used for the conducted experiments.

2.2. Instrumentation

The lyophilization of apple peels was performed using a Biobase BK-FD10P freeze-dryer (Jinan, China). After extraction, the centrifugation was carried out using NEYA 16R (Remi Elektrotechnik Ltd., Palghar, India) to separate the supernatant from the solid residue. For PEF processing of the samples, we used a UPG100 mode/arbitrary waveform generator (ELV Elektronik AG, Leer, Germany), a Leybold high-voltage power generator (LD Didactic GmbH, Huerth, Germany), a Rigol DS1052E digital oscilloscope (Beaverton, OR, USA), and two custom stainless-steel chambers (Val-Electronic, Athens, Greece). To conduct US pretreatment, an Elmasonic P70H ultrasonication bath (Elma Schmidbauer GmbH, Singen, Germany) was utilized. Spectrophotometric analyses were conducted with a Shimadzu UV-1900i PharmaSpec spectrophotometer (Kyoto, Japan). Finally, to perform chromatographic analyses and quantify the identified individual polyphenols, we used a Shimadzu CBM-20A liquid chromatograph with a Shimadzu SPD-M20A diode array detector (DAD) (Shimadzu Europa GmbH, Duisburg, Germany). The separation of targeted compounds occurred in a Phenomenex Luna C18(2) column (Phenomenex Inc. Torrance, CA, USA) maintained at 40 °C (100 Å, 5 μm, 4.6 mm × 250 mm).

2.3. Sample and Extract Preparation

Red apple (*Malus domestica*) fruits from the 'Starking Delicious' cultivar were purchased in Karditsa, Greece, at a neighborhood store. The apples were thoroughly rinsed with running tap water and then dried using paper towels. At first, the apple peels were cut off with a blade, and then manually chopped into small pieces and lyophilized. The peels were finely ground to a diameter of less than 400 μm and then stored in the freezer at −40 °C, pending further analysis.

2.4. Plant Extraction Procedure

Various extraction combinations of different pre-treatment techniques were employed to determine the best conditions for recovering antioxidant compounds from apple peels. For each experiment, an amount of 1 g of dried apple peel powder was mixed with solvent at a solid-to-liquid ratio of 1:20 g/mL. The solvents used ranged from 0% to 100% v/v aqueous ethanol. PEF and US techniques were employed to enhance the conventional stirring extraction process. Once these techniques were used, the peel powder was hydrated by including the suitable solvent and allowed to sit for 10 min. The sample underwent a 20-min treatment with PEF or US individually. When both techniques were used together, the sample received a 20-min PEF treatment, then a 20-min US treatment. Ultimately, all samples underwent a conventional stirring extraction process with the use of a stirring hotplate. For the samples that demanded PEF treatment, an amount of 1.0 kV/cm of electric field strength was applied, with a pulse length of 10 µs within a pulse period of 1 ms (frequency: 1 kHz). For the US treatment, the equipment operated at a nominal frequency of 37 kHz with the temperature being stable at 30 °C. PEF and US conditions were chosen according to our previous methodology [19].

The mixtures comprising the appropriate solvent and apple peels were placed in screw-capped glass bottles and subjected to temperatures that ranged from 20 to 80 °C for a duration varying between 30 and 150 min while being stirred continuously at 500 rpm. The sample extracts were centrifuged for 10 min at 10,000× g after the extraction process was finished. The supernatants were gathered and kept refrigerated at −40 °C until further analysis. The extraction was performed using various combinations of the analyzed parameters, which are represented by the coded levels in Table 1.

Table 1. The independent variables with the respective actual and coded levels were utilized for the optimization process.

Independent Variables	Code Units	Coded Variable Level				
		1	2	3	4	5
Technique	X_1	ST	PEF + ST	US + ST	PEF + US + ST	–
C (%, v/v)	X_2	0	25	50	75	100
t (min)	X_3	30	60	90	120	150
T (°C)	X_4	20	35	50	65	80

2.5. Experiment Design and Optimization Using Response Surface Methodology (RSM)

The RSM method was utilized to assess the antioxidant capacity and quantity of bioactive compounds in apple peel extracts. The extraction of targeted bioactive compounds (i.e., individual polyphenols and ascorbic acid) could be enhanced through an optimization process made by RSM. This optimization could probably increase the antioxidant activity of the extracts. The optimization involved refining key extraction parameters including the extraction method (i.e., conventional stirring with or without pretreatment green techniques such as PEF and US), adjusting the ethanol-to-water ratio (C, % v/v), and adjusting the extraction duration (t, min) in several temperature ranges (T, °C). The Main Effect Screening design of the experiment was employed for that reason, with 20 design points being examined. The process variables were set at five levels according to the experimental design. A minimum confidence level of 95% using analysis of variance (ANOVA) and summary-of-fit tests was used to assess the overall model significance (R^2, p-value) and the significance of the model coefficients. The following second-order polynomial model was employed to predict the response variable as a function of the independent variables under investigation:

$$Y_k = \beta_0 + \sum_{i=1}^{2} \beta_i X_i + \sum_{i=1}^{2} \beta_{ii} X_i^2 + \sum_{i=1}^{2}\sum_{j=i+1}^{3} \beta_{ij} X_i X_j \qquad (1)$$

where the intercept and regression coefficients for the initial, linear, quadratic, and interaction terms are denoted by β_0, β_i, β_{ii}, and β_{ij}, respectively. Terms X_i and X_j represent the independent variables, while Y_k denotes the expected response variable.

The largest peak area was determined through RSM wherein the effect of a significant independent variable on the response was also assessed. The results from the above model equation were visually represented through 3D plots, which illustrate surface response graphs.

2.6. Bioactive Compounds Quantification

2.6.1. Determination of Total Polyphenol Content (TPC)

The TPC determination procedure was based on a previously established methodology [22]. In summary, 0.2 mL of sample volume was combined with 0.2 mL of Folin–Ciocalteu reagent. Right after 2 min, 1.6 mL of a sodium carbonate aqueous solution (5% w/v) was inserted into the mixture in a 2 mL Eppendorf tube. The incubation process at 40 °C for 20 min was followed for the mixture with its absorbance being measured at 740 nm. A calibration curve of gallic acid was used to determine the concentration of total polyphenols (C_{TP}), in which TPC was expressed as mg of gallic acid equivalents (GAE)/g dw, calculated using the following equation:

$$\text{TPC (mg GAE/g dw)} = \frac{C_{TP} \times V}{w} \quad (2)$$

where V denotes the total volume (in L) of the extraction medium, and w represents the dry weight of the sample (in g).

2.6.2. Chromatographic Polyphenol Quantification

An established methodology from our previous research [23] involving high-performance liquid chromatography (HPLC) was used to identify and quantify individual polyphenols from the apple peel extracts. Mixtures of formic acid constituted the mobile phase, since 0.5% aqueous formic acid (A) and 0.5% formic acid in acetonitrile (B) were used. The gradient program was set to start from 0 to 40% B, increase to 50% B over 10 min, then to 70% B in another 10 min, and hold steady for 10 min. We kept the flow rate stable at 1 mL/min. The identification of targeted compounds was made through the comparison of their absorbance spectra and their respective retention times with pure analytical standards, using calibration curves ranging from 0 to 50 mg/L of excellent linearity (>0.99).

2.6.3. Determination of Ascorbic Acid Content (AAC)

AAC analysis was carried out following a protocol previously described in reference [24]. Briefly, a mixture of 0.1 mL of extract volume and 0.5 mL of aqueous Folin–Ciocalteu reagent (10% v/v) was combined with 0.9 mL of aqueous trichloroacetic acid (10% w/v) within an Eppendorf tube. After 10 min intervals, the absorbance at 760 nm was recorded. Ascorbic acid was used as a calibration standard.

2.7. Antioxidant Assays

2.7.1. Ferric-Reducing Antioxidant Power (FRAP) Assay

The protocol published by Shehata et al. [25] was utilized for the evaluation of FRAP. A 0.1 mL aliquot of an appropriately diluted sample was combined with 0.1 mL of 4 mM $FeCl_3$ solution, which was prepared in 0.05 M HCl. This mixture underwent incubation for 30 min at 37 °C. Subsequently, 1.8 mL of TPTZ solution (1 mM in 0.05 M HCl) was added, with the mixture being vortexed. The absorbance was recorded at 620 nm after 5 min. The antioxidant activity through ferric-reducing power (P_R) was determined using an ascorbic acid calibration curve in 0.05 M HCl, with its concentrations (i.e., C_{AA}) ranging from 50 to 500 µM. The P_R was expressed as µmol of ascorbic acid equivalents (AAE) per g of dw, calculated using Equation (3) as follows:

$$P_R \text{ (µmol AAE/g dw)} = \frac{C_{AA} \times V}{w} \quad (3)$$

where V denotes the total volume of the extraction medium (in L), and w represents the dry weight of the sample (in g).

2.7.2. DPPH• Antiradical Activity Assay

The antioxidant activity of apple peel extracts was also evaluated through radical inhibition activity, employing a modified DPPH• methodology as described by Shehata et al. [25]. In this method, 0.1 mL of the extract was combined with 3.9 mL of a 100 µM DPPH• solution in methanol. The final mixture was then incubated for 30 min in the dark at ambient temperature. Subsequently, the absorbance at 515 nm was recorded. Additionally, a blank sample containing the DPPH• solution and methanol was used, with its absorbance being immediately recorded in the same wavelength. The percentage of inhibition was computed using Equation (4):

$$\text{Inhibition (\%)} = \frac{A_{515(i)} - A_{515(f)}}{A_{515(i)}} \times 100 \quad (4)$$

Different ascorbic acid concentrations (C_{AA}) were used in the calibration curve, represented by Equation (5), which was utilized to assess the antiradical activity (A_{AR}), expressed as µmol AAE per g of dw as follows:

$$A_{AR}(\mu\text{mol AAE/g dw}) = \frac{C_{AA} \times V}{w} \quad (5)$$

where V denotes the total volume of the extraction medium (in L), and w represents the dry weight of the sample (in g).

2.8. Statistical Analysis

A statistical study focusing on distribution analysis and response surface methodology was carried out using JMP® Pro 16 software (SAS, Cary, NC, USA). Quantitative analyses were conducted in triplicate, with each batch of apple peel extracts undergoing the extraction process at least twice. To ensure data normality, the Kolmogorov–Smirnov test was used. One-way analysis of variance (ANOVA) was utilized to identify statistically significant differences, which was followed by post hoc Tukey HSD (honestly significant difference) test calculations applying the Tukey–Kramer method. Results are presented as means ± standard deviations. Additionally, Pareto plot and partial least squares (PLS) analyses were conducted for extraction optimization, whereas correlation analyses (i.e., principal component analysis (PCA), and multivariate correlation analysis (MCA)) were used to interpret any correlations between the variables under investigation. All mentioned analyses were performed using JMP® Pro 16 software.

3. Results and Discussion

The apple fruits weighed 171.35 ± 12.85 g. The total soluble solid value was 14.17 ± 0.91 °Brix, the pH value was 4.08 ± 0.06, and the titratable acidity (given as % malic acid) was measured at 0.39 ± 0.01. These findings fall within the typical range for fruits, indicating a balance between acidity and sweetness. Also, these values are vital in determining the ripeness, flavor profile, and overall quality of the fruit, which are essential for consumer satisfaction and processing requirements.

Despite similarities in the methodology with our previously mentioned study [19], it is worth mentioning that apples create a vast amount of waste annually. Apples have a significantly higher global annual production (~86 million tons) [26] than mandarins (~38 million tons) [27]. To that end, this research was concerned with the extraction optimization of polyphenols from apple peels. The conditions and combinations of extraction techniques were examined through RSM in order to identify the optimal model for generating extracts that are abundant in bioactive compounds (i.e., polyphenolic compounds and ascorbic acid) and possess significant antioxidant activity. Additionally, previous studies have demonstrated that incorporating PEF and US into the extraction process can enhance its effectiveness [28]. Ethanol has the potential to be mixed with water to provide an extraction solvent that is well suited for application within the food industry [29]. However, it is crucial to take into account that polyphenols are thermolabile chemicals [30] for which the ideal temperature range for conventional extraction procedures to attain the maximum polyphenol recovery is typically between 50 and 80 °C [31,32]. An extensive examination is required to evaluate the influence of duration on extraction, considering the established effectiveness of both brief and extended extraction times in prior investigations [33]. PEF and the US are environmentally sustainable as they require reduced extraction time and energy consumption [34].

3.1. Extraction Parameters Optimization

The recovery of bioactive compounds may encounter challenges that stem from changes in the solubility and polarity of their chemical structures [35]. Hence, it is imperative to enhance the efficiency of this procedure [36]. The solvent's composition is critical, as its features significantly influence chemical extraction [37]. Moderately polar compounds, such as polyphenols, are challenging to extract using highly polar solvents like water. Therefore, solvents with moderate polarity such as organic solvents are frequently utilized to further enhance the extraction process. Ethanol, for example, is a unique organic solvent (i.e., an alcohol) that can be mixed with water to create a extraction solvent appropriate for use in the food sector [29]. The extracts underwent a screening process that included performing spectrophotometric analysis for the specified assays. The assessment of polyphenols was carried out using the widely known Folin–Ciocalteu spectrophotometric method, which is a rapid, cost-effective, and highly sensitive methodology for measuring the total amount of polyphenols [36]. This method is widely recognized for its significant correlation with the liquid chromatographic method of measurement [38,39]. The obtained model from partial least squares analysis determined the optimal sample with high levels of polyphenols and strong antioxidant properties. Subsequently, liquid chromatography was employed to identify the specific polyphenols found in apple peel. In Table 2, the experimental findings out of the four independent variables are displayed. The TPC values of the extract ranged from 1.57 to 15.93 mg GAE/g dw, with design points 16 and 13 representing the lower and the higher TPC values, respectively. These values are consistent with the literature, considering that Villamil-Galindo and Piagentini [40] reported a similar range, 6.33 to 11.90 mg GAE/g dw, on 'Granny Smith' apple peels. As for the antioxidant activity of the extracts, the higher values are reported on design point 8 on both FRAP and DPPH assays, and the reported values are 128.06 and 105.66 μmol AAE/g dw, respectively, while the lowest antioxidant activity is on design point 1 for FRAP with a value of 16.98 μmol AAE/g dw, and 16 for DPPH, with a value of 16.71 μmol AAE/g dw. The highest ascorbic acid content resulted on design point 20, where the AAC is 4.16 mg/g dw, and the lower value is observed in design point 1, with a value of 0.56 mg/g dw.

The statistical parameters, including second-order polynomial equations (models) and their respective coefficients ($R^2 > 0.97$) derived from each model, are displayed in Table 3. Moreover, the adjustment to the R^2 model is provided for each equation, and all values are higher than ~0.87, indicating a good fit of the equations to the model. An excellent fit between the produced models and the observed data was observed based on the data. Additionally, all responses exhibit low p-values (ranging from 0.0005 to 0.0090), enhancing the above statement. Figure S2–S5 present plots illustrating the relationship between the actual and the predicted response for each parameter under examination and the corresponding desirability functions. Figure 1 displays 3D response plots for TPC, whereas Figure S6–S8 display 3D response plots for the other responses under investigation. In Figure S2, it is evident that only factor X_2 (i.e., solvent composition) significantly affects TPC, as it is the only factor that has a different influence on polyphenol recovery, while the other factors seem to lead to maximum recovery at each value, with little variation between them, with a desirability function of ~0.87. In Figure 1, it is profound that the TPC is maximized when only ST is employed at high temperatures and for a short duration of time. Regarding the antioxidant capacity through the FRAP assay, in Figure S3 it is noted that the antioxidant capacity of the extracts is maximized when ST and low temperatures are employed for relatively long times, with a desirability function of ~0.89. The suitable solvent composition for this assay is relatively non-polar. These conclusions are also enhanced by the 3D plots in Figure S6. Similar results are drawn regarding Figures S4 and S7 for the DPPH assay, with a desirability function of ~0.91. Concerning the AAC of the extracts, Figures S5 and S8 imply that the most favorable extraction technique is the combination of PEF, US, and ST for an intermediate temperature and at a short time, and the desirability function is ~0.84.

Table 2. Findings from the experiment on the four independent variables and their corresponding responses of the dependent variables.

Design Point	Independent Variables				Responses			
	X_1	X_2	X_3	X_4	TPC [1]	FRAP [2]	DPPH [3]	AAC [4]
1	3	1	3	4	3.04 ± 0.17 [j]	16.98 ± 0.51 [j]	27.63 ± 1.66 [k,l]	0.56 ± 0.04 [n]
2	3	2	1	3	10.42 ± 0.58 [e,f]	76.39 ± 4.66 [f,g]	59.71 ± 3.58 [e,f,g]	2.77 ± 0.06 [c,d,e,f]
3	2	3	4	3	11.17 ± 0.78 [d,e]	101.98 ± 2.45 [c,d]	67.89 ± 3.06 [c,d,e]	2.13 ± 0.04 [i,j,k]
4	2	4	5	4	10.78 ± 0.57 [d,e,f]	102.24 ± 3.78 [c,d]	73.29 ± 4.10 [c,d]	2.45 ± 0.18 [f,g,h,i]
5	3	5	4	2	6.53 ± 0.23 [i]	55.76 ± 2.23 [i]	38.38 ± 1.65 [i,j]	2.53 ± 0.16 [e,f,g,h]
6	4	1	4	5	3.99 ± 0.12 [j]	28.42 ± 1.02 [j]	22.95 ± 0.80 [l,m]	1.49 ± 0.06 [l,m]
7	4	2	3	1	10.36 ± 0.71 [e,f]	71.18 ± 1.85 [f,g,h]	51.63 ± 1.14 [g,h]	2.36 ± 0.13 [g,h,i,j]
8	1	3	3	2	14.40 ± 0.71 [b]	128.06 ± 4.74 [a]	105.66 ± 2.54 [a]	2.50 ± 0.11 [e,f,g,h]
9	1	4	4	1	13.96 ± 0.82 [b]	127.88 ± 6.52 [a]	90.74 ± 2.63 [b]	2.85 ± 0.08 [c,d,e]
10	1	5	1	4	8.89 ± 0.30 [g,h]	82.22 ± 5.51 [e,f]	64.57 ± 1.68 [d,e,f]	2.18 ± 0.14 [h,i,j]
11	1	1	2	3	2.70 ± 0.18 [j,k]	19.17 ± 0.40 [j]	32.07 ± 2.34 [j,k]	1.14 ± 0.09 [m]
12	1	2	5	5	13.65 ± 0.30 [b]	126.73 ± 9.25 [a,b]	97.62 ± 5.37 [a,b]	2.96 ± 0.12 [c,d]
13	4	3	2	4	15.93 ± 0.56 [a]	114.91 ± 4.14 [a,b,c]	71.30 ± 4.06 [c,d]	2.94 ± 0.09 [c,d]
14	3	4	2	5	13.37 ± 0.27 [b]	113.82 ± 3.41 [b,c]	73.18 ± 5.34 [c,d]	3.53 ± 0.23 [b]
15	2	5	3	5	7.48 ± 0.18 [h,i]	68.24 ± 2.80 [g,h,i]	42.96 ± 2.23 [h,i]	2.01 ± 0.10 [j,k]
16	2	1	1	1	1.57 ± 0.06 [k]	21.43 ± 0.96 [j]	16.71 ± 1.17 [m]	0.62 ± 0.04 [n]
17	2	2	2	2	9.41 ± 0.37 [f,g]	79.61 ± 4.06 [f,g]	57.17 ± 3.49 [f,g]	1.80 ± 0.12 [k,l]
18	3	3	5	1	11.86 ± 0.34 [c,d]	94.23 ± 2.17 [d,e]	73.87 ± 1.63 [c]	3.05 ± 0.09 [c]
19	4	4	1	2	13.15 ± 0.55 [b,c]	105.28 ± 3.26 [c,d]	58.90 ± 2.47 [e,f,g]	2.68 ± 0.10 [d,e,f,g]
20	4	5	5	3	9.49 ± 0.21 [f,g]	57.57 ± 2.88 [h,i]	29.71 ± 1.57 [j,k,l]	4.16 ± 0.11 [a]

Values are calculated as the mean values of triplicates (±standard deviation). Lowercase letters (e.g., a–n) within each column denote the statistically significant differences ($p < 0.05$); [1] values in mg GAE/g dw; [2] values in μmol AAE/g dw; [3] values in μmol AAE/g dw; [4] values in mg AA/g dw.

Table 3. Mathematical models employing RSM have been utilized to optimize the extraction process of apple peels.

Responses	Second-Order Polynomial Equations (Models)	R^2 Predicted	R^2 Adjusted	p-Value	Equation
TPC	$Y = -4.03 - 5.13X_1 + 15.79X_2 - 2.39X_3 + 0.14X_4 + 1.05X_1^2 - 2.39X_2^2 - 0.003X_3^2 + 0.47X_4^2 - 0.0004X_1X_2 + 0.18X_1X_3 - 0.27X_1X_4 + 0.48X_2X_3 - 0.47X_2X_4 - 0.04X_3X_4$	0.9779	0.9159	0.0033	(6)
FRAP	$Y = -24.83 - 33.24X_1 + 140.01X_2 - 18.61X_3 - 16.43X_4 + 4.79X_1^2 - 19.82X_2^2 + 0.14X_3^2 + 4.68X_4^2 + 0.04X_1X_2 + 1.06X_1X_3 - 0.61X_1X_4 + 1.6X_2X_3 - 3.84X_2X_4 + 1.99X_3X_4$	0.9896	0.9606	0.0005	(7)
DPPH	$Y = -25.99 - 22.48X_1 + 97.17X_2 - 13.59X_3 + 4.52X_4 + 3.5X_1^2 - 13.76X_2^2 - 0.14X_3^2 + 3.16X_4^2 - 1.14X_1X_2 + 1.82X_1X_3 - 2.55X_1X_4 + 2.33X_2X_3 - 4.24X_2X_4 - 0.15X_3X_4$	0.9662	0.8716	0.0090	(8)
AAC	$Y = 1.17 - 1.5X_1 + 3.96X_2 - 2.62X_3 + 0.21X_4 + 0.2X_1^2 - 0.6X_2^2 + 0.03X_3^2 + 0.33X_4^2 + 0.06X_1X_2 + 0.38X_1X_3 - 0.26X_1X_4 + 0.39X_2X_3 - 0.37X_2X_4 - 0.006X_3X_4$	0.9766	0.9109	0.0038	(9)

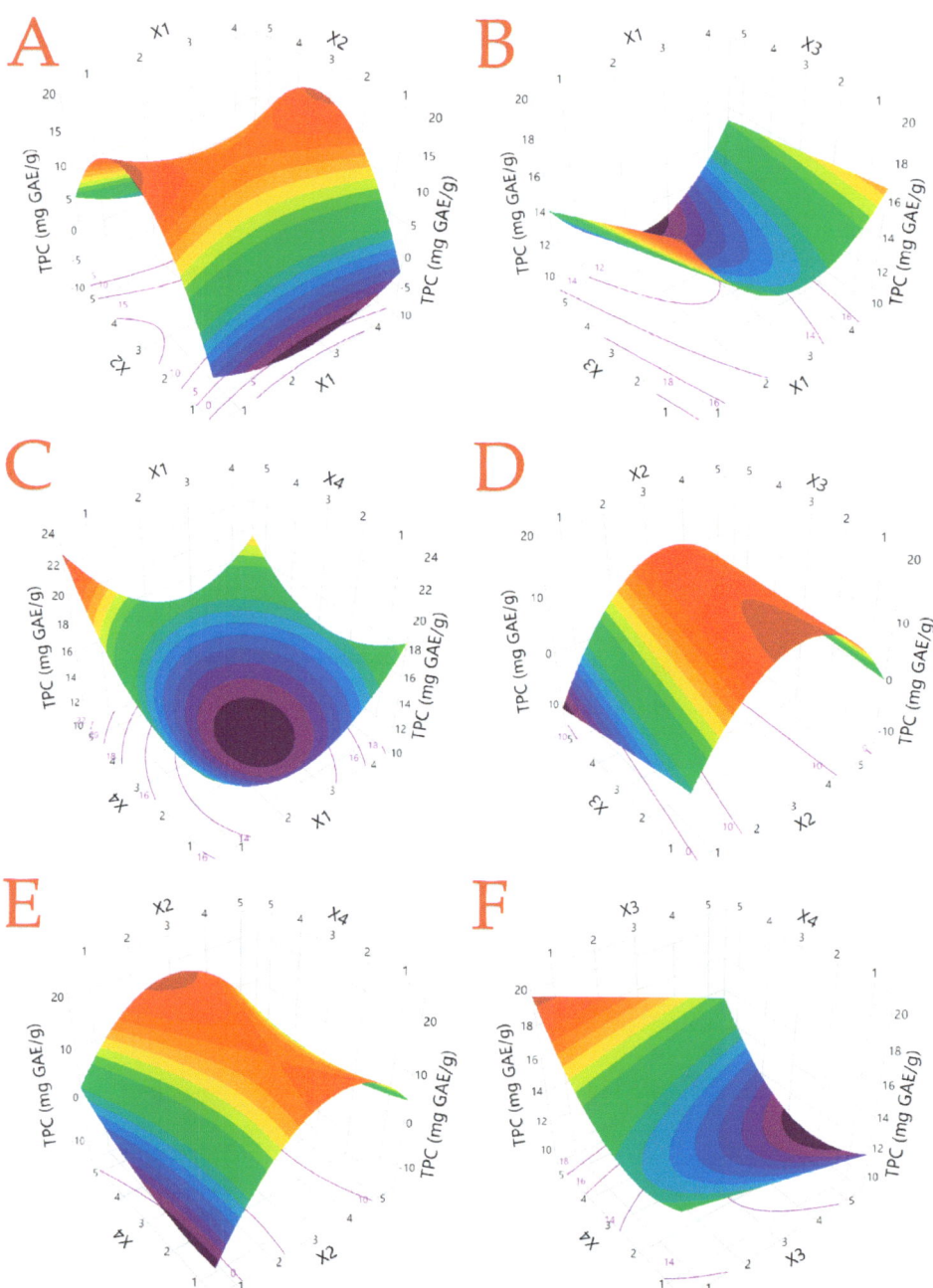

Figure 1. The optimized extraction conditions of apple peel extracts are depicted in 3D plots that highlight the impact of the process variables considered in the response (total polyphenol content—TPC, mg GAE/g). Plot (**A**), covariation of X_1: extraction technique and X_2: ethanol concentration; C, % v/v; plot (**B**), covariation of X_1 and X_3: extraction duration; t, min; plot (**C**), covariation of X_1 and X_4: extraction temperature; T, °C; plot (**D**), covariation of X_2 and X_3; plot (**E**), covariation of X_2 and X_4; plot (**F**), covariation of X_3 and X_4.

3.2. Impact of Extraction Parameters to Assays Through Pareto Plot Analysis

On the notion of statistical significance ($p < 0.05$), the main effects and their interactions were assessed through a standardized Pareto plot. The independent variables (extraction technique, X_1; solvent composition, X_2; extraction duration, X_3; temperature, X_4) and their interactions that impacted TPC, FRAP, DPPH, and AAC are illustrated in Figure 2. The depicted values also include the orthogonal coded estimates, which are obtained through the orthogonalization of the transformation estimates. Concerning all assays, it is obvious that factor X_2*X_2 has a negative impact on all TPC, FRAP, DPPH, and AAC. On the contrary, factor X_2 positively affects these values. As the solvent has a high level of efficacy in the recovery of polar compounds, it is now ascertained that a binary solvent is necessary to enhance the efficacy of the extraction [41]. It is also important to mention that factor X_1 has a negative impact on both FRAP and DPPH, but it positively impacts AAC. This could lead to the conclusion that the pre-treatment of the extracts with PEF or US led to the degradation of some polyphenols [12], and so FRAP and DPPH were negatively affected, whereas AA was not affected at all by these techniques.

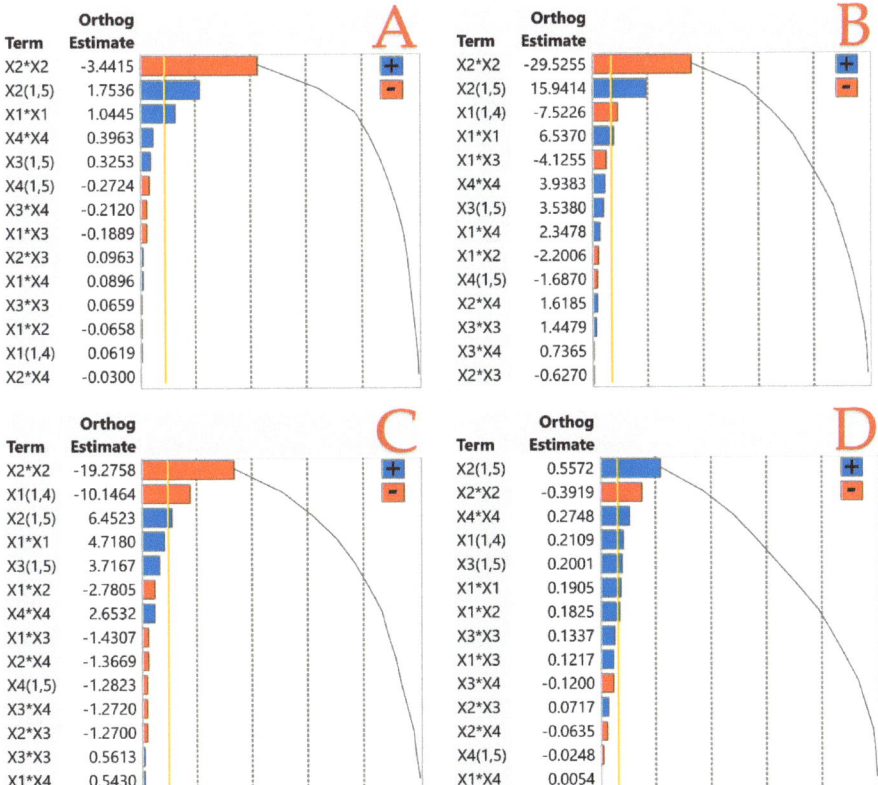

Figure 2. Pareto plots display transformed estimates for TPC (**A**), FRAP (**B**), DPPH (**C**), and AAC (**D**) assays. The significance level ($p < 0.05$) is denoted by a gold reference line, which is included on the plot.

3.3. Optimized Extraction Conditions

To achieve the maximum yields of TPC, FRAP, DPPH, and AAC, the desirability function was employed, and the results are displayed in Table 4. It can be noted that TPC and AAC are favored when PEF, US, and ST are combined, with their predicted responses at 15.45 mg GAE/g dw and 3.92 mg/g, respectively, while the antioxidant activity of the extracts is maximized only through ST, with predicted responses 127.68 µmol AAE/g dw for FRAP and 103.06 µmol AAE/g dw for DPPH. The solvent composition ranges from intermediate polarity to low polarity, while a long extraction duration seems to favor all responses. The extraction temperature varies from 20 to 65 °C. Since these conditions are fundamentally distinct, RSM, ANOVA, and PLS statistical analyses are necessary to determine the general best settings for all four responses to simultaneously extract the greatest possible quantities.

Table 4. Optimal extraction conditions and maximum predicted responses for dependent variables.

Responses	Maximum Predicted Response	Optimal Conditions			
		Technique (X_1)	C (%, v/v) (X_2)	t (min) (X_3)	T (°C) (X_4)
TPC (mg GAE/g dw)	15.45 ± 2.33	PEF + US + ST (4)	50 (3)	60 (2)	65 (4)
FRAP (µmol AAE/g dw)	127.68 ± 17.57	ST (1)	75 (4)	120 (4)	20 (1)
DPPH (µmol AAE/g dw)	103.06 ± 18.63	ST (1)	50 (3)	120 (4)	65 (4)
AAC (mg/g dw)	3.92 ± 0.63	PEF + US + ST (4)	100 (5)	150 (5)	50 (3)

3.4. Correlation Analyses

PCA (Figure 3) and MCA (Table 5) were utilized to further examine the responses more thoroughly and provide information about the correlations among the examined responses. The objective of MCA is to identify patterns and relationships among multiple variables concurrently, while PCA is employed to represent a multivariate data table as a reduced set of variables (summary indices) to facilitate the observation of trends, anomalies, clusters, and outliers. PCA is particularly advantageous for examining complex datasets, enabling a more profound comprehension of the data and their correlation. AAC, TPC, FRAP, and DPPH have a positive correlation with both factors X_2 and X_3 but a negative correlation with factors X_1 and X_4. As for MCA analysis, TPC seems to have a high correlation with all three responses and, more specifically, it correlates with FRAP at ~0.96, at ~0.88 with DPPH, and ~0.78 with AAC. FRAP and DPPH are also in line with one another. However, a relatively low correlation is observed between AAC and FRAP (~0.68) and between AAC and DPPH (~0.53).

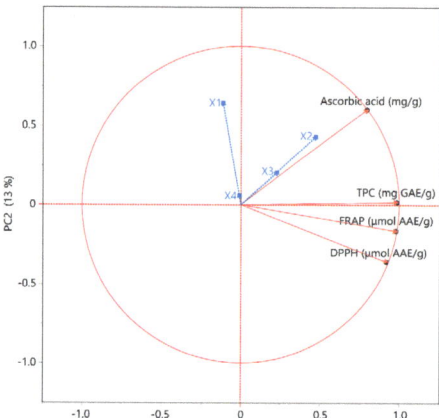

Figure 3. PCA plot of the measured variables, along with the four X variables that are represented in blue.

Table 5. Multivariate correlation analysis of the measured variables.

Responses	TPC	FRAP	DPPH	AAC
TPC	–	0.9616	0.8757	0.7759
FRAP		–	0.9474	0.6755
DPPH			–	0.5269
AAC				–

3.5. Partial Least Squares (PLS) Analysis

The impact of the four extraction parameters was assessed by applying a PLS model. Figure 4 illustrates the PLS model used to create a correlation loading plot in which the impact of extraction conditions of apple peels is visually displayed. The projection factor with a value over 0.8 indicates that this variable exhibits a high contribution. It is obvious that the X_1 variable, just stirring, was far from enough to yield maximum polyphenol recovery, the antioxidant activity of the extracts, and ascorbic acid content simultaneously; however, this became feasible when PEF, US, or coupled PEF and US pretreated extractions were executed. This phenomenon can be attributed to the electroporation processes associated with PEF, which has proven efficacy, as the current disrupts the membrane of cells in apple peels, facilitating the recovery of targeted bioactive compounds [12]. The cavitation effect of ultrasound, which induces the creation of microcracks on solid surfaces due to the release of high energy upon bubble collapse, also results in increased yields [42]. Regarding the X_2 variable, it was observed that a moderate-polarity solvent comprising 75% v/v ethanol was ideal in all assays. The explanation for this may be in the moderate-polarity polyphenols recovered in comparison to water [43]. Extraction duration (i.e., variable X_3) had the most effective impact when short periods were applied in all assays. Finally, in all cases, it was observed that temperatures as high as 80 °C (variable X_4) favored the maximization of the extraction yields. The desirability of this model is ~0.92, which implies a good fit of the model in the data.

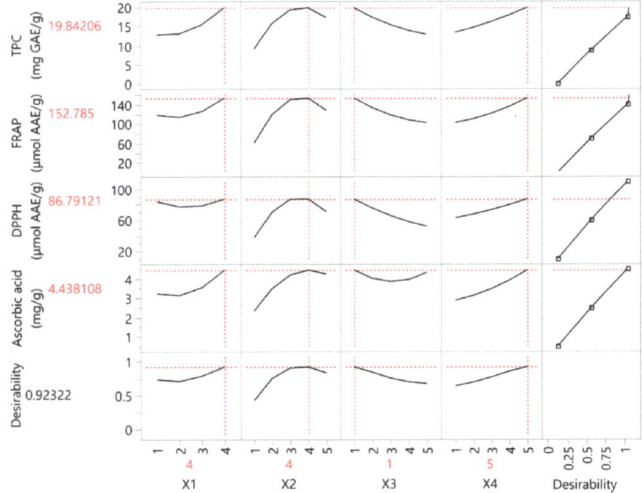

Figure 4. Partial least squares (PLS) prediction profiler for each variable and the desirability function, including extrapolation control, aimed at optimizing apple peel extracts. Variables include X_1 (extraction technique), X_2 (ethanol concentration, % v/v), X_3 (extraction duration, min), and X_4 (extraction temperature, °C).

After comparing the PLS model values with those derived from experimental investigations, a correlation of 0.9845 between them is observed and they exhibit no significant deviations, possessing a p-value < 0.0001.

3.6. Analysis of the Optimal Extract

The optimal extraction conditions were determined as a combination of PEF, US, and ST, utilizing 75% v/v aqueous ethanol as a solvent, for 30 min at 80 °C. There is a plethora of research in the literature where the combination of PEF, US, and ST is shown to be effective for the recovery of bioactive compounds [12]. The combination of electroporation and cavitation effect, as pretreatment techniques, appear to considerably enhance the performance of conventional extraction through stirring, and thus it is possible to obtain higher yields in shorter times (about 30 min), and thus less energy is expended. The polyphenols recovery, along with the antioxidant capacity (FRAP, DPPH, and AAC values) are exhibited in Table 6. Moreover, in Table 7, the individual polyphenolic compounds determined through HPLC-DAD are displayed, and in Figure S9 a representative chromatograph of these polyphenolic compounds is depicted. The TPC recovered under the optimal conditions, which is 17.23 mg GAE/g dw, is very close to the value predicted by the PLS program. This result is almost double the one reported by Tian et al. [15], who determined 9.22 mg GAE/g dw on apple peel flesh. Kunradi Vieira et al. [44] subjected apple fruit and peels in the US for 15 min and 80% v/v acetone and reported a TPC on apple peels that is ~199% lower than ours. It is evident that the selection of a suitable solvent is crucial for the recovery of bioactive compounds, and that an ethanol–water mixture, which produces a solvent of moderate polarity, results in much greater yields compared to acetone, a non-polar solvent. Yue et al. [21] reported a TPC of ~13 mg GAE/g dw from apple peels when 50% ethanol was utilized with 30 min of ultrasonication at 50 °C. It is important to highlight that in our study, the TPC obtained under the optimal conditions utilized lower temperature and ultrasonication duration. Wang et al. [20] studied how PEF affects the extraction yield and the degradation of bioactive compounds from apple peels, and the results indicated that all PEF parameters play a crucial role in the TPC yield. Sethi et al. [16] determined the AAC on several cultivars of apple peels, and the reported content is ~1187 lower than the one reported in this study. It is once again evident that the nature of the raw material, in this case, the different varieties of the same fruit, clearly affects the extraction yield. The most abundant polyphenolic compound in apple peels identified in this research is pelargonin chloride. Pelargonin is an anthocyanin to which the red color of the apple peels is attributed [45]. Several pelargonidin pigments have been found in red apples by other researchers [46]. The second most abundant polyphenolic compound is catechin. Kondo et al. [47] also determined catechin in three different apple cultivars, and the amounts are from ~147% to ~394% lower than ours. Moreover, Karaman et al. [48] explored the polyphenolic compounds in apple flesh and peels from different cultivars, and they also determined catechin and chlorogenic acid. The highest amount of catechin they found was in the peels of the 'Amasya' cultivar and of chlorogenic acid in the 'King Luscious' cultivar, but these results are ~888% and ~663% lower than ours, respectively. The same research team also determined epicatechin in their apple samples, and the highest amount in apple peels was in 'Ervin Spur' for epicatechin, which is in line with our value.

Table 6. Optimal extraction conditions (X_1:4, X_2:4, X_3:1, and X_4:5) and maximum desirability for all variables using the partial least squares (PLS) prediction profiler.

Variables	PLS Model Values	Experimental Values
TPC (mg GAE/g dw)	19.84	17.23 ± 0.65
FRAP (μmol AAE/g dw)	152.79	130.87 ± 6.15
DPPH (μmol AAE/g dw)	86.79	95.38 ± 3.05
AAC (mg/g dw)	4.44	3.99 ± 0.13

Table 7. Polyphenolic compounds identified under optimal extraction conditions (X_1:4, X_2:4, X_3:1, and X_4:5).

Polyphenolic Compound	Optimal Extract (mg/g dw)
Pelargonin chloride	2.56 ± 0.19
Catechin	1.58 ± 0.03
Chlorogenic acid	0.61 ± 0.04
Homovanillic acid	0.01 ± 0
Epicatechin	0.11 ± 0
Quercetin 3-*D*-galactoside	0.60 ± 0.02
Hesperidin	0.59 ± 0.04
Total identified	6.06 ± 0.34

4. Conclusions

By conducting thorough research and carefully analyzing various conditions, this study sought to identify the most effective method for extracting bioactive compounds from apple peels. Utilizing the RSM approach, we were able to identify the key parameters for extraction. However, the PLS model provided us with the opportunity to further refine and optimize these parameters. The significance of incorporating PEF and US into traditional extraction methods was discovered. It is worth mentioning that additional research could explore alternative green techniques or extraction conditions to enhance the recovery of bioactive compounds. The findings of this study highlight the effectiveness of green extraction techniques for specific byproducts that can be utilized to develop food additives, animal feed, dietary supplements, and cosmetics. Additionally, they offer an effective environmentally friendly alternative for minimizing waste in the food sector.

Supplementary Materials: The following supporting information can be downloaded at: https://www.mdpi.com/article/10.3390/eng5040150/s1, Figure S1 illustrates the primary polyphenol classes present in apple peels. The comparisons between the predicted and actual responses for each parameter being investigated are depicted in Figures S2–S5, which also include the desirability functions. Three-dimensional response diagrams for the remaining responses are illustrated in Figures S6–S8. A representative chromatograph of the optimal extract at 280 and 320 nm is provided in Figure S9.

Author Contributions: Conceptualization, V.A., T.C., and S.I.L.; methodology, V.A. and T.C.; software, V.A.; validation, T.C. and V.A.; formal analysis, D.K., E.B., V.A., and T.C.; investigation, M.M. and B.-C.B.; resources, S.I.L.; data curation, D.K., M.M., and E.B.; writing—original draft preparation, B.-C.B., M.M., and E.B.; writing—review and editing, V.A., T.C., D.K., K.K., E.B., M.M., and S.I.L.; visualization, D.K., K.K., and M.M.; supervision, V.A., T.C., and S.I.L.; project administration, S.I.L.; funding acquisition, S.I.L. All authors have read and agreed to the published version of the manuscript.

Funding: This research received no external funding.

Institutional Review Board Statement: Not applicable.

Informed Consent Statement: Not applicable.

Data Availability Statement: All related data and methods are presented in this paper. Additional inquiries should be addressed to the corresponding author.

Conflicts of Interest: The authors declare no conflicts of interest.

References

1. Vasile, M.; Bunea, A.; Ioan, C.R.; Ioan, B.C.; Socaci, S.; Viorel, M. Phytochemical Content and Antioxidant Activity of *Malus Domestica* Borkh Peel Extracts. *Molecules* **2021**, *26*, 7636. [CrossRef] [PubMed]
2. Patocka, J.; Bhardwaj, K.; Klimova, B.; Nepovimova, E.; Wu, Q.; Landi, M.; Kuca, K.; Valis, M.; Wu, W. *Malus Domestica*: A Review on Nutritional Features, Chemical Composition, Traditional and Medicinal Value. *Plants* **2020**, *9*, 1408. [CrossRef]
3. Davies, T.; Watts, S.; McClure, K.; Migicovsky, Z.; Myles, S. Phenotypic Divergence between the Cultivated Apple (*Malus Domestica*) and Its Primary Wild Progenitor (*Malus Sieversii*). *PLoS ONE* **2022**, *17*, e0250751. [CrossRef]

4. FAOSTAT. Food and Agriculture Association of the United States. Available online: http://www.fao.org/faostat/en/ (accessed on 27 March 2024).
5. Fernandes, P.A.R.; Wessel, D.F.; Coimbra, M.A.; Cardoso, S.M. Apple (*Malus Domestica*) By-Products: Chemistry, Functionality and Industrial Applications. In *Mediterranean Fruits Bio-Wastes: Chemistry, Functionality and Technological Applications*; Ramadan, M.F., Farag, M.A., Eds.; Springer International Publishing: Cham, Switzerland, 2022; pp. 349–373. ISBN 978-3-030-84436-3.
6. Shehzadi, K.; Rubab, Q.; Asad, L.; Ishfaq, M.; Shafique, B.; Ali, M.M.; Ranjha, N.; Mahmood, S.; Mueen-Ud-Din, G.; Javaid, T.; et al. A Critical Review on Presence of Polyphenols in Commercial Varieties of Apple Peel, Their Extraction and Health Benefits. *Open Access J. Biol. Sci. Res.* 2020, 6, 18. [CrossRef]
7. Acquavia, M.A.; Pascale, R.; Foti, L.; Carlucci, G.; Scrano, L.; Martelli, G.; Brienza, M.; Coviello, D.; Bianco, G.; Lelario, F. Analytical Methods for Extraction and Identification of Primary and Secondary Metabolites of Apple (*Malus Domestica*) Fruits: A Review. *Separations* 2021, 8, 91. [CrossRef]
8. Shafi, A.; Ahmad, F.; Mohammad, Z.H. Effect of the Addition of Banana Peel Flour on the Shelf Life and Antioxidant Properties of Cookies. *ACS Food Sci. Technol.* 2022, 2, 1355–1363. [CrossRef]
9. Gupta, N.; Singh, S.; Chauhan, D.; Srivastava, R.; Singh, V.K. Exploring the Anticancer Potentials of Polyphenols: A Comprehensive Review of Patents in the Last Five Years. *Recent Pat. Anti-Cancer Drug Discov.* 2023, 18, 3–10. [CrossRef]
10. Bashmil, Y.M.; Ali, A.; Bk, A.; Dunshea, F.R.; Suleria, H.A.R. Screening and Characterization of Phenolic Compounds from Australian Grown Bananas and Their Antioxidant Capacity. *Antioxidants* 2021, 10, 1521. [CrossRef] [PubMed]
11. Tzima, K.; Brunton, N.P.; Lyng, J.G.; Frontuto, D.; Rai, D.K. The Effect of Pulsed Electric Field as a Pre-Treatment Step in Ultrasound Assisted Extraction of Phenolic Compounds from Fresh Rosemary and Thyme by-Products. *Innov. Food Sci. Emerg. Technol.* 2021, 69, 102644. [CrossRef]
12. Chatzimitakos, T.; Athanasiadis, V.; Kalompatsios, D.; Mantiniotou, M.; Bozinou, E.; Lalas, S.I. Pulsed Electric Field Applications for the Extraction of Bioactive Compounds from Food Waste and By-Products: A Critical Review. *Biomass* 2023, 3, 367–401. [CrossRef]
13. Arshad, R.N.; Abdul-Malek, Z.; Roobab, U.; Qureshi, M.I.; Khan, N.; Ahmad, M.H.; Liu, Z.-W.; Aadil, R.M. Effective Valorization of Food Wastes and By-Products through Pulsed Electric Field: A Systematic Review. *J. Food Process Eng.* 2021, 44, e13629. [CrossRef]
14. González-Centeno, M.R.; Comas-Serra, F.; Femenia, A.; Rosselló, C.; Simal, S. Effect of Power Ultrasound Application on Aqueous Extraction of Phenolic Compounds and Antioxidant Capacity from Grape Pomace (*Vitis vinifera* L.): Experimental Kinetics and Modeling. *Ultrason. Sonochemistry* 2015, 22, 506–514. [CrossRef]
15. Tian, J.; Wu, X.; Zhang, M.; Zhou, Z.; Liu, Y. Comparative Study on the Effects of Apple Peel Polyphenols and Apple Flesh Polyphenols on Cardiovascular Risk Factors in Mice. *Clin. Exp. Hypertens.* 2018, 40, 65–72. [CrossRef] [PubMed]
16. Sethi, S.; Joshi, A.; Arora, B.; Bhowmik, A.; Sharma, R.R.; Kumar, P. Significance of FRAP, DPPH, and CUPRAC Assays for Antioxidant Activity Determination in Apple Fruit Extracts. *Eur. Food Res. Technol.* 2020, 246, 591–598. [CrossRef]
17. Fotirić Akšić, M.; Nešović, M.; Ćirić, I.; Tešić, Ž.; Pezo, L.; Tosti, T.; Gašić, U.; Dojčinović, B.; Lončar, B.; Meland, M. Polyphenolics and Chemical Profiles of Domestic Norwegian Apple (*Malus* × *Domestica* Borkh.) Cultivars. *Front. Nutr.* 2022, 9, 941487. [CrossRef] [PubMed]
18. Carpentieri, S.; Soltanipour, F.; Ferrari, G.; Pataro, G.; Donsì, F. Emerging Green Techniques for the Extraction of Antioxidants from Agri-Food By-Products as Promising Ingredients for the Food Industry. *Antioxidants* 2021, 10, 1417. [CrossRef]
19. Kalompatsios, D.; Ionescu, A.-I.; Athanasiadis, V.; Chatzimitakos, T.; Mantiniotou, M.; Kotsou, K.; Bozinou, E.; Lalas, S.I. Maximizing Bioactive Compound Extraction from Mandarin (*Citrus Reticulata*) Peels through Green Pretreatment Techniques. *Oxygen* 2024, 4, 307–324. [CrossRef]
20. Wang, L.; Boussetta, N.; Lebovka, N.; Vorobiev, E. Cell Disintegration of Apple Peels Induced by Pulsed Electric Field and Efficiency of Bio-Compound Extraction. *Food Bioprod. Process.* 2020, 122, 13–21. [CrossRef]
21. Wang, L.; Boussetta, N.; Lebovka, N.; Vorobiev, E. Selectivity of Ultrasound-Assisted Aqueous Extraction of Valuable Compounds from Flesh and Peel of Apple Tissues. *LWT* 2018, 93, 511–516. [CrossRef]
22. Athanasiadis, V.; Chatzimitakos, T.; Mantiniotou, M.; Kalompatsios, D.; Bozinou, E.; Lalas, S.I. Investigation of the Polyphenol Recovery of Overripe Banana Peel Extract Utilizing Cloud Point Extraction. *Eng* 2023, 4, 3026–3038. [CrossRef]
23. Chatzimitakos, T.; Athanasiadis, V.; Makrygiannis, I.; Kalompatsios, D.; Bozinou, E.; Lalas, S.I. An Investigation into *Crithmum Maritimum* L. Leaves as a Source of Antioxidant Polyphenols. *Compounds* 2023, 3, 532–551. [CrossRef]
24. Athanasiadis, V.; Chatzimitakos, T.; Mantiniotou, M.; Bozinou, E.; Lalas, S.I. Exploring the Antioxidant Properties of *Citrus Limon* (Lemon) Peel Ultrasound Extract after the Cloud Point Extraction Method. *Biomass* 2024, 4, 202–216. [CrossRef]
25. Shehata, E.; Grigorakis, S.; Loupassaki, S.; Makris, D.P. Extraction Optimisation Using Water/Glycerol for the Efficient Recovery of Polyphenolic Antioxidants from Two Artemisia Species. *Sep. Purif. Technol.* 2015, 149, 462–469. [CrossRef]
26. Asma, U.; Morozova, K.; Ferrentino, G.; Scampicchio, M. Apples and Apple By-Products: Antioxidant Properties and Food Applications. *Antioxidants* 2023, 12, 1456. [CrossRef] [PubMed]
27. Gormez, E.; Golge, O.; González-Curbelo, M.Á.; Kabak, B. Pesticide Residues in Mandarins: Three-Year Monitoring Results. *Molecules* 2023, 28, 5611. [CrossRef]

28. Jintawiwat, R.; Punamorntarakul, N.; Hirunyasiri, R.; Jarupoom, P.; Pankasemsuk, T.; Supasin, S.; Kawee-ai, A. Testing the Efficacy of a Prototype That Combines Ultrasound and Pulsed Electric Field for Extracting Valuable Compounds from *Mitragyna Speciosa* Leaves. *AgriEngineering* **2023**, *5*, 1879–1892. [CrossRef]
29. Chemat, F.; Rombaut, N.; Meullemiestre, A.; Turk, M.; Perino, S.; Fabiano-Tixier, A.-S.; Abert-Vian, M. Review of Green Food Processing Techniques. Preservation, Transformation, and Extraction. *Innov. Food Sci. Emerg. Technol.* **2017**, *41*, 357–377. [CrossRef]
30. Antony, A.; Farid, M. Effect of Temperatures on Polyphenols during Extraction. *Appl. Sci.* **2022**, *12*, 2107. [CrossRef]
31. Siddiqui, S.A.; Ali Redha, A.; Salauddin, M.; Harahap, I.A.; Rupasinghe, H.P.V. Factors Affecting the Extraction of (Poly)Phenols from Natural Resources Using Deep Eutectic Solvents Combined with Ultrasound-Assisted Extraction. *Crit. Rev. Anal. Chem.* **2023**, 1–22. [CrossRef] [PubMed]
32. Osorio-Tobón, J.F. Recent Advances and Comparisons of Conventional and Alternative Extraction Techniques of Phenolic Compounds. *J. Food Sci. Technol.* **2020**, *57*, 4299–4315. [CrossRef]
33. Chatzimitakos, T.; Athanasiadis, V.; Kotsou, K.; Mantiniotou, M.; Kalompatsios, D.; Makrygiannis, I.; Bozinou, E.; Lalas, S.I. Optimization of Pressurized Liquid Extraction (PLE) Parameters for Extraction of Bioactive Compounds from Moringa Oleifera Leaves and Bioactivity Assessment. *Int. J. Mol. Sci.* **2024**, *25*, 4628. [CrossRef] [PubMed]
34. Chatzimitakos, T.; Athanasiadis, V.; Kalompatsios, D.; Kotsou, K.; Mantiniotou, M.; Bozinou, E.; Lalas, S.I. Sustainable Valorization of Sour Cherry (*Prunus Cerasus*) By-Products: Extraction of Antioxidant Compounds. *Sustainability* **2024**, *16*, 32. [CrossRef]
35. Thoo, Y.; Ng, S.Y.; Khoo, M.; Mustapha, W.; Ho, C. A Binary Solvent Extraction System for Phenolic Antioxidants and Its Application to the Estimation of Antioxidant Capacity in *Andrographis Paniculata* Extracts. *Int. Food Res. J.* **2013**, *20*, 1103–1111.
36. Awad, A.M.; Kumar, P.; Ismail-Fitry, M.R.; Jusoh, S.; Ab Aziz, M.F.; Sazili, A.Q. Green Extraction of Bioactive Compounds from Plant Biomass and Their Application in Meat as Natural Antioxidant. *Antioxidants* **2021**, *10*, 1465. [CrossRef] [PubMed]
37. Dirar, A.I.; Alsaadi, D.H.M.; Wada, M.; Mohamed, M.A.; Watanabe, T.; Devkota, H.P. Effects of Extraction Solvents on Total Phenolic and Flavonoid Contents and Biological Activities of Extracts from Sudanese Medicinal Plants. *S. Afr. J. Bot.* **2019**, *120*, 261–267. [CrossRef]
38. Šeruga, M.; Novak, I.; Jakobek, L. Determination of Polyphenols Content and Antioxidant Activity of Some Red Wines by Differential Pulse Voltammetry, HPLC and Spectrophotometric Methods. *Food Chem.* **2011**, *124*, 1208–1216. [CrossRef]
39. Alessandri, S.; Ieri, F.; Romani, A. Minor Polar Compounds in Extra Virgin Olive Oil: Correlation between HPLC-DAD-MS and the Folin-Ciocalteu Spectrophotometric Method. *J. Agric. Food Chem.* **2014**, *62*, 826–835. [CrossRef]
40. Villamil-Galindo, E.; Piagentini, A. Green Solvents for the Recovery of Phenolic Compounds from Strawberry (*Fragaria* x *Ananassa* Duch) and Apple (*Malus Domestica*) Agro-Industrial Bio-Wastes. *Rev. Fac. Cienc. Agrar. UNCuyo* **2024**, *56*, 1. [CrossRef]
41. Plaskova, A.; Mlcek, J. New Insights of the Application of Water or Ethanol-Water Plant Extract Rich in Active Compounds in Food. *Front. Nutr.* **2023**, *10*, 1118761. [CrossRef] [PubMed]
42. Geow, C.H.; Tan, M.C.; Yeap, S.P.; Chin, N.L. A Review on Extraction Techniques and Its Future Applications in Industry. *Eur. J. Lipid Sci. Technol.* **2021**, *123*, 2000302. [CrossRef]
43. Gil-Martín, E.; Forbes-Hernández, T.; Romero, A.; Cianciosi, D.; Giampieri, F.; Battino, M. Influence of the Extraction Method on the Recovery of Bioactive Phenolic Compounds from Food Industry By-Products. *Food Chem.* **2022**, *378*, 131918. [CrossRef] [PubMed]
44. Kunradi Vieira, F.G.; da Silva Campelo Borges, G.; Copetti, C.; Valdemiro Gonzaga, L.; da Costa Nunes, E.; Fett, R. Activity and Contents of Polyphenolic Antioxidants in the Whole Fruit, Flesh and Peel of Three Apple Cultivars. *Arch. Latinoam. De Nutr.* **2009**, *59*, 101–106.
45. Liu, H.; Liu, Z.; Wu, Y.; Zheng, L.; Zhang, G. Regulatory Mechanisms of Anthocyanin Biosynthesis in Apple and Pear. *Int. J. Mol. Sci.* **2021**, *22*, 8441. [CrossRef]
46. Zhang, X.; Xu, J.; Xu, Z.; Sun, X.; Zhu, J.; Zhang, Y. Analysis of Antioxidant Activity and Flavonoids Metabolites in Peel and Flesh of Red-Fleshed Apple Varieties. *Molecules* **2020**, *25*, 1968. [CrossRef] [PubMed]
47. Kondo, S.; Tsuda, K.; Muto, N.; Ueda, J. Antioxidative Activity of Apple Skin or Flesh Extracts Associated with Fruit Development on Selected Apple Cultivars. *Sci. Hortic.* **2002**, *96*, 177–185. [CrossRef]
48. Karaman, Ş.; Tütem, E.; Başkan, K.S.; Apak, R. Comparison of Antioxidant Capacity and Phenolic Composition of Peel and Flesh of Some Apple Varieties. *J. Sci. Food Agric.* **2013**, *93*, 867–875. [CrossRef]

Disclaimer/Publisher's Note: The statements, opinions and data contained in all publications are solely those of the individual author(s) and contributor(s) and not of MDPI and/or the editor(s). MDPI and/or the editor(s) disclaim responsibility for any injury to people or property resulting from any ideas, methods, instructions or products referred to in the content.

Article

Gasification of Lignocellulosic Waste in Supercritical Water: Study of Thermodynamic Equilibrium as a Nonlinear Programming Problem

Julles Mitoura dos Santos Junior * and Adriano Pinto Mariano

Faculdade de Engenharia Química, Universidade Estadual de Campinas (UNICAMP), Av. Albert Einstein 500, Campinas 13083-852, Brazil; adpm@unicamp.br
* Correspondence: j264216@dac.unicamp.com.br; Tel.: +55-(98)-99106-0212

Abstract: As one of the main industrial segments of the current geoeconomics scenario, agro-industrial activities generate excessive amounts of waste. The gasification of such waste using supercritical water (SCWG) has the potential to convert the waste and generate products with high added value, hydrogen being the product of greatest interest. Within this context, this article presents studies on the SCWG processes of lignocellulosic residues from cotton, rice, and mustard husks. The Gibbs energy minimization (minG) and entropy maximization (maxS) approaches were applied to evaluate the processes conditioned in isothermal and adiabatic reactors, respectively. The thermodynamic and phase equilibria were written as a nonlinear programming problem using the *Peng–Robinson* state solution for the prediction of fugacity coefficients. As an optimization tool, TeS (Thermodynamic Equilibrium Simulation) software v.10 was used with the help of the *trust-constr* algorithm to search for the optimal point. The simulated results were validated with experimental data presenting surface coefficients greater than 0.99, validating the use of the proposed modeling to evaluate reaction systems of interest. It was found that increases in temperature and amounts of biomass in the process feed tend to maximize hydrogen formation. In addition to these variables, the H_2/CO ratio is of interest considering that these processes can be directed toward the production of synthesis gas (syngas). The results indicated that the selected processes can be directed to the production of synthesis gas, including the production of chemicals such as methanol, dimethyl ether, and ammonia. Using an entropy maximization approach, it was possible to verify the thermal behavior of reaction systems. The maxS results indicated that the selected processes have a predominantly exothermic character. The initial temperature and biomass composition had predominant effects on the equilibrium temperature of the system. In summary, this work applied advanced optimization and modeling methodologies to validate the feasibility of SCWG processes in producing hydrogen and other valuable chemicals from agro-industrial waste.

Keywords: lignocellulosic biomass; gasification in supercritical water; hydrogen

Citation: dos Santos Junior, J.M.; Mariano, A.P. Gasification of Lignocellulosic Waste in Supercritical Water: Study of Thermodynamic Equilibrium as a Nonlinear Programming Problem. *Eng* **2024**, *5*, 1096–1112. https://doi.org/10.3390/eng5020060

Academic Editor: Antonio Gil Bravo

Received: 29 April 2024
Revised: 5 June 2024
Accepted: 7 June 2024
Published: 12 June 2024

Copyright: © 2024 by the authors. Licensee MDPI, Basel, Switzerland. This article is an open access article distributed under the terms and conditions of the Creative Commons Attribution (CC BY) license (https:// creativecommons.org/licenses/by/ 4.0/).

1. Introduction

Fossil fuels are limited, and the urgent need to increase the use of renewable energy sources, along with the alarming daily release of CO_2 into the atmosphere, is a major concern. Consequently, various economic and environmental factors are driving a renewed interest in developing and improving manufacturing alternatives to reduce greenhouse gas (GHG) emissions and support modern society [1].

In this context, using agro-industrial waste, especially in Brazil, presents an appealing alternative due to the country's high production rates and the large amount of waste generated. Lignocellulosic residues, in particular, are produced in significant volumes through various processes. Examples include sugarcane bagasse from the sugar and ethanol industries, as well as residues from other production chains like rice, cotton, and mustard [2,3].

These wastes can be used in various value-added processes, including thermochemical processes, especially those aimed at hydrogen production. Hydrogen has a higher energy density compared to common and widely used fossil fuels, such as natural gas, with a key difference that hydrogen can be obtained from renewable sources and has water as its only oxidation product—thus contributing to avoiding climate change and reducing GHG emissions. A variety of such processes have been under investigation over the years, such as the thermochemical conversion of agro-industrial residues into hydrogen [4].

One pathway for producing hydrogen using thermochemical processes from biomass materials involves supercritical water (SCW). This is a promising reaction medium for organic components due to its unique physicochemical properties that at temperatures above 374 °C and pressures above 22.1 MPa allow overlap with the critical point of water, weakening the hydrogen bonds and resulting in a high dielectric constant and strong solubility of organic compounds such as biomass [5].

The composition of gasified biomass is primarily influenced by the reaction temperature [6]. There are three key reactions involved in the gasification of lignocellulosic biomass, as shown in Equations (1)–(3):

$$C + H_2O \rightarrow CO + H_2 \quad \text{Endothermic} \tag{1}$$

$$CO + H_2O \rightarrow CO_2 + H_2 \quad \text{Athermic} \tag{2}$$

$$C + 3H_2 \rightarrow CH_4 + H_2O \quad \text{Exothermic} \tag{3}$$

The thermal characteristics of Equations (1)–(3) were evaluated at a reference temperature of 298 K. Under SCWG conditions, the system exhibits low viscosity, which enhances mass transfer. This improved mixing of biomass with water hinders polymerization reactions, leading to the formation of simpler compounds, such as H_2, and increases gasification efficiency [7,8]. In biomass conversion processes, water can be said to fulfill all possible functions it can act as a solvent, catalyst, catalyst precursor, and reagent [9].

In Freitas and Guirardello [10], sugarcane bagasse was converted into syngas using SCW media with CO_2 as a co-reactant. This study was performed using the Gibbs energy minimization method in combination with the virial Equation of State (EoS). In Mitoura et al. [11], the methane cracking process was verified using ideal gas consideration in a thermodynamic study based on Gibbs energy minimization and entropy maximization methods. In Freitas and Guirardello [12], various glycerol (residue of biodiesel production) reforming technologies were studied to produce hydrogen. In all of these studies, the systems were thermodynamically considered using the Gibbs energy minimization method in combination with the virial EoS. Furthermore, the systems were thermodynamically evaluated using simple thermodynamic models (ideal gas and virial EoS); meanwhile, in other studies in the literature more complex thermodynamic models were used, such as Peng Robinson's state research. But in these cases, a stoichiometric approach (which does not consider all possible reactions) was used, as in Castello and Fiori [13].

This work addresses gasification processes using agro-industrial waste (composed of cotton, mustard and rice husk) using supercritical water as a reaction medium. Thermodynamic models based on Gibbs energy minimization and entropy maximization methods combined with the Peng–Robinson (PR) equation of state (EoS) were used to represent the phase behaviors of these complex systems. The models were formatted as optimization routines as non-linear programming problems using a non-stoichiometric approach (considering all possible reactions).

Process variables, such as temperature, pressure, and inlet composition were also evaluated to maximize hydrogen production and study the thermal behavior of all processes. The results of this work provide a complete thermodynamic analysis of this type of system, including the development of a robust, reliable, and efficient thermodynamic model. Moreover, its computational time is low despite the use of complex state equations such as the PR EoS.

2. Methodology

2.1. Thermodynamic Approach

To predict the biomass gasification process in isothermal reactors, the thermodynamic Gibbs energy minimization method was employed. In this method, a thermodynamic equilibrium is reached when the system's total Gibbs free energy is minimized, making it a common objective function for evaluating equilibrium processes [6,14]. To verify the thermal behavior of the reaction process, the entropy maximization method was utilized [15].

The use of Gibbs energy minimization and entropy maximization methodologies is crucial for evaluating the thermodynamic behavior of SCWG processes. MinG predicts equilibrium states by minimizing Gibbs energy, offering insights into isothermal reactor performance, while maxS assesses thermal behavior by maximizing entropy, key for adiabatic reactor dynamics. Both methods were formulated as nonlinear programming tasks with appropriate constraints. The following sections detail the thermodynamic approach for lignocellulosic biomass SCWG processes.

2.1.1. Chemical and Phase Equilibrium Formulated as a Gibbs Energy Minimization Problem: Calculation of Isothermal Reactor

The Gibbs energy of a system at constant temperature (T) and pressure (P) is described by Equation (4). In this context, the index i represents the different chemical species present, and k represents the different phases present:

$$G = \sum_{i=1}^{NC} \sum_{k=1}^{NF} n_i^k \mu_i^k \qquad (4)$$

According to Sandler [16], the chemical potential can be defined as shown in Equation (5):

$$\mu_i^k = \mu_i^o + RT \ln(\hat{f}_i^k / f_i^o) \qquad (5)$$

Fugacity in a multicomponent and multiphase system can be determined using the Phi–Phi (φ–φ) method, which is more suitable for handling cubic state equations. Thus, the objective function for minimizing the total Gibbs energy of the system, at constant pressure and temperature, can be rewritten as Equation (6):

$$G = \sum_{i=1}^{NC} \sum_{k=1}^{NF} n_i^k \mu_i^k \qquad (6)$$

By directly minimizing Equation (6), while taking into account mass balance and stoichiometry constraints, a unified chemical and phase equilibrium point is attained. To ensure the system reaches a suitable solution, two additional constraints must be incorporated. The first constraint involves ensuring the non-negativity of the number of moles, as outlined in Equation (7), for each component within each phase [10].

Gasification using supercritical water as a reactive medium occurs under conditions of high pressure and temperature. It is assumed that no components form in the liquid phase; however, both phases are still considered in the modeling process. To simplify the thermodynamic modeling, the solid phase is treated as ideal, as indicated by Equation (7), thus eliminating the need to estimate non-idealities. This approach seems reasonable given the significant amounts of water introduced into the reactive system during supercritical water gasification, which hinders the formation of solid phase components [10,12,17].

$$\mu_i^s = \mu_i^o \qquad (7)$$

In contrast to the assumption of ideality for the solid phase, the vapor and liquid phases cannot be considered ideal due to the process conditions, which make such an assumption impractical. Equations (8) and (9) describe the chemical potentials of the vapor and liquid phases, respectively:

$$\mu_i^v = \mu_i^0 + RT(\ln \hat{\phi}_i^v + ln y_i + \ln P) \quad (8)$$

$$\mu_i^l = \mu_i^0 + RT(\ln \hat{\phi}_i^l + ln x_i + \ln P_i^{sat}) \quad (9)$$

For predicting fugacity coefficients in reaction systems under supercritical conditions, the selection of equations of state (EoS) must be done carefully. There are reports of successful applications of the Peng–Robinson (PR) and Peng–Robinson Boston Mathias (PR-BM) EoS in studying supercritical biomass-in-water gasification reaction systems [6,18,19]. In this study, the Peng–Robinson Palatino Linotype EoS was applied to calculate the fugacity coefficients of the verified system, using simplified mixing rules as reported by Downling et al. [20].

2.1.2. Chemical and Phase Equilibrium Formulated as an Entropy Maximization Problem: Calculation of an Adiabatic Reactor

The second stage of this work involves applying the entropy maximization methodology, as described in Equation (10), for the thermodynamic evaluation of gasification reactions with supercritical water from various biomass sources. These methodologies have been applied to a wide range of systems in the field of chemical reaction engineering, including systems that transform biomass into fuels [10] and methane cracking processes [11].

$$\max S = \sum_{i=1}^{NC} \sum_{k=1}^{NF} n_i^k \bar{S}_i^k \quad (10)$$

These problems are subject to the restrictions of Equations (8) and (9), which pertain to the non-negativity of the number of moles and the balance of atoms, respectively, as well as the minimization of Gibbs energy. Additionally, they must comply with the enthalpy balance restriction, as outlined in Equation (11):

$$\sum_{i=1}^{NC} \sum_{k=1}^{NF} n_i^k \bar{H}_i^k = \sum_{i=1}^{NC} n_i^0 \bar{H}_i^0 = H^0 \quad (11)$$

2.2. Calculation of Fugacity Coefficients Using the Peng-Robinson Equation

The Peng–Robinson (PR) equation of state was utilized to assess the non-ideality of the system. This equation is noted for its balance of simplicity and accuracy. It was developed to perform as well as or better than the Soave–Redlich–Kwong (SRK) equation by altering the attractive pressure term of the semi-empirical Van der Waals equation [21]. Equations of state can be formulated as cubic equations in terms of the compressibility factor Z, commonly represented by Equation (12):

$$f(Z) = Z^3 - (1 + B - uB)Z^2 + (A + wB^2 - uB - uB^2)Z - AB - wB^2 - wB^3 \quad (12)$$

where A and B are dimensional coefficients that vary with temperature, pressure, and phase composition according to the classical mixing rule described by Downling et al. [20]. The parameters u and w are set to 2 and -1, respectively, as specified in the Peng–Robinson equation of state. Equation (12) approximates the real behavior of the liquid and vapor phases for various fluids. Solving this equation yields either one or three real roots, which are then used to calculate the fugacity coefficients using the phi–phi method applied in this study [16].

2.3. Mathematical Formulation and Solution of the Equilibrium Problem

In their research, Kamath, Biegler, and Grossmann [22] established criteria for the cubic equation of state to ensure accurate root selection. They found that the first derivative with respect to Z must be positive to avoid selecting the root's mean value. Additionally, the second derivative plays a crucial role in determining the roots of the liquid and vapor phases. The larger root corresponds to the vapor phase, and its second derivative must be greater than or equal to zero. Conversely, the smaller root, representing the liquid phase, must have a second derivative less than or equal to zero. Equations (13)–(16) present these constraints specifically for the Peng–Robinson equation:

$$f'(Zg) = 3Zg^2 - 2(1-B)Zg + A - 2B - 3B^2 \geq 0 \qquad (13)$$

$$f'(Zl) = 3Zl^2 - 2(1-B)Zl + A - 2B - 3B^2 \geq 0 \qquad (14)$$

$$f''(Zg) = 6Zg + 2B - 2 \geq -M\sigma^g \qquad (15)$$

$$f''(Zl) = 6Zl + 2B - 2 \geq -M\sigma^l \qquad (16)$$

To prevent the selection of a root lacking physical significance, Kamath, Biegler, and Grossmann [22] introduced slack variables (σ^v and σ^l) into Equations (15) and (16). This strategy mitigates discontinuities caused by the disappearance of one of the system phases, leaving only either the gaseous or liquid phase.

A total of 12 potential components were taken into account during the biomass SCWG process. These components, along with their thermodynamic properties, are detailed in Table 1. They were chosen to represent the main compounds expected to emerge during the supercritical water biomass gasification. This selection stems from extensive literature findings, which suggest that these components are commonly present in notable quantities during gasification processes involving biomass from diverse origins [6,10,12,17,23,24].

The lignocellulosic residues were written as pseudocomponents. Table 2 shows the analysis for each biomass evaluated in this work, as well as their key parameters such as hydrogen/carbon and oxygen/carbon ratios. Kang et al. [25] and Vassilev et al. [26] reported the ultimate analysis for many agroindustrial residues.

Table 1. Critical properties and formation properties reported by Poling et al. [27].

Components	T_c (K)	P_c (bar)	V_c (m^3/kmol)	ω	ΔH_f (cal/mol)	ΔG_f (cal/mol)
H_2O	647.14	220.64	0.056	0.344	-5.78×10^4	-5.46×10^4
H_2	32.98	12.93	0.064	-0.217	0	0
CH_4	190.56	45.99	0.099	0.011	-1.78×10^4	-1.21×10^4
CO_2	304.15	73.74	0.094	0.225	-9.41×10^4	-9.43×10^4
CO	132.85	34.94	0.093	0.045	-2.64×10^4	-3.28×10^4
O_2	154.58	50.43	0.073	0.022	0	0
N_2	126.20	33.98	0.090	0.037	0	0
CH_4O	512.64	80.97	0.118	0.565	-4.80×10^4	-3.88×10^4
C_2H_6	305.32	48.72	0.146	0.099	-2.00×10^4	-7.61×10^3
C_3H_8	369.83	42.48	0.200	0.152	-2.50×10^4	-5.81×10^3
NH_3	405.40	113.53	0.072	0.257	-1.10×10^4	-3.92×10^3
C_2H_4	282.34	50.41	0.131	0.087	1.25×10^4	1.64×10^4

Table 2. Mass composition of different types of biomass.

Biomass	C	H	N	O	H/C *	O/C *	Reference
Rice husk	49.3	6.1	0.8	43.7	1.48	0.59	[25]
Soy husk	45.4	6.7	0.9	46.9	1.77	0.69	[25]
Mustard husk	45.8	9.2	0.4	44.4	2.41	0.65	[25]
Cotton husk	50.4	8.4	1.4	39.8	2.00	0.53	[26]

* The ratios of hydrogen to carbon and oxygen to carbon are represented on a molar basis.

The formulated nonlinear programming (NLP) problems were addressed using the TeS—Thermodynamic Equilibrium Simulation software and the *trust-constr* solver. This choice is informed by the numerous advantages offered by the *trust-constr* solver for the type of approach employed in this study. Specifically, it is suitable for models characterized by highly nonlinear constraints, designed to handle large-scale models, and can be applied to models without differentiable functions [28]. This approach has consistently demonstrated high levels of accuracy and efficiency and has yielded excellent results in a variety of systems, particularly those involving chemical equilibrium and mixed phases [12,17,29,30].

3. Results and Discussion

3.1. Methodology Validation

This section aims to present results that support the application of the proposed methodology. The methodologies presented in Section 2 were validated using experimental data reported by Chakinala et al. [31] and Gomes et al. [6] who verified the SCWG processes of microalgal biomass in isothermal and adiabatic systems.

3.1.1. Methodology Validation for Isothermal Systems Using Gibbs Energy Minimization

Chakinala et al. [31] investigated the SCWG process of *Chlorella vulgaris* microalgal biomass using Ru/TiO_2 catalysts. This study was conducted at 240 bar, with the biomass feed set at 7.3% wt. Figure 1 presents a comparison between the results obtained by applying the previously described Gibbs energy minimization methodology, using the Peng–Robinson equation of state and the ideal gas model. The calculated results are plotted alongside those reported by Chakinala et al. [31] at 973.15 K.

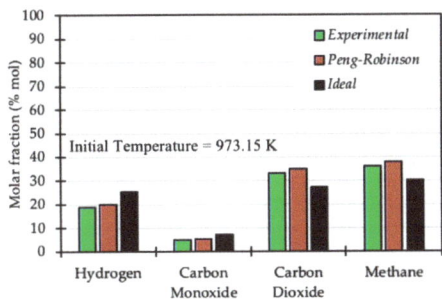

Figure 1. Comparison between molar composition results at the thermodynamic equilibrium condition calculated by the Gibbs energy minimization methodology (Peng–Robinson and Ideal Gas) with experimental data reported by Chakinala et al. [31].

The results in Figure 1 indicate a noticeable divergence between the calculated data and the literature when using the ideal gas model, which is expected given the nature of the reaction system and the thermodynamic limitations of the ideal gas model.

When the gas is considered real, by using the Peng–Robinson (PR) equation of state to calculate the fugacity coefficients the calculated results align satisfactorily with those reported by Chakinala et al. [31], achieving a correlation coefficient (R^2) greater than 0.999. This outcome underscores the necessity of employing more robust models to predict non-idealities. The PR equation offers better predictions than ideal models due to its high precision within the utilized temperature and pressure range and its accurate calculation of fugacity for the components present [32]. Furthermore, the proposed methodology proved effective for verifying the SCWG process of the *C. vulgaris* microalgal biomass. In terms of process design, robust models like the one presented in this work are essential for the development and scale-up of gasification processes with water in a supercritical state.

3.1.2. Methodology Validation for Adiabatic Systems Using Entropy Maximization

To validate this methodology, we used experimental data reported by Gomes et al. [6] who verified the SCWG process of *Nannochloropsis* sp. at 250 bar with 15% wt of biomass in the process feed. Figure 2 presents a comparison between data reported by Gomes et al. [6] and data calculated using the methodology described in this study for calculating adiabatic systems.

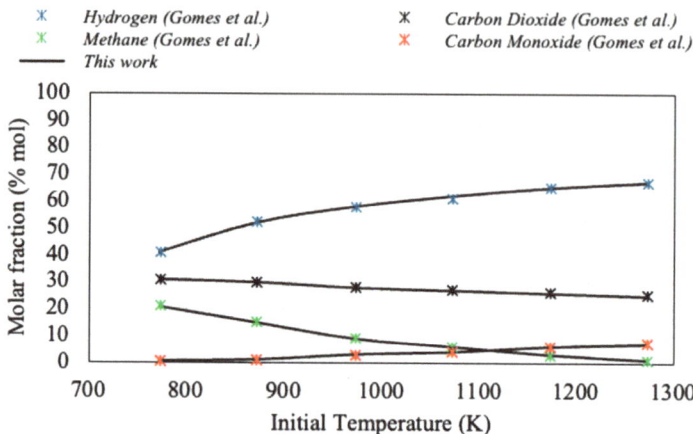

Figure 2. Comparisons between data from Gomes et al. [6] and simulated data for product composition as a function of initial temperature, pressure at 250 bar and initial composition of *Nannochloropsis* sp. 15% wt.

The results reported in Figure 2 indicate that the methodology proposed in this study has a good correlation with the data used for validation ($R^2 > 0.99$). Thus, considering both results from the verified literature, the proposed methodologies produced satisfactory results. The next section presents a more detailed study of the behavior of the SCWG process of the selected lignocellulosic residues.

3.2. Study of Equilibrium Compositions of SCWG Processes of Lignocellulosic Waste

The approach presented in this paper for Gibbs energy minimization mimics the conditions of an ideal isothermal reactor. Therefore, hydrogen production rates are anticipated to be high given that hydrogen generation reactions are mainly endothermic and benefit from the thermal stability provided by the isothermal reactor [6,11,33]. The thermodynamic study of this process was conducted under controlled conditions, with temperatures ranging from 723 to 1223 K and pressures from 200 to 300 bar. The proportions of biomass in the process input were maintained between 26% and 53%, with a constant amount of water of 2.0 mol.

Figure 3 shows the formation behavior of the main components throughout the SCWG processes of the verified lignocellulosic waste (hydrogen, methane, carbon monoxide, and carbon dioxide). However, other components were formed in smaller quantities ($<5 \times 10^{-4}$ mol) for all verified reaction conditions, such as NH_3, N_2, and C_2H_6. As the model is based on chemical and phase equilibrium, the molar fractions of these components are orders of magnitude lower than the majority of products, regardless of the conditions used. Likewise, the impact of these components on the formation reactions of the main gaseous products is low, even though the biomass obtained from food production contains relative concentrations of elements such as nitrogen or sulfur [18].

The results in Figure 3 show the combined effects of temperature and biomass mass composition in the process feed on the formation of the majority of components at 257.14 bar. As expected, increases in temperature and amounts of biomass in the system feed tend to maximize the formation of hydrogen and carbon monoxide for both processes. This fact is justified by the endothermic nature of the gasification reaction. These results follow what was expected taking into account previous results already reported in the literature [6,13,18,34–36].

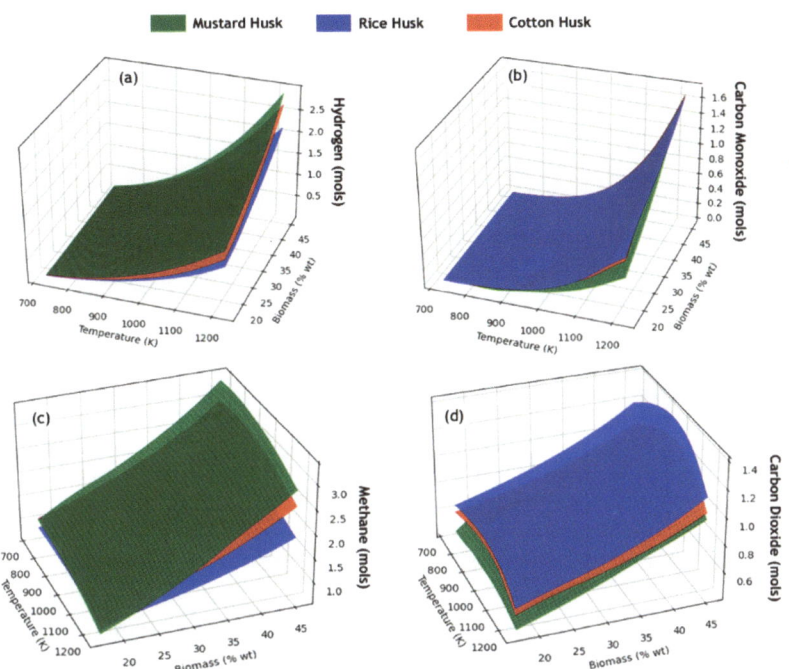

Figure 3. Behavior of the equilibrium compositions of hydrogen (**a**), carbon monoxide (**b**), methane (**c**), and carbon dioxide (**d**) as a function of temperature and mass composition of biomass in the reaction process feed at 257.14 bar.

In addition to the formation of hydrogen and carbon monoxide, throughout the SCWG processes verified, the formation of methane and carbon dioxide is significant. Contrary to the behavior of hydrogen formation from carbon monoxide, the formation of methane is disfavored by increases in temperature, and this result is justified by the fact that the methanation reaction is exothermic, as observed by Jin et al. [37] and Guo et al. [36]. Similarly to the formation of methane, the formation of carbon dioxide is favored with the addition of biomass in the feed of the reaction system. However, the formation of carbon dioxide presents maximum points at intermediate temperatures (800–900 K) for both processes.

It can be seen that the SCWG process of mustard husks presents higher rates of hydrogen and methane formation, this result is justified by the fact that this substrate presents a higher ratio of H/C atoms than the other substrates verified (H/C = 2.41), thus allowing more hydrogen and methane to be formed throughout the reaction process. Conversely, the rice husk SCWG process presents greater formation of carbon monoxide and dioxide due to its higher ratio of oxygen atoms to carbon atoms (O/C = 0.59) compared to other substrates.

Cotton husks present intermediate values for the O/C and H/C ratios, and, for this reason, the SCWG process for this substrate presents intermediate indices for the formation of both components among the processes verified.

In addition to the combined effects of temperature and the amount of substrate in feeding the reaction process, the influence of temperature on the behavior of the reactions is of great importance [17]. Figure 4 presents the combined effects of temperature and pressure on the behavior of the equilibrium compositions for SCWG processes of lignocellulosic waste, fixing the mass composition of the biomass in the process feed at 31.19 %wt.

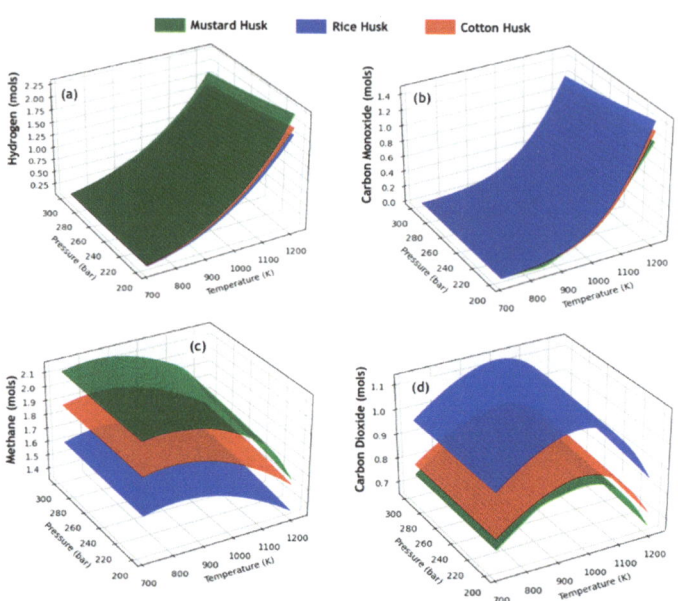

Figure 4. Behavior of the equilibrium compositions of hydrogen (**a**), carbon monoxide (**b**), methane (**c**), and carbon dioxide (**d**) as a function of temperature and pressure with the biomass composition set at 31.19 %wt in the reaction system feed.

For both conditions verified in Figure 4, the SCWG process of mustard husks presents higher rates of hydrogen and methane formation. The rice husk SCWG process presents higher rates of carbon monoxide and carbon dioxide formation. Quite apparent visually, pressure increases tend to minimize the formation of hydrogen and carbon monoxide. At 1223 K and 31.19%wt of biomass in the process feed, the pressure variation from 200 to 300 bar minimizes the formation of hydrogen by 17.21% and by 16.92% for the formation of carbon monoxide. Applying the same verification to the formation of methane and carbon dioxide, an increase in the formation of these components of 7.83% and 1.43%, respectively, is noted. This result was predicted by Whitag et al. [38], who described that the increase in pressure disfavors the formation of products of interest during the process. This is justified by the fact that the increase in pressure disfavors the water–gas displacement reactions and the methanation reaction is favored following *Le Chatelier's* principle. Thus, hydrogen is largely consumed, forming methane and carbon dioxide.

In short, from the results reported in Figures 3 and 4, it is concluded that an increase in both temperature and biomass in the feed of the reaction process tends to increase the number of moles of hydrogen and carbon monoxide. Although this behavior is observed, it is important to highlight that, in terms of molar composition, the hydrogen fractions decrease as a function of increasing feed concentration [39]. Furthermore, an increase in pressure tends to minimize the formation of these components. Higher rates of hydrogen formation are observed for the SCWG process of mustard husks due to its higher ratio of H/C atoms; considering that this is the component of greatest interest, among the lignocellulosic residues verified, mustard husks would be more suitable due to their higher rates of hydrogen formation.

In addition to hydrogen, the SCWG process can be characterized by the production of synthesis gas (syngas), which is predominantly composed of H_2 and CO [40–42]. An interesting parameter to consider when studying the potential for syngas formation is the relationship between the amounts of H_2 and CO (molar ratio H_2/CO) [43]. When the H_2/CO molar ratio is close to one, the generated syngas is favorable for the synthesis of methanol (CH_3OH) and the production of light hydrocarbons such as methane (CH_4) and ethylene (C_2H_4) using Fischer–Tropsch-type synthesis reactions. For H_2/CO molar ratios close to two, the generated synthesis gas is considered ideal for the production of methanol (CH_3OH) and dimethyl ether (DME), which are important as fuels and as intermediates in the chemical industry; when the H_2 ratio/CO is equal to three, the generated syngas is favorable to produce ammonia (NH_3) through the Haber–Bosch synthesis reaction [44]. Figure 5 shows the behavior of the H_2/CO molar ratio as a function of temperature and the amount of biomass in the process feed at 200 bar for both processes verified.

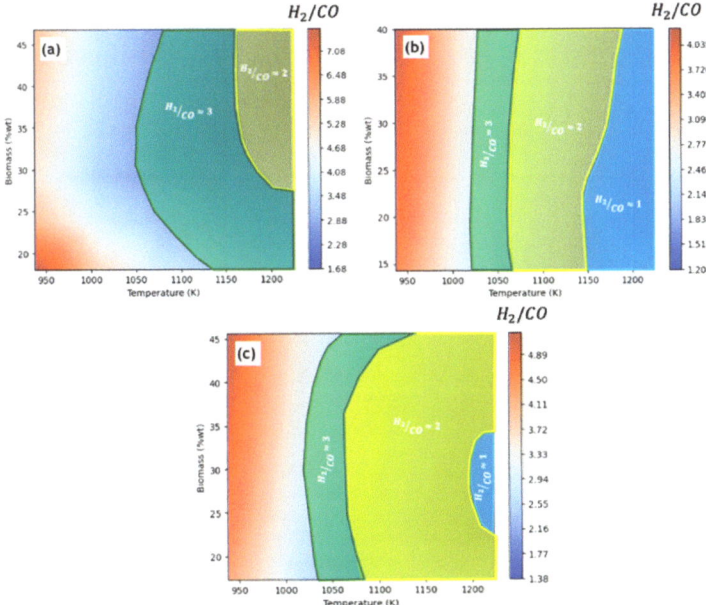

Figure 5. Behavior of the H_2/CO ratio as a function of temperature and biomass composition for the SCWG processes verified at 200 bar (**a**): mustard husks, (**b**): rice husks, (**c**): cotton husks.

The results of the H_2/CO molar ratios for both verified processes are presented in Figure 5 as a function of temperature and biomass composition in the process feed (%wt) at 200 bar.

The results in Figure 5 indicate a significant difference in the behavior of the H_2/CO ratio depending on the composition of the substrate used throughout the SCWG processes. The mustard husk SCWG process presents H_2/CO ratios greater than two, indicating that this process may be more suitable for producing synthesis gas as a substrate for DME production processes. This result is similar to the SCWG process of cotton husks. The rice husk SCWG process presents higher rates of synthesis gas formation with H_2/CO ratios close to one. A H_2/CO ratio close to one is ideal for producing long-chain hydrocarbons such as paraffins, naphthas, and oils. This is because this balanced proportion of hydrogen and carbon monoxide is necessary for the efficient synthesis of products through processes like the Fischer–Tropsch synthesis [45].

As observed, biomass processing using water as a reaction medium under supercritical conditions shows good hydrogen formation rates, and the H_2:CO ratios under conditions that maximize the formation of these products allow their application as synthesis gas. Despite the good formation rates of the desired products, operating reactions under supercritical conditions involves various associated costs. Alternative routes for biomass processing, such as pyrolysis, are conducted under reduced temperature and pressure conditions; however, they result in high CO_2 formation and the production of large quantities of solid components [46].

The different behaviors of the SCWG processes of the residues verified result from the different compositions of each substrate. Figure 6 shows the maximum formation of the majority of components formed throughout the SCWG processes as a function of the H/C and O/H ratios. The results presented correspond to conditions that maximize the formations of the respective verified components. Following what was presented in Figures 4 and 5, increases in the H/C ratio tend to maximize the formation of hydrogen and methane. Increases in the O/C ratio maximize the formation of carbon dioxide and carbon monoxide. However, this result is barely noticeable for the formation of carbon monoxide because, for the conditions that maximize the formation of this component, the differences between the quantities formed for the processes verified are minimal. Both results seen in Figure 6 make sense if analyzed by taking into account the main reactions throughout the biomass SCWG processes.

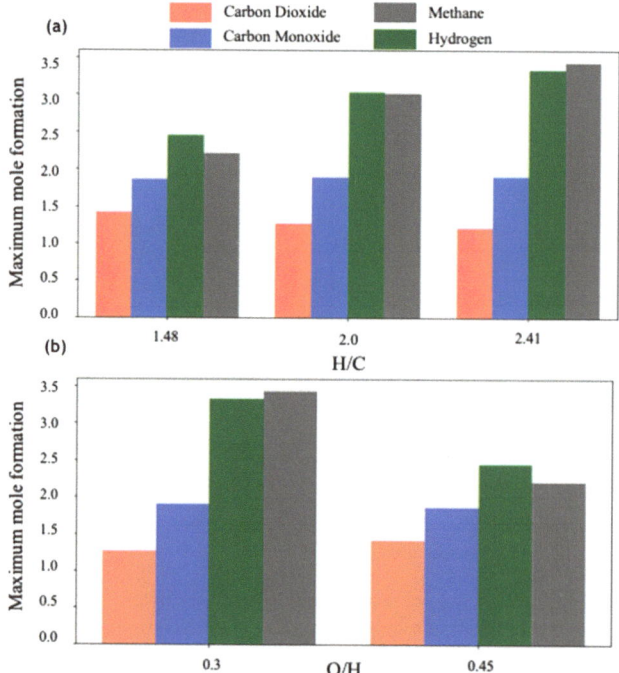

Figure 6. Behavior of the maximum formation of the main products of the SCWG processes of the waste verified as a function of the H/C (**a**) and O/H ratios (**b**).

3.3. Thermal Behavior of SCWG Processes of Lignocellulosic Waste

In addition to equilibrium compositions, the study of the thermal behavior of reaction systems is of fundamental importance for project development. To verify the equilibrium temperatures in non-isothermal systems, the entropy maximization methodology associated with the Peng–Robinson equation of state was used to calculate non-idealities. This methodology has already been used by Gomes et al. [6] for verifying the thermal behavior of reaction systems under supercritical conditions with satisfactory results validated with experimental data.

Figure 7 illustrates the thermal behavior of SCWG processes for lignocellulosic waste. The effects studied are: initial temperature (Figure 7a) with pressure at 220 bar and 10% wt biomass feed; pressure (Figure 7b) with initial temperature at 900 K and 10% wt biomass feed; and biomass composition (Figure 7c) with initial temperature at 900 K and pressure at 220 bar.

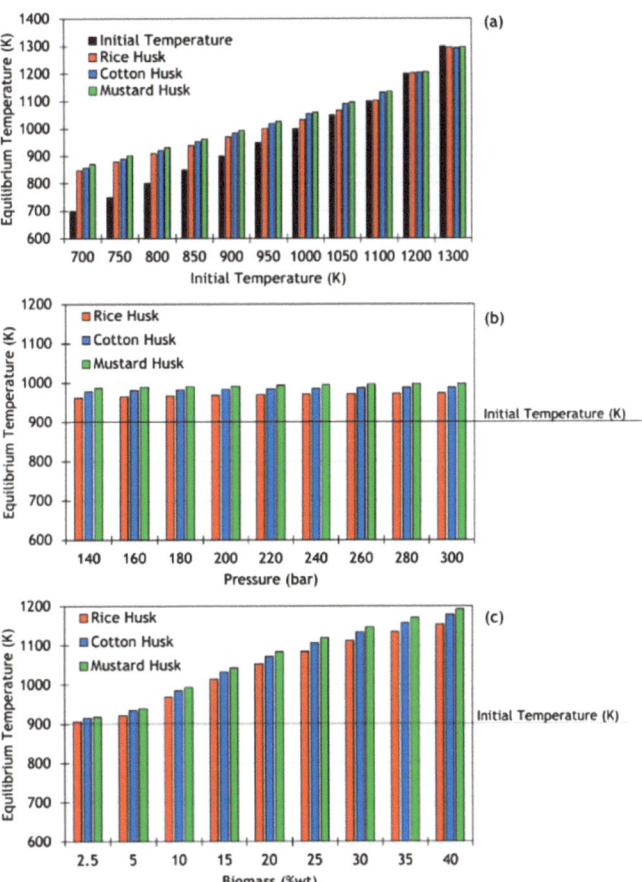

Figure 7. Thermal behavior of reaction systems as a function of initial temperature (**a**), pressure (**b**), and initial biomass composition (**c**).

From the results presented in Figure 7, it is clear that the initial temperature and the biomass composition in the feed of the reaction system have predominant effects on the behavior of the equilibrium temperature. This verification is in accordance with the results shown in Figures 3 and 4, considering that pressure has less significant effects on the development of the reactions.

Checking the effect of the initial temperature on the behavior of the equilibrium temperature (Figure 7a), it is clear that, in general, the SCWG process of mustard husks presents higher equilibrium temperatures than the other processes. This result is justified by the fact that higher H/C ratios tend to favor methanation reactions, which are exothermic, thus increasing the equilibrium temperature of the reaction system. Similarly, the rice husk SCWG process presents lower values for equilibrium temperatures, since this substrate has the lowest H/C ratio (1.48) among the substrates studied.

For both cases, exothermic effects are predominant. However, it can be seen that the differences between the equilibrium temperatures and the initial temperature of the reaction system tend to decrease with increases in temperature at the beginning of the process, indicating that the reactions initially have an exothermic character that is minimized with an increase in the initial temperature of the reaction system. At 1200 K, the differences between the initial temperature and the equilibrium temperatures for both reaction processes are minimal, as can be confirmed in Table 3. At 1300 K, it is clear that the equilibrium temperatures become lower than the initial temperature, which indicates the endothermic effect becomes predominant.

Table 3. Variation of the equilibrium temperature as a function of the initial temperature for the SCWG processes of lignocellulosic waste at 220 bar with biomass composition in the feed set at 10% wt.

Initial Temperature (K)	Temperature Deviation (%)		
	Rice Husk	Cotton Husk	Mustard Husk
700	+17.5	+18.4	+19.6
750	+14.7	+15.7	+16.8
800	+12.0	+13.1	+14.1
850	+9.6	+10.8	+11.7
900	+7.2	+8.6	+9.4
950	+5.1	+6.8	+7.4
1000	+3.2	+5.2	+5.7
1050	+1.6	+3.8	+4.2
1100	+0.27	+2.8	+3.0
1200	+0.25	+0.34	+0.57
1300	−0.33	−0.56	−0.28

In addition to the effect of initial temperature, the results in Figure 7 indicate that pressure has a small, significant effect on the behavior of the equilibrium temperature. This result follows what was expected taking into account the behavior of the equilibrium compositions as a function of the system pressure. Other authors verified similar results when studying the effect of reaction system pressure on the thermal behavior of the process when conditioned in adiabatic reactors [6,11,33].

4. Conclusions

This study reveals the effectiveness of the proposed methodologies for evaluating the thermodynamic behavior of supercritical water gasification (SCWG) processes, both in validating the results and in analyzing the thermodynamic equilibrium and thermal behavior of the reaction systems. By utilizing more robust models such as the Peng–Robinson equation of state, it was possible to achieve a satisfactory correlation between the calculated and experimental data ($R^2 > 0.99$), highlighting the importance of considering gas non-idealities for more precise predictions.

The analysis of the thermodynamic behavior of the SCWG processes of lignocellulosic residues revealed consistent patterns in the formation of gaseous products such as hydrogen, methane, carbon monoxide, and carbon dioxide as a function of temperature, pressure, and biomass composition.

Significantly, biomass composition significantly influenced the composition of the formed products, with substrates like mustard husk exhibiting higher hydrogen and methane formation due to their higher H/C ratio. For the processes verified, the mustard husk SCWG process showed higher rates of hydrogen and methane formation, which is justified by the greater number of hydrogen atoms in its composition. Conversely, the rice SCWG process presents lower rates of hydrogen formation, which also presents higher rates of carbon monoxide and carbon dioxide formation due to its greater content of carbon atoms. The biomass of rice husks presents intermediate values with respect to the compositions of H and C, and for this reason it presents intermediate values for the compositions of the verified components (H_2, CH_4, CO and CO_2).

Moreover, the analysis of the H_2/CO molar ratio indicated the potential of SCWG processes for synthesis gas production, with important implications for the production of chemicals such as methanol, dimethyl ether, and ammonia, depending on the temperature conditions and biomass composition. The SCWG process of lignocellulosic waste showed good rates of synthesis gas formation with H_2/CO ratios between two and three, indicating that this can be directed to the production of ammonia and DME. The SCWG processes of rice and cotton husks also have H_2/CO indices close to one under specific conditions of temperature, pressure and biomass composition, indicating that the syngas generated from this process can be directed to the production of methanol.

Lastly, the investigation of the thermal behavior of SCWG processes highlighted the importance of initial temperature and biomass composition in determining the equilibrium temperature of the system, with pressure exerting a less significant effect. These insights are crucial for the development and scaling of supercritical water gasification processes, underscoring the potential of utilizing lignocellulosic residues for the production of high-value-added products.

5. Patents

The results presented in this article were developed using the TeS—Thermodynamic Equilibrium Simulation software. This software was developed by the authors of this text and this article marks the first publication using this tool. The TeS was registered by the National Institute of Industrial Property with registration number BR512024000275-8.

Author Contributions: J.M.d.S.J.: project proposal and methodology development. J.M.d.S.J.: research and validation. J.M.d.S.J.: development of results. J.M.d.S.J. and A.P.M.: constant evaluation of results. A.P.M.: supervision and guidance throughout the development of the article. All authors have read and agreed to the published version of the manuscript.

Funding: The authors of this text thank Fundação de Amparo à Pesquisa do Estado de São Paulo (FAPESP) (grant number 23/01072-7).

Institutional Review Board Statement: Not applicable.

Informed Consent Statement: Not applicable.

Data Availability Statement: The data will be made available if requested by readers. However, we provide free access to the thermodynamic simulation software applied in this study. TeS—Thermodynamic Equilibrium Simulation: https://drive.google.com/drive/folders/1z-5jygS4syg8-_W6xIUT_7sraYoKtF9z?usp=share_link accessed on 10 June 2024.

Acknowledgments: The authors extend their gratitude to the dedicated faculty at the University of Campinas and the Federal University of Maranhão for their invaluable contributions to the personal and professional growth of countless people, as well as to all educators who actively promote the advancement of society. A.P. Mariano thanks Fundação de Amparo à Pesquisa do Estado de São Paulo (FAPESP) (grant number 23/01072-7). Finally, J.M. Santos Junior expresses his gratitude to Radix Engineering and Software for generously providing the time and essential tools needed to create these results.

Conflicts of Interest: The authors declare no conflicts of interest.

Nomenclature

G	Total Gibbs energy	P	Pressure
l	Liquid phase	n	Number of moles
s	Solid phase	NC	Number of components
v	Vapor phase	NF	Number of phases
NC	Number of components	T	Temperature
NF	Number of phases	R	Universal gas constant
$\hat{f}_i^k$	Fugacity of component i in phase k	a_{mi}	Number of atoms of element i in component m
f_i^o	Fugacity of pure species i in a standard reference state	n_i^o	Number of moles in standard state
H_i^k	Enthalpy of component i in phase k	H^0	Total enthalpy
H_i^0	Enthalpy of component i in the standard state	μ_i^0	Chemical potential of component i in a standard reference state
y_i	Mole fraction of component i in the vapor phase	P_i^{sat}	Component saturation pressure i
x_i	Molar fraction of component i in the liquid phase	Z_i	Compressibility factor
a_m	Attraction parameter for mixtures	k_{ij}	Binary interaction parameter
b_m	Repulsion parameter for mixtures	A, B, u, w	Parameters of the cubic equation of state

M	Constant for Kamath, Biegler, and Grossmann constraints
σ^k	Slack variables for Kamath, Biegler, and Grossmann constraints
$T_{c,I}$	Critical component temperature i
$P_{c,i}$	Critical component pressure i
w_i	Acentric factor
n_i^k	Number of moles of component i in phase k; $i = [1, 2, 3, \ldots, NC]$; $k = [v, l, s]$
μ_i^k	Chemical potential of component i in phase k; $i = [1, 2, 3, \ldots, NC]$; $k = [v, l, s]$
$\hat{\varnothing}_i^k$	Coefficient of fugacity of component i in phase k; $i = [1, 2, 3, \ldots, NC]$; $k = [v, l]$

References

1. Dechamps, P. The IEA World Energy Outlook 2022—A brief analysis and implications. *Eur. Energy Clim. J.* **2023**, *11*, 100–103. [CrossRef]
2. Virmond, E.; De Sena, R.F.; Albrecht, W.; Althoff, C.A.; Moreira, R.F.P.M.; José, H.J. Characterisation of agroindustrial solid residues as biofuels and potential application in thermochemical processes. *Waste Manag.* **2012**, *32*, 1952–1961. [CrossRef] [PubMed]
3. Araújo, D.J.C.; Machado, A.V.; Vilarinho, M.C.L.G. Availability and Suitability of Agroindustrial Residues as Feedstock for Cellulose-Based Materials: Brazil Case Study. *Waste Biomass Valorization* **2019**, *10*, 2863–2878. [CrossRef]
4. Saravanan, A.; Kumar, P.S.; Aron, N.S.M.; Jeevanantham, S.; Karishma, S.; Yaashikaa, P.R.; Chew, K.W.; Show, P.L. A review on bioconversion processes for hydrogen production from agro-industrial residues. *Int. J. Hydrogen Energy* **2022**, *47*, 37302–37320. [CrossRef]
5. Khandelwal, K.; Nanda, S.; Boahene, P.; Dalai, A.K. Conversion of biomass into hydrogen by supercritical water gasification: A review. *Environ. Chem. Lett.* **2023**, *21*, 2619–2638. [CrossRef]
6. Gomes, J.G.; Mitoura, J.; Guirardello, R. Thermodynamic analysis for hydrogen production from the reaction of subcritical and supercritical gasification of the *C. vulgaris* microalgae. *Energy* **2022**, *260*, 125030. [CrossRef]
7. Osada, M.; Sato, T.; Watanabe, M.; Shirai, M.; Arai, K. Catalytic gasification of wood biomass in subcritical and supercritical water. *Combust. Sci. Technol.* **2006**, *178*, 537–552. [CrossRef]
8. Guo, L.; Jin, H. Boiling coal in water: Hydrogen production and power generation system with zero net CO2 emission based on coal and supercritical water gasification. *Int. J. Hydrogen Energy* **2013**, *38*, 12953–12967. [CrossRef]
9. Thiruvenkadam, S.; Izhar, S.; Yoshida, H.; Danquah, M.K.; Harun, R. Process application of Subcritical Water Extraction (SWE) for algal bio-products and biofuels production. *Appl. Energy* **2015**, *154*, 815–828. [CrossRef]
10. Freitas, A.C.D.; Guirardello, R. Use of CO_2 as a co-reactant to promote syngas production in supercritical water gasification of sugarcane bagasse. *J. CO2 Util.* **2015**, *9*, 66–73. [CrossRef]
11. Mitoura dos Santos Junior, J.; Gomes, J.G.; de Freitas, A.C.D.; Guirardello, R. An Analysis of the Methane Cracking Process for CO_2-Free Hydrogen Production Using Thermodynamic Methodologies. *Methane* **2022**, *1*, 243–261. [CrossRef]
12. Freitas, A.C.D.; Guirardello, R. Comparison of several glycerol reforming methods for hydrogen and syngas production using Gibbs energy minimization. *Int. J. Hydrogen Energy* **2014**, *39*, 17969–17984. [CrossRef]
13. Castello, D.; Fiori, L. Supercritical water gasification of biomass: Thermodynamic constraints. *Bioresour. Technol.* **2011**, *102*, 7574–7582. [CrossRef] [PubMed]
14. Ciuffi, B.; Chiaramonti, D.; Rizzo, A.M.; Frediani, M.; Rosi, L. A critical review of SCWG in the context of available gasification technologies for plastic waste. *Appl. Sci.* **2020**, *10*, 6307. [CrossRef]
15. Rossi, C.C.R.S.; Berezuk, M.E.; Cardozo-Filho, L.; Guirardello, R. Simultaneous calculation of chemical and phase equilibria using convexity analysis. *Comput. Chem. Eng.* **2011**, *35*, 1226–1237. [CrossRef]
16. Sandler, S.I. *Chemical, Biochemical, and Engineering Thermodynamics*; John Wiley & Sons: Hoboken, NJ, USA, 2017.
17. Barros, T.V.; Carregosa, J.D.; Wisniewski, A.W., Jr.; Freitas, A.C.; Guirardello, R.; Ferreira-Pinto, L.; Bonfim-Rocha, L.; Jegatheesan, V.; Cardozo-Filho, L. Assessment of black liquor hydrothermal treatment under sub- and supercritical conditions: Products distribution and economic perspectives. *Chemosphere* **2022**, *286*, 131774. [CrossRef] [PubMed]
18. Louw, J.; Schwarz, C.E.; Knoetze, J.H.; Burger, A.J. Thermodynamic modelling of supercritical water gasification: Investigating the effect of biomass composition to aid in the selection of appropriate feedstock material. *Bioresour. Technol.* **2014**, *174*, 11–23. [CrossRef] [PubMed]
19. Hantoko, D.; Yan, M.; Prabowo, B.; Susanto, H.; Li, X.; Chen, C. Aspen Plus Modeling Approach in Solid Waste Gasification. *Current Developments in Biotechnology and Bioengineering*; Elsevier: Amsterdam, The Netherlands, 2019; pp. 259–281. [CrossRef]
20. Dowling, A.W.; Balwani, C.; Gao, Q.; Biegler, L.T. Optimization of sub-ambient separation systems with embedded cubic equation of state thermodynamic models and complementarity constraints. *Comput. Chem. Eng.* **2015**, *81*, 323–343. [CrossRef]
21. Peng, D.-Y.; Robinson, D.B. A New Two-Constant Equation of State. *Ind. Eng. Chem. Fundam.* **1976**, *15*, 59–64. [CrossRef]
22. Kamath, R.S.; Biegler, L.T.; Grossmann, I.E. An equation-oriented approach for handling thermodynamics based on cubic equation of state in process optimization. *Comput. Chem. Eng.* **2010**, *34*, 2085–2096. [CrossRef]
23. Tang, H.; Kitagawa, K. Supercritical water gasification of biomass: Thermodynamic analysis with direct Gibbs free energy minimization. *Chem. Eng. J.* **2005**, *106*, 261–267. [CrossRef]
24. Basu, P.; Mettanant, V. Biomass Gasification in Supercritical Water—A Review. *Int. J. Chem. React. Eng.* **2009**, *7*. [CrossRef]

25. Kang, Q.; Appels, L.; Tan, T.; Dewil, R. Bioethanol from Lignocellulosic Biomass: Current Findings Determine Research Priorities. *Sci. World J.* **2014**, *2014*, 298153. [CrossRef]
26. Vassilev, S.V.; Baxter, D.; Andersen, L.K.; Vassileva, C.G. An overview of the chemical composition of biomass. *Fuel* **2010**, *89*, 913–933. [CrossRef]
27. Poling, B.E.; Prausnitz, J.M.; O'connell, J.P. *Properties of Gases and Liquids*; McGraw-Hill Education: New York, NY, USA, 2001.
28. Ćalasan, M.P.; Nikitović, L.; Mujović, S. CONOPT solver embedded in GAMS for optimal power flow. *J. Renew. Sustain. Energy* **2019**, *11*, 046301. [CrossRef]
29. Rocha, S.A.; Guirardello, R. An approach to calculate solid–liquid phase equilibrium for binary mixtures. *Fluid. Phase Equilib.* **2009**, *281*, 12–21. [CrossRef]
30. Voll, F.; Rossi, C.; Silva, C.; Guirardello, R.; Souza, R.; Cabral, V.; Cardozo-Filho, L. Thermodynamic analysis of supercritical water gasification of methanol, ethanol, glycerol, glucose and cellulose. *Int. J. Hydrogen Energy* **2009**, *34*, 9737–9744. [CrossRef]
31. Chakinala, A.G.; Brilman, D.W.F.; van Swaaij, W.P.; Kersten, S.R.A. Catalytic and Non-catalytic Supercritical Water Gasification of Microalgae and Glycerol. *Ind. Eng. Chem. Res.* **2010**, *49*, 1113–1122. [CrossRef]
32. Lopez-Echeverry, J.S.; Reif-Acherman, S.; Araujo-Lopez, E. Peng-Robinson equation of state: 40 years through cubics. *Fluid. Phase Equilib.* **2017**, *447*, 39–71. [CrossRef]
33. Dos Santos, J.M.; De Sousa, G.F.B.; Vidotti, A.D.S.; De Freitas, A.C.D.; Guirardello, R. Optimization of glycerol gasification process in supercritical water using thermodynamic approach. *Chem. Eng. Trans.* **2021**, *86*, 847–852. [CrossRef]
34. De Blasio, C.; Järvinen, M. *Supercritical Water Gasification of Biomass*. Encyclopedia of Sustainable Technologies; Elsevier: Amsterdam, The Netherlands, 2017; pp. 171–195. [CrossRef]
35. Rodriguez Correa, C.; Kruse, A. Supercritical water gasification of biomass for hydrogen production—Review. *J. Supercrit. Fluids* **2018**, *133*, 573–590. [CrossRef]
36. Yan, Q.; Guo, L.; Lu, Y. Thermodynamic analysis of hydrogen production from biomass gasification in supercritical water. *Energy Convers. Manag.* **2006**, *47*, 1515–1528. [CrossRef]
37. Ding, W.; Shi, J.; Wei, W.; Cao, C.; Jin, H. A molecular dynamics simulation study on solubility behaviors of polycyclic aromatic hydrocarbons in supercritical water/hydrogen environment. *Int. J. Hydrogen Energy* **2021**, *46*, 2899–2904. [CrossRef]
38. Withag, J.A.M.; Smeets, J.R.; Bramer, E.A.; Brem, G. System model for gasification of biomass model compounds in supercritical water—A thermodynamic analysis. *J. Supercrit. Fluids* **2012**, *61*, 157–166. [CrossRef]
39. Ferreira-Pinto, L.; Silva Parizi, M.P.; Carvalho de Araújo, P.C.; Zanette, A.F.; Cardozo-Filho, L. Experimental basic factors in the production of H_2 via supercritical water gasification. *Int. J. Hydrogen Energy* **2019**, *44*, 25365–25383. [CrossRef]
40. Luyben, W.L. Control of parallel dry methane and steam methane reforming processes for Fischer–Tropsch syngas. *J. Process Control* **2016**, *39*, 77–87. [CrossRef]
41. Cheng, C.K.; Foo, S.Y.; Adesina, A.A. H_2-rich synthesis gas production over Co/Al_2O_3 catalyst via glycerol steam reforming. *Catal. Commun.* **2010**, *12*, 292–298. [CrossRef]
42. Lundgren, J.; Ekbom, T.; Hulteberg, C.; Larsson, M.; Grip, C.-E.; Nilsson, L.; Tunå, P. Methanol production from steel-work off-gases and biomass based synthesis gas. *Appl. Energy* **2013**, *112*, 431–439. [CrossRef]
43. Saad, J.M.; Williams, P.T. Manipulating the H_2/CO ratio from dry reforming of simulated mixed waste plastics by the addition of steam. *Fuel Process. Technol.* **2017**, *156*, 331–338. [CrossRef]
44. Rostrup-Nielsen, J.; Christiansen, L.J. *Concepts in Syngas Manufacture*; World Scientific: Singapore, 2011; Volume 10.
45. Rahimpour, M.R.; Arab Aboosadi, Z.; Jahanmiri, A.H. Synthesis gas production in a novel hydrogen and oxygen perm-selective membranes tri-reformer for methanol production. *J. Nat. Gas. Sci. Eng.* **2012**, *9*, 149–159. [CrossRef]
46. Jia, D.; Liang, J.; Liu, J.; Chen, D.; Evrendilek, F.; Wen, T.; Cao, H.; Zhong, S.; Yang, Z.; He, Y. Insights into pyrolysis of ginger via TG-FTIR and Py-GC/MS analyses: Thermochemical behaviors, kinetics, evolved gas, and products. *J. Anal. Appl. Pyrolysis* **2024**, *179*, 106442. [CrossRef]

Disclaimer/Publisher's Note: The statements, opinions and data contained in all publications are solely those of the individual author(s) and contributor(s) and not of MDPI and/or the editor(s). MDPI and/or the editor(s) disclaim responsibility for any injury to people or property resulting from any ideas, methods, instructions or products referred to in the content.

Article

Screening of Azo-Dye-Degrading Bacteria from Textile Industry Wastewater-Activated Sludge

Grazielly Maria Didier de Vasconcelos [1], Isabela Karina Della-Flora [1,2], Maikon Kelbert [1], Lidiane Maria de Andrade [3], Débora de Oliveira [1], Selene Maria de Arruda Guelli Ulson de Souza [1], Antônio Augusto Ulson de Souza [1] and Cristiano José de Andrade [1,*]

[1] Department of Chemical Engineering and Food Engineering, Federal University of Santa Catarina, Florianópolis 88040-900, SC, Brazil; graziellydidier@gmail.com (G.M.D.d.V.); isa_dellaflora@hotmail.com (I.K.D.-F.); maikonkelbert@gmail.com (M.K.); debora.oliveira@ufsc.br (D.d.O.); selene.souza@ufsc.br (S.M.d.A.G.U.d.S.); antonio.augusto.souza@ufsc.br (A.A.U.d.S.)

[2] Institute of Biomaterials, Department of Materials Science and Engineering, University of Erlangen-Nuremberg, Cauerstrasse 6, 91058 Erlangen, Germany

[3] Dempster MS Lab, Chemical Engineering Department of Polytechnic School of University of São Paulo (USP), R. Do Lago, 250–Butantã, São Paulo 05338-110, SP, Brazil; lidiane.andrade@gmail.com

* Correspondence: eng.crisja@gmail.com; Tel.: +55-19-98154-3393

Abstract: This study investigates the biodegradation of Reactive Red 141 (RR 141), an azo dye prevalent in the textile industry, by bacteria isolated from activated sludge in a textile effluent treatment plant. RR 141, characterized by nitrogen–nitrogen double bonds (-N=N-), contributes to environmental issues when improperly disposed of in textile effluents, leading to reduced oxygen levels in water bodies, diminished sunlight penetration, and the formation of potentially carcinogenic and mutagenic aromatic amines. This research focuses on identifying bacteria from activated sludge with the potential to decolorize RR 141. Microbiological identification employs MALDI-TOF-MS, known for its precision and rapid identification of environmental bacteria, enhancing treatment efficiency. Results highlight *Bacillus thuringiensis* and *Kosakonia radicincitans* as the most promising strains for RR 141 decolorization. Analysis of micro-organisms in activated sludge and database exploration suggests a correlation between these strains and the decolorization process. It is worth noting that this is the first report on the potential use of *K. radicincitans* for azo dye decolorization. Three distinct culture media—BHI, MSG, and MS—were assessed to investigate their impact on RR 141 decolorization. Notably, BHI and MSG media, incorporating a carbon source, facilitated the bacterial growth of both tested species (*B. thuringiensis* and *K. radicincitans*), a phenomenon absent in the MS medium. This observation suggests that the bacteria exhibit limited capability to utilize RR 141 dye as a carbon source, pointing towards the influence of the culture medium on the discoloration process. The study evaluates performance kinetics, decolorization capacity through UV-VIS spectrophotometry, potential degradation pathways via HPLC-MS analysis, phytotoxicity, and enzymatic activity identification. *B. thuringiensis* and *K. radicincitans* exhibit potential in decolorizing RR141, with 38% and 26% removal individually in 120 h. As a consortium, they achieved 36% removal in 12 h, primarily through biosorption rather than biodegradation, as indicated by HPLC-MS analyses. In conclusion, the research emphasizes the importance of exploring bacteria from activated sludge to optimize azo dye degradation in textile effluents. *B. thuringiensis* and *K. radicincitans* emerge as promising candidates for bioremediation, and the application of MALDI-TOF-MS proves invaluable for rapid and precise bacteria identification.

Keywords: biodegradation; Reactive Red 141; MALDI-TOF MS; toxicity; *Bacillus thuringiensis*; *Kosakonia radicincitans*

1. Introduction

The global textile trade is expected to grow by 4.4% between 2021 and 2028 [1]. This continuous growth in textile products relates to a wastewater production rate above 130 L per kg of produced fabric [2]. The diversity and recalcitrance of contaminants are associated with the toxicity of some substances, such as dyes, resulting in an effluent with high biological oxygen demand (BOD), chemical oxygen demand (COD), intensive color, and salinity [3,4]. Azo dyes are recalcitrant and highly water-soluble compounds widely used by the textile industry. They are chemically composed of nitrogen-to-nitrogen double bonds (-N=N-) [5]. The incorrect disposal of textile effluents impacts the environment holistically, for instance, reducing the oxygen concentration in water bodies as well as the sunlight penetration, with a consequent reduction in the photosynthetic activity of aquatic algae and plants, the formation of aromatic amines—potentially carcinogenic and mutagenic, among others [6,7]. An example is Reactive Red 141 (RR 141), an azo dye commonly used in the textile industry [8]. It is a compound with high stability and limited biodegradability, presenting a challenge in its remediation processes [8].

Usually, textile industry effluent treatment is composed of physical–chemical steps (coagulation/flocculation and decantation) followed by biological processes, mainly activated sludge [9]. In this context, the microbiome of activated sludge from wastewater treatment plants is being explored to identify isolated species with dye decolorization potential [7,10,11]. These environments indicate the presence of bacteria groups that have significantly adapted to a diverse range of azo dyes, thereby supporting the possibility of discovering novel and more effective bacteria for azo dye decolorization. [12]. Exploring activated sludges as a microbial consortia is a vital perspective to optimize biodegradation. Also, studying isolated strains from activated sludges enables the elucidation of biodegradation pathways. MALDI-TOF-MS (matrix-assisted laser desorption/ionization—time of flight—mass spectrometry) is highlighted as a promising methodology for activated sludge microbiome identification with consequent treatment-yield enhancement [2]. MALDI-TOF has high accuracy and precision for identifying environmental bacteria, thereby improving time and cost savings [13,14]. With enhanced reference databases, it should be routinely applied in environmental studies [13]. The technique demonstrates high accuracy in identifying various bacterium types, aiding rapid bacterial identification in diverse environmental settings [13,15].

Therefore, this study aimed to isolate and identify bacterial strains from activated sludge of a treatment textile wastewater plant capable of decolorizing RR-141 (Reactive Red 141) azo dye. The two most promising strains were used to investigate kinetic performance, decolorization ability through UV-VIS spectrophotometry, possible pathways via HPLC-MS (High Performance Liquid Chromatography—Mass Spectrometry) analysis, phytotoxicity, and enzyme activity identification. The scanning electron microscope of the activated sludge sample was also executed.

2. Materials and Methods

2.1. Chemicals

Reactive Red 141 (RR-141/RHE7B), acetonitrile (CH_3CN, Sigma-Aldrich, St. Louis, MO, USA), brain heart infusion (BHI, KASVI, Milan, Italy), dibasic potassium phosphate (K_2HPO_4, VETEC, anhydrous, Duque de Caxias, Brazil), ethanol (C_2H_6O, Sigma-Aldrich, St. Louis, MO, USA), magnesium sulfate ($MgSO_4 \cdot 7H_2O$, NUCLEAR, Angra dos Reis, Brazil), sodium chloride (NaCl, VETEC, Duque de Caxias, Brazil), formic acid (CH_2O_2, BIOTEC, São Paulo, Brazil), α-cyano-4-hydroxycinnamic acid ($C_{10}H_7NO_3$, Sigma-Aldrich, St. Louis, MO, USA), trifluoroacetic acid (CF_3COOH, ÊXODO CIENTÍFICA, Sumaré, Brazil), barium chloride ($BaCl_2 \cdot 2H_2O$, DINÂMICA, São Paulo, Brazil), methanol (CH_3OH, UV-IR-HPLC-HPLC isocratic Panreac, Castellar del Vallés, Spain), calcium chloride ($CaCl_2 \cdot 2H_2O$, NEON, São Paulo, Brazil), magnesium sulfate ($MgSO_4 \cdot 7H_2O$, VETEC, Duque de Caxias, Brazil), sodium bicarbonate ($NaHCO_3$, NUCLEAR, Angra dos Reis, Brazil), potassium chloride (KCl, LAFAN, São Paulo, Brazil), dipotassium phosphate

(K_2HPO_4, Sigma-Aldrich, São Paulo, Brazil), monopotassium phosphate (KH_2PO_4, NUCLEAR, Angra dos Reis, Brazil), ammonium sulfate ((NH_4)$_2SO_4$, NUCLEAR, Angra dos Reis, Brazil), and glucose ($C_6H_{12}O_6$, MERCK, São Paulo, Brazil) were used as chemicals, without any pre-treatment.

2.2. Bacterial Isolation

Activated sludge from the secondary settling tank of a textile industry treatment plant located in the city of Blumenau, Brazil was collected and transported to the laboratory at room temperature. The solid fraction of an activated sludge sample was separated from the liquid fraction through centrifugation ($10,000 \times g$ for 5 min). Homogenized biological sludge (4 g L^{-1}) was added in a 37 g L^{-1} of Brain Heart Infusion (BHI) solution and incubated at 30 °C and 150 rpm for 24 h. Later, the samples were diluted in a 0.85% NaCl solution, employing a dilution factor of 10^{-6}. From this solution, 1 mL was spread in BHI agar and grown at 30 °C for 24 h in an incubator. The different colonies were transferred to BHI agar tubes and grown at 30 °C for 24 h. After, the isolated bacteria were maintained on BHI agar slants and 1.5 mL microcentrifuge tubes using a cryopreservative liquid (BHI with glycerol, 2:8) and preserved at -20 °C for further assays.

2.3. Microbiome Identification

The incubated cells (Petri plate—BHI at 30 °C for 24 h) were transferred, using a pipette tip, to a 1.5 mL screw-cap extraction tube (Eppendorf, Hamburg, Germany) and wholly mixed with 0.3 mL of double-distilled water. Absolute ethanol (0.9 mL) was added and cautiously mixed, and the tubes were centrifuged for 2 min at $20,000 \times g$. The supernatant was rejected. The precipitate was air-dried and mixed thoroughly with 50 µL of formic acid (70%) and 50 µL of acetonitrile. The mixture was submitted to centrifugation ($20,000 \times g$, 2 min). The supernatant (1 µL) was dried at room temperature on a ground steel MALDI target plate. The samples received an extra layer of 2 µL of a saturated solution of α-cyano-4-hydroxycinnamic acid in 50% acetonitrile and 2.5% trifluoroacetic acid and were dried at room temperature [16]. An UltrafleXtreme MALDI-TOF mass spectrometer (Bruker Daltonics, Bremen, Germany) performed the mass spectrometry analysis on the linear positive ion mode. Mass spectra were obtained in a range from 2 to 20 kDa with ions generated via the irradiation of a smart beam using a frequency of 2000 Hz, PIE 100 ns, 7 kV lens [17]. The voltages were 25 and 23 kV for the first and second ion sources, respectively.

MALDI Biotyper CA System software (Bruker Daltonics, Bremen, Germany) was used to identify bacteria with cut-off values higher than 1.7 for species identification [18]. The values labeled as "score" in Table 1 signify the resemblance of the identified species to the database employed in the identification technique. A score > 2.3 implies a "highly probable identification", while a score between 2 and 2.299 indicates a "confident identification of the genus and probable identification of the species". Scores falling between 1.7 and 1.999 suggest a "probable identification of the genus", and a score < 1.7 denotes an "unreliable identification" [19]. Lastly, the micro-organisms identified were compared with several databases (DOAJ, JSTOR, Science Direct, Scopus, Springer, and Google Scholar) to prospect potential azo-dye-degrading strains, and the used keywords were "strain name" AND "decolorization".

The dominant micro-organisms within the biological sludge, harnessed as a microbial reservoir for dye remediation, are documented in Table 1. Additionally, the documental database used to identify scientific trends in the context of biological dye remediation is shown in Table 1. The most abundant strain in the activated sludge is *Bacillus thuringiensis*, which presents a moderate number of reports associated with decolorization than *E. coli* and *B. cereus* species (widely used). Moreover, certain identified bacteria, such as *Kosakonia* sp., display subtle yet discernible correlations with the decolorization process.

Table 1. Micro-organisms identified in the biological sludge and its association studies with decolorization detected using different journal directories.

Strain	Score	DOAJ *	JSTOR *	Science Direct *	Scopus *	Springer *	Google Scholar *
Bacillus cereus	2.16	0	1	92	7	335	11,600
Klebsiella oxytoca	1.74	0	0	8	1	57	605
Bacillus thuringiensis	2.21	0	1	31	2	112	1190
Kosakonia cowanii	1.96	0	0	1	0	1	11
Lysinibacillus fusiformis	1.81	0	0	4	1	17	198
Acinetobacter baumannii	2.16	0	0	10	1	43	676
Kosakonia radicincitans	1.86	0	0	0	0	0	7
Escherichia coli	1.75	7	1	318	29	1028	17,200

* Journal directories.

Acinetobacter baumannii, *Klebsiella oxytoca*, and *Escherichia coli* are gram-negative bacteria that are widely related to dye decolorization. *A. baumannii* aerobically decolorized two textile azo dyes—Reactive Blue and Reactive Black 5—with 90% and 87% efficiency after 48 h [20] and were also tested to decolorize Reactive black 5, Reactive blue 19, Reactive red 120, and Reactive Red 198 reaching yields above 96% [21,22]. *K. oxytoca* promoted the highest decolorization potential of 69.68% for vat brown dye [23] and achieved simultaneous decolorization (83.8% within 24 h) and biohydrogen production (2.47 mL h^{-1}) [24]. *E. coli* was used to biodegrade methylene blue [25]. The authors reported 92.9% of dye removal. *E. coli* spp. can also be applied simultaneously with other micro-organisms, such as *Pseudomonas putida* [26], *Enterobacter asburiae*, *E. ludwigii*, and *B. thuringiensis*, with an excellent yield of over 96% [27].

2.4. Prospection of Potential Azo-Dye-Degrading Bacteria in Solid and Liquid Mediums

The azo dye decolorization potential in a solid medium was evaluated using the streak plate method. Preculture broth (100 µL) of each culture was streaked on a solid medium composed of BHI (37 g L^{-1}) and RR-141 (60 mg L^{-1}) and incubated at 30 °C for 168 h [28].

Three distinct growth mediums were employed for the investigation conducted in the liquid environment: the first medium consisted of BHI (37 g L^{-1}); the second medium, referred to as mineral salt media (MS), was composed of the following components per liter: NaCl (5 g), MgSO$_4$·7H$_2$O (0.1 g), K$_2$HPO$_4$ (10 mg), KH$_2$PO$_4$ (1 g), and (NH$_4$)$_2$SO$_4$ (2 g). For the third cultivation medium, the same composition as MS was used, but glucose (3 g L^{-1}) was added as a carbon source. This last medium will be referred to as MSG. All liquid mediums had the addition of the dye RR-141 (30 mg L^{-1}). The liquid culture mediums were inoculated with 10% (v/v) of each bacteria strain (approximately 1×10^9 cells mL^{-1}) and incubated at 30 °C and 100 rpm for 7 days.

Following the incubation period, each culture sample was subjected to centrifugation at 10,000× *g* for 10 min. The resulting supernatant was then analyzed using a UV/Vis spectrometer (Femto Cirrus 80, São Paulo, Brazil) with measurements taken at a wavelength of 516 nm. The extent of color removal was quantified using the equation defined as Equation (1):

$$Decolorization(\%) = \frac{(ABS_0 - ABS_f)}{ABS_0} \times 100 \quad (1)$$

where ABS_f is the sample absorbance after 7 days and ABS_0 is the initial system absorbance. Two strains (*B. thuringiensis* and *K. radicincitans*) that revealed the highest decolorizing potential in all assays were selected for the subsequent assays.

2.5. Evaluation of Carbon Sources on the Kinetic Degradation of Azo Dye

Batch assays were performed to confirm the azo-dye-decolorizing capability of the selected strains. The decolorization rate and their performance as isolated strains or a consortium were performed. The consortium was standardized in the same work volume, considering 10% (v/v) of inoculum (CFU (Colony formed unit) $\simeq 1 \times 10^9$ cells mL^{-1} to each bacteria species). The decolorization kinetics were evaluated in three culture media: BHI, MS, and MSG (composition described in Section 2.4). All liquid mediums had the addition of the dye RR-141 (30 mg L^{-1}).

The pH, microbial growth (assessed via optical density at 600 nm using a UV-Vis spectrophotometer), and decolorization (quantified at 516 nm using a UV-Vis spectrophotometer as outlined in Section 2.4) were monitored over 120 h.

2.6. Phytotoxicity Assay

Due to the rapid and uniform germination, *Lactuca sativa* seeds were used as standard assays [29]. Before inoculation, the seeds were cleaned, and the surface was sterilized using 99% ethanol solution for 5 min and then washed several times using sterilized distilled water. The assays were performed on sterile Petri dishes (Ø 90 mm) covered with qualitative filter paper (Unifil®, Curitiba, Brazil, 80 g m^{-2}). In each plate, 3 mL of the solution to be tested were added, and 10 seeds were equally spaced on the filter paper. Tap water (TW) was a positive control [30]. The Petri dishes were sealed and incubated (TECNAL TE-371, type BOD, Piracicaba, Brazil) at 30 °C. The germination and growth rates were analyzed daily for 7 days. The experiment was carried out in duplicate.

2.7. Detection of Azo Dye

The azo dye was detected using an HPLC coupled to mass spectrometry detection (HPLC–MS) equipped via a C18 column (Shimpack XR-ODS 50 x 2.0 mm I.D.). The samples were prepared via precipitation with $BaCl_2$ followed by filtration. The samples were eluted at a flow rate of 0.05 L.min^{-1} and monitored at 370 nm. The eluents A (ultrapure water containing 1% formic acid) and B (methanol) served as mobile phases in an isocratic mode (30% A and 70% B). Nitrogen was used as the nebulizing gas (1.50 L.min^{-1}), heated sheath gas, and drying gas (3 L.min^{-1}, 250 °C).

2.8. Enzymatic Azo Dye Degradation

Isolated colonies of each bacterial culture were inoculated (needle) in Petri dishes containing BHI agar and RR-141 (60 mg L^{-1}) and then incubated at 30 °C for 24 h. The differences in the halo formation diameters were calculated considering the total and CFU diameter differences. The diameters resulted from two perpendicular axis measurement averages [31].

3. Results and Discussion

3.1. Screening of Dye Decolourization in Solid and Liquid Mediums

In biological processes, the bioavailability of enzymes interferes directly with the dye transformation, which can be performed extracellularly and intracellularly. Nevertheless, the most effective strategy involves extracellular degradation [32]. Since azo dyes have complex structures, their diffusion through cell membranes is hampered. Therefore, the assays performed in the solid medium are strictly related to degradation by extracellular enzymes. The visual analysis of Petri plates (Figure 1) indicated that *K. oxytoca*, *B. thuringiensis*, *K. cowanii*, and *K. radicincitans* colonies have the highest potential for decolorizing RR-141 (60 mg L^{-1}) after 168 h of incubation at 30 °C. Moreover, the decolorization occurred primarily within 48 h of incubation.

A similar trend was observed by Kiayi et al. (2019) [28], where the solid-plate test promoted total decolorization of carmoisine (50 mg L^{-1}) within 4 days by *S. cerevisiae* colonies, with no visual changes in the fifth and sixth days.

Biofilm-producing bacteria is an important factor since biofilm is an excellent means to retain micro-organisms and improve their performance in environmental biotechnologies [33,34]. Proteins and carbohydrates from EPS allow binds between the microbial biomasses and substrates, favoring their activities [33,34]. In this context, it is worth noting that the diffusion effects are essential to reach the high yields of biodegradability. Considering SEM micrographs of activated sludge (Appendix A, Figure A1), no biofilm was produced as an alternative to enhance the degradation process; the dye diffusion to the micro-organism is hampered in this experiment.

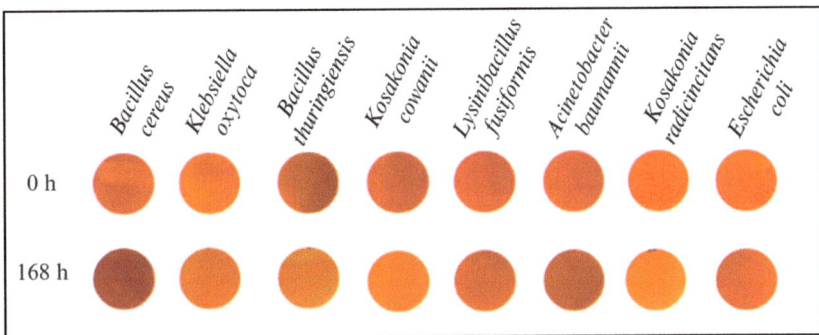

Figure 1. Decolorization potential study in the solid medium at 0 h and 168 h.

For the tests in a liquid medium, BHI and MSG media aimed to elucidate the optimal pathway for strain performance. This was achieved by evaluating their behavior in a nutrient-rich and opaque medium (BHI) and a less enriched and more translucid medium (MSG).

The analysis of the results confirmed *B. thuringiensis* (Figure 2) as the most promising dye-degrading species. It also indicated *K. radicincitans*, *B. cereus*, and *A. baumannii* as potential strains for RR-141 degradation.

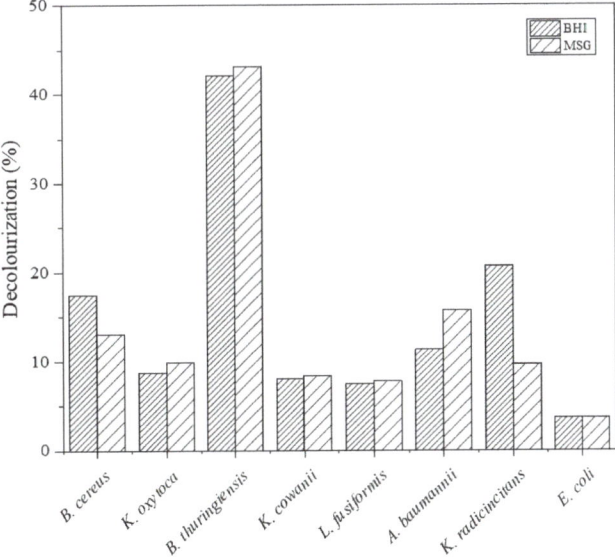

Figure 2. Decolorization yields of each isolated culture in BHI and MSG media.

Hence, building upon the outcomes of the solid and liquid mediums and considering the frequency of reports associated with each species (as depicted in Table 1), *B. thuringiensis* and *K. radicincitans* were chosen as candidates for subsequent investigations.

3.2. Effect of Carbon Sources and Dye-Decolorizing Kinetics by B. thuringiensis or K. radicincitans

The potential correlation between the carbon source provided by the culture medium and its impact on bacterial dye decolorization capacity was explored. To investigate this, the decolorization of the dye was examined while concurrently monitoring microbial growth and pH variation (Figure 3). This investigation was conducted using three distinct culture media (BHI, MS, and MSG), each within the context of the two chosen bacterial strains, *B. thuringiensis* and *K. radicincitans*.

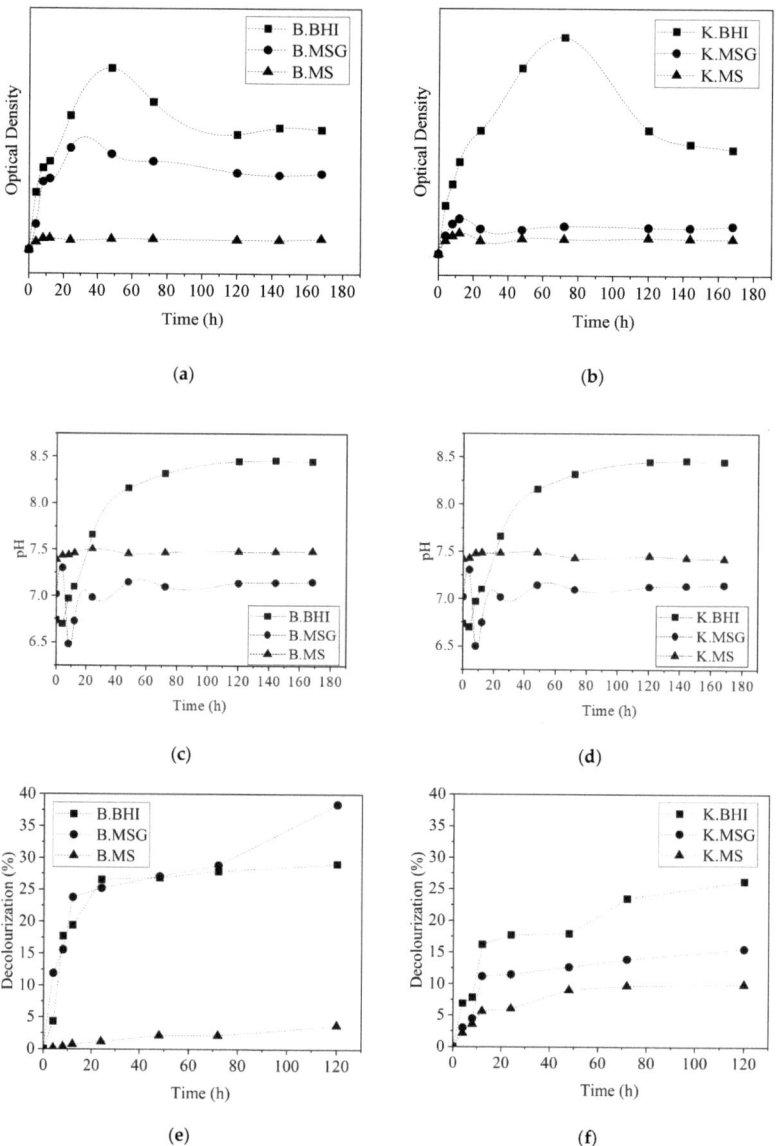

Figure 3. Temporal evolution of bacterial growth ((**a**) *B. thuringiensis*; (**b**) *K. radicincitans*), pH variation ((**c**) *B. thuringiensis*; (**d**) *K. radicincitans*), and RR-141 decolorization ((**e**) *B. thuringiensis*; (**f**) *K. radicincitans*) in three culture media (BHI, MSG, and MS).

Bacterial growth was more significant in the BHI medium, followed by MSG, for both strains (Figure 3a,b). No substantial changes in the optical density were identified in the MS medium (absence of carbon source, except azo dye RR-141), which could indicate the lack of ability of the strains to use the dye as a primary carbon source (Figure 4). Most micro-organisms are not capable of utilizing dyes as a carbon source for growth and require a carbohydrate source, such as peptone (present in BHI medium) or glucose (present in the MSG medium) [35].

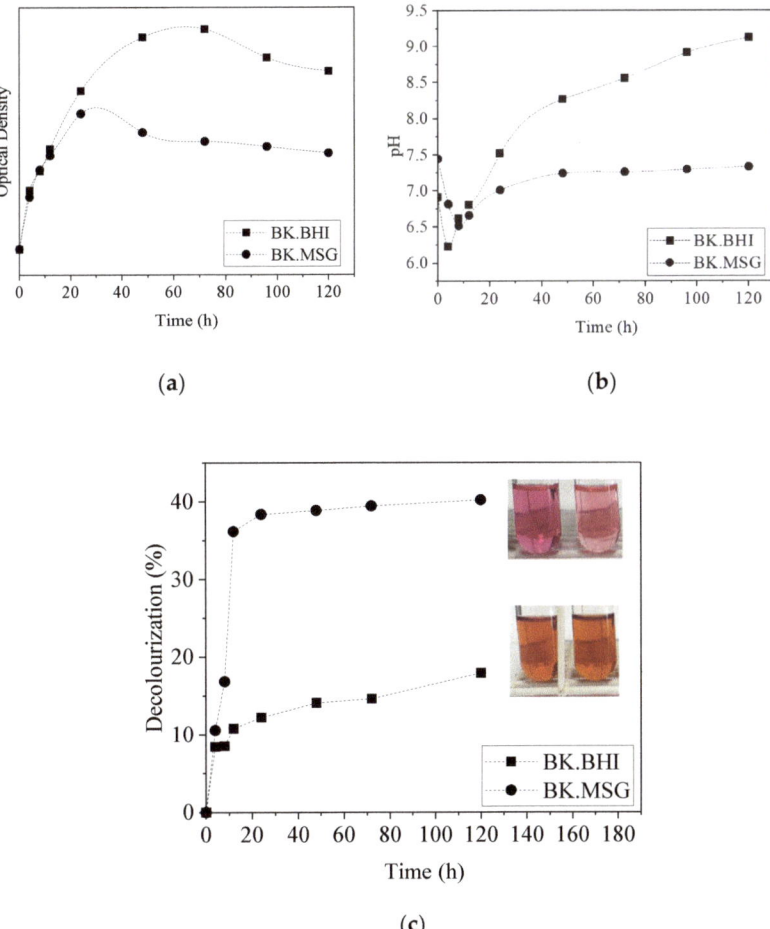

Figure 4. Temporal evolution of bacterial growth (**a**), pH variation (**b**), and decolorization rate of the consortia system in BHI and MSG media over 120 h, with pictures of the initial and end states of both medium (**c**).

In the system with the BHI medium, a more pronounced alkalinization of the medium is observed over time than in the MSG system (Figure 3c,d). In the MS system, where no bacterial growth was observed, there were no significant changes in pH. The decolorization efficiency of the azo dye is significantly influenced by pH. This could be attributed to the transportation of dye molecules across the cell membrane, a step recognized as the limiting factor in the decolorization process [35]. The optimal pH for decolorizing azo dyes varies according to the bacterial species but is often reported to be between 6 and 10 [35].

The decolorization of the azo dye RR-141 is depicted in Figure 3e,f. The *B. thuringiensis* strain performed better in MSG (38%), while *K. radicincitans* achieved the highest yield in the BHI medium (26%). This behavior could be linked to how these bacteria perform under specific conditions, including carbon and nutrient sources and pH levels.

The discoloration observed in this experiment may result from two combined or isolated mechanisms: (i) The process of biosorption and bioaccumulation of the azo dye by the bacterial biomass; (ii) The biodegradation of the azo dye through the bacteria's metabolic/enzymatic machinery [36].

In the first case, the dye is removed from the liquid medium and stored inside the bacterial cell, but the azo dye does not undergo any chemical modification. This mechanism is especially interesting for applications in wastewater treatment processes contaminated by dyes when the objective is to reuse the dye, which can be separated from the effluent through the biosorption process.

On the other hand, in the biodegradation process, the azo dye is biodegraded into secondary metabolites or completely mineralized to H_2O and CO_2 [37]. The biodegradation process carried out by bacteria is generally an enzymatic process [36,38]. Enzymes such as azoreductaze, laccase, peroxidase, and phenoloxidase have been reported in the biodegradation processes of azo dyes [36,39–42]. Enzyme action comprehension and identification at each metabolism stage are essential to improve the degradation process and are indicated for further studies.

3.3. Enzymatic Azo Dye Degradation

The test in a solid medium was carried out to investigate the presence of enzymes in the azo dye decolorization process [43]. *B. thuringiensis* and *K. radicincitans* showed, after 24 h, a halo formation represented by a more translucid area surrounding the CFU that could be associated with extracellular enzyme production [31]. Considering the halo's measurement (Table 2), *K. radicincitans* presented the best results, which indicate good development (Figure 4, OD graphic) and consequent enzymatic activity. In addition, the cell growth on dye-supplemented BHI agar medium with white colonies aspect can be associated with dye-decolorizing potential. Since cell mat coloring results from dye biosorption, maintaining the original mat color indicates biodegradation via an enzymatic process [44].

Table 2. Measurements of the colony-forming unit and its halo for each bacteria species in the enzyme activity assay.

	Bacillus thuringiensis	*Kosakonia radicincitans*
Øcolony (cm)	0.250 ± 0.056	0.549 ± 0.031
Øcolony + halo (cm)	0.339 ± 0.052	0.675 ± 0.051
Øhalo (cm)	0.091 ± 0.017	0.126 ± 0.031

Oxidative and reductive enzymes are vital to azo dye biodegradation [45]. Azo dye remediation intermediated by enzymes can be intra or extracellular. However, the high complexity of azo dyes hampers their diffusion through cell membranes, so the preferable route is via enzyme release in the extracellular environment [46]. Microbial strains used in the decolorization process must have efficient enzymes and a transport system to permit the absorption of dyes in cells [45]. Enzymes improve the reductive cleavage of azo bonds, producing intermediate metabolites posteriorly degraded by aerobic or anaerobic mechanisms [47].

Azoreductases, laccases, and peroxidases are enzymes often related to azo dye discoloration. The action of azoreductase occurs through reduction mechanisms mediated by a flavoprotein in the microbial electron transport chain [48], which converts azo dyes into colorless products (aromatic amines) [47]. The laccases act either via direct or indirect oxidation [49] through an unspecific free radical mechanism that results in phenolic products and prevents the formation of aromatic amines [50]. The peroxidase mechanism comprises the oxidation of the phenolic group and the production of a radical close to the azo bonds [51].

3.4. Dye-Decolorizing Kinetics by B. thuringiensis and K. radicincitans as a Consortium

Degradation of azo dyes is frequently accomplished by employing microbial consortia. The cooperative metabolic interactions within these microbial communities contribute to a more extensive biodegradation and mineralization process. In this study, the consortium evaluation revealed a good development in both media, with MSG promoting better conditions for a higher decolorization yield (Figure 5). Compared to the isolated kinetic study, the decolorization process achieved better results in a minor period, with approximately 36% of decolorization yielded within 12 h, associated with a synergic effect.

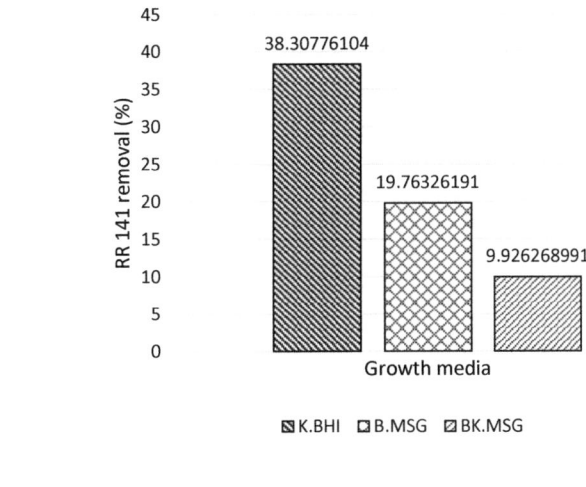

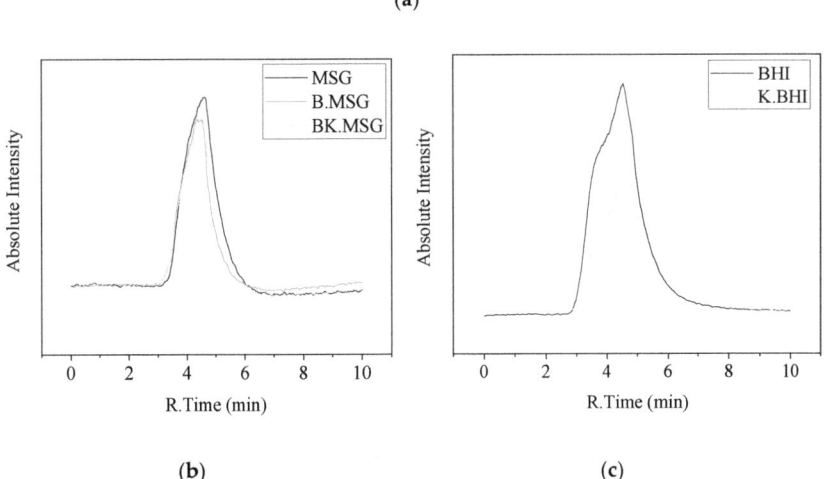

Figure 5. Removal percentage of RR 141 dye for the tests K.BHI [1], B.MSG [2], and BK.MSG [3] (**a**) along with their respective chromatograms (MSG (**b**) and BHI (**c**)). [1] *K. radicincitans* in BHI medium; [2] *B. thuringiensis* in MSG medium; [3] the consortium in MSG medium.

Some microbial consortia had achieved yields and objectives that no individual strain could successfully reach in biodegradation studies [52–54]. Additionally, mixed culture evaluations are more similar to practical situations since aseptic conditions can add more cost and higher stability over environmental stress such as composition, pH, or temperature variations. The synergistic action can occur via different pathways: (i) A micro-organism causes dye biotransformation, rendering it more reachable to another organism that would not be able to act on this dye in its original state; (ii) Only the micro-organism promotes some solution decolorization via the modification of the chromophore. However, the complete degradation is not achieved, and the metabolic products may have a toxic nature, as in an anaerobic reduction of azo dyes. Once another micro-organism takes this unwished metabolite as a nutrient source, the complete degradation, leading to carbon dioxide, ammonia, and water, can be achieved (only by mixed populations). This mineralization reaction ensures that no potentially harmful degradation products are released into the environment [32].

It is not easy to reproduce and effectively interpret the results when using mixed cultures because it only provides a wide view of what is happening in the system; it is difficult to identify and quantify individual culture growth and it hampers the elucidation of the degradation mechanism [2]. For these reasons, the application and deep comprehension of color removal via single bacterial cultures are essential since they promote a more straightforward interpretation of experimental observations and reproducibility [55]. Comparing the two carbon sources tested, better discoloration was obtained in the MSG medium (40%) than in the BHI medium (18%). Studies on the decolorization of azo dyes have reported higher removal rates, but this removal may depend on the dye dosage [56]. Eslami et al. (2019) [56] showed a 98% removal of the RR198 dye by a bacterial consortium (*Enterococcus faecalis* and *Klebsiella variicola*) at concentrations of 10–25 mg L^{-1}. Although, the removal efficiency was reduced at higher contractions (50, 75, and 100 mg L^{-1}) (55.62%, 25.82%, and 15.42%, respectively). This indicates that the dye can be toxic to bacteria at higher concentrations and reduce the decolorization efficiency. *B. thuringiensis* and *K. radicincitans* may perform better in decolorizing R141 at lower concentrations than those used in this study (30 mg L^{-1}). This hypothesis can be investigated in future studies.

3.5. Biodegradation Analysis

HPLC-MS analysis was conducted to verify the reduction in dye quantity within both individual bacterial systems and the consortium. This assessment was conducted in the culture medium where the most significant discoloration occurred. Specifically, we examined the removal of dye by *K. radicincitans* in BHI medium (K.BHI) and by *B. thuringiensis* in MSG medium (B.MSG) and evaluated the removal of dye in MSG medium by the bacterial consortium (BK.MSG).

The chromatograms of dye solutions at the initial and after each studied media treatment were obtained (Figure 5). The HPLC profile of the RR-141 solution in both media exhibited a single peak at retention times of 4.53 and 4.63 min to BHI and MSG, respectively (matrix effect—chromatography). The assay results of the solution after treatment exposed peaks corresponding to RR 141 with reduced intensity. The highest dye removal was observed in the K.BHI system, reaching 38%, followed by the B.MSG system with 19%, and the BK.MSG bacterial consortium system achieved a 10% reduction (Figure 5a).

While UV-Vis analysis (Sections 3.2 and 3.4) and HPLC-MS measurements revealed a decrease in dye concentration following the biological treatments, no new peaks were observed in the chromatograms. The emergence of new peaks in chromatograms after biological treatment typically signifies the formation of byproducts resulting from bacterial metabolism, a common occurrence during the biodegradation process of RR 141 [57–59]. The absence of new peaks in the chromatograms of treated solutions contributes to a possible adsorption strand via the biomass generated [60–62]. Once adsorption significantly influences the decolorization phenomena, another speculation can be made about the effect of culture media and biomass composition and adsorption.

Table 3 summarizes the percentages of decolorization and removal of the RR 141 dye, measured via UV-Vis spectroscopy and HPLC techniques. While both methods aim to comprehend the process of dye decolorization, their outcomes may not necessarily align or exhibit the same trend. The optimal decolorization condition for *B. thuringiensis*, identified through UV-Vis spectroscopy analysis, demonstrated a 38% reduction, whereas HPLC analysis only indicated a 19% decolorization. Likewise, the bacterial consortium exhibited an optimal 40% decolorization via UV-Vis spectroscopy, while HPLC analysis revealed a lower 10% decolorization. In contrast, for *K. radicincitans*, HPLC analysis indicated a higher optimal decolorization (38%) than UV-Vis analysis (26%).

Table 3. Percentages of decolorization (UV-Vis spectroscopy analysis) and removal (HPLC analysis) of RR 141 in liquid media.

Micro-Organisms	Decolorization Percentage of RR 141 Dye			Removal Percentage of RR 141 Dye	
	BHI	MSG	MS	BHI	MSG
B. thuringiensis	29%	38%	4%	-	19%
K. radicincitans	26%	15%	10%	38%	-
Consortium	18%	40%	-	-	10%

The decrease in UV-Vis spectroscopy absorbance is commonly misconstrued as dye biodegradation, primarily indicating decolorization rather than the actual degradation of the dye molecule [63]. UV-Vis spectroscopy, relying on a single wavelength, may misinterpret dye concentration in solutions due to factors like intermediate formation, overall solution changes, and pH-dependent dye peaks [63]. Because of this, UV-Vis responses cannot be directly compared with responses from techniques such as HPLC; they must complement each other. Therefore, UV-Vis analysis can be used to assess discoloration and HPLC to investigate the mechanisms of discoloration and the removal percentage [64].

3.6. Phytotoxicity Assay

A phytotoxicity study was carried out to evaluate the toxic or non-toxic nature of the dye. The analysis of phytotoxicity (Figure 6) revealed the toxic nature of Reactive Red 141 to the *Lactuca sativa* seeds. The germination rate was lower with Reactive Red 141 (60%, i.e., six of ten seeds germinated, on average) compared to tap water (80%, i.e., eight of ten seeds germinated, on average).

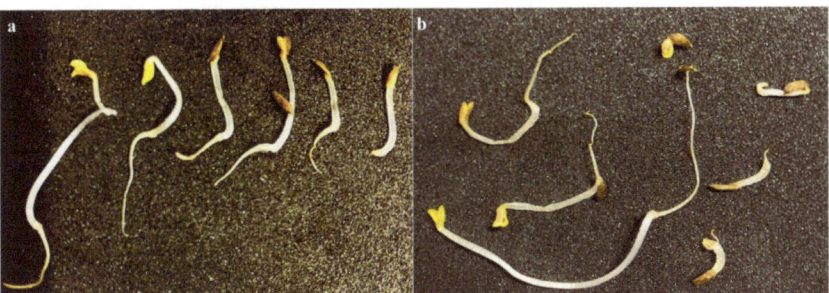

Figure 6. Phytotoxicity experiments: (**a**) In RR-141 solution; (**b**) Tap water.

These findings align with previous studies highlighting the phytotoxic of azo dyes. An investigation by Sompark et al. (2021) revealed that the germination rate of mung bean seeds exposed to RR 141 at a concentration of 0.5 g/L was 62.50%, in contrast to the 100% germination observed when the seeds were subjected solely to distilled water treatment [65].

RR 141 was assessed for its impact on the germination of beetroot, cabbage, and tomato seeds, resulting in notably reduced germination rates of 36.7%, 40.0%, and 33.3%, respectively [59]. In contrast, when these identical seed varieties were germinated in a control group with pure water, a significantly higher germination rate exceeding 85% was observed [59].

The Reactive Red 141 is significantly inhibitory for the plants' germination, indicating the presence of phytotoxic compounds. Azo dyes and some metabolites present carcinogenic, toxic, and mutagenic properties to the environment and humankind [66]. Bacterial azo dye degradation comprises the cleavage of azo bonds and the intermediates breakdown. However, three mechanism routes promote the carcinogenic activation of azo dyes: (i) Direct oxidation of the azo linkage to diazonium salts with highly reactive electrophilic behavior; (ii) Oxidation of azo dyes in the presence of structures formed by free aromatic amine groups; (iii) Reduction and cleavage of the azo bond with consequent formation of aromatic amines [67,68].

4. Conclusions

This study explores the potential of various bacteria for azo dye remediation in industrial wastewater. *Bacillus thuringiensis* and *Kosakonia radicincitans* show promise in decolorizing the azo dye RR141, achieving around 38% and 26% decolorization in 120 h, respectively. When used together as a bacterial consortium, they remove 36% of the dye in 12 h. The dye removal analyses conducted via HPLC-MS suggest that the primary removal mechanisms are associated with the biosorption of the dye into the bacterial biomass rather than its biodegradation. The study also investigates the impact of carbon sources and pH on decolorization, emphasizing the importance of nutrient-rich media. It suggests that microbial consortia can lead to more efficient decolorization but underscores the need to study individual bacterial cultures to understand their capabilities and mechanisms. Overall, the research offers valuable insights into bioremediation, highlighting the potential of specific bacterial strains and microbial consortia for addressing dye pollution in industrial wastewater. Further studies in this area could lead to more efficient and eco-friendly solutions.

Author Contributions: Conceptualization, G.M.D.d.V. and C.J.d.A.; methodology, G.M.D.d.V. and C.J.d.A.; formal analysis, G.M.D.d.V. and L.M.d.A.; investigation, G.M.D.d.V.; Vasconcelos, I.K.D.-F. and L.M.d.A.; writing—original draft preparation, G.M.D.d.V. and I.K.D.-F.; writing—review and editing, C.J.d.A., D.d.O., S.M.d.A.G.U.d.S., A.A.U.d.S. and M.K.; supervision, C.J.d.A. All authors have read and agreed to the published version of the manuscript.

Funding: This research was funded by Conselho Nacional de Desenvolvimento Científico e Tecnológico (CNPq 133874/2019-2) and by Fundação de Amparo a Pesquisa e Inovação do Estado de Santa Catarina (FAPESC).

Institutional Review Board Statement: Not applicable.

Informed Consent Statement: Not applicable.

Data Availability Statement: Data are contained within the article.

Conflicts of Interest: The authors declare no conflicts of interest. The funders had no role in the study's design; in the collection, analyses, or interpretation of data; in the writing of the manuscript; or in the decision to publish the results.

Appendix A. Scanning Electron Microscope (SEM)

The analysis of SEM micrographs of activated sludge indicated the presence of a biofilm matrix, very likely to beflocculating bacteria. Some essential characteristics are related to the structure and function of biofilm, for example, organic compounds such as extracellular polymeric substances (EPS), which play a significant role in modifying the surface (charge and hydrophobicity) to give suitable conditions for bacterial connection [21–23]. Large amounts of EPS and different bacterial species can be found in biofilms [24]. EPS is defined as a complex combination of high-molecular-weight microbial biopolymers. Its composition is based on humic substances, lipids, polysaccharides, proteins, and uranic acids. Its liquid anionic composition enables the effective sequestration of positively charged species, such as some dyes [25]. There are two primary forms of EPS: as a capsule covalently bounded to the cell surface or as slime polysaccharides roughly associated with the cell surface, as detected in Figure A1.

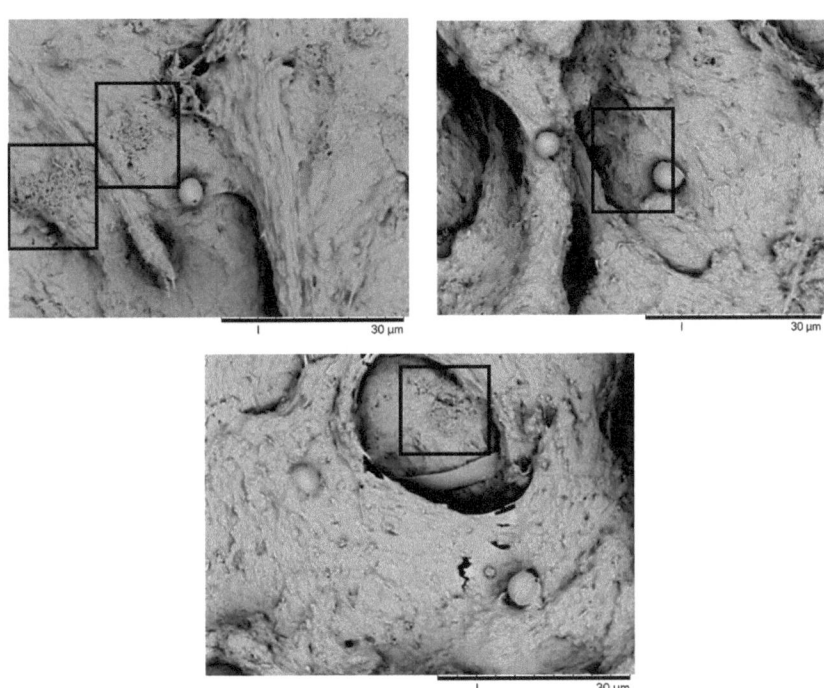

Figure A1. Scanning electron microscope of the activated sludge—biofilm formation.

References

1. *Grand View Research Textile Market Size, Share & Trends Analysis Report by Raw Material (Wool, Chemical, Silk), by Product (Natural Fibers, Polyester), by Application (Household, Technical), by Region, and Segment Forecasts, 2021–2028*; Grand View Research: San Francisco, CA, USA, 2021.
2. de Vasconcelos, G.M.D.; Mulinari, J.; Souza, S.M.d.A.G.U.d.; de Souza, A.A.U.; de Oliveira, D.; de Andrade, C.J. Biodegradation of Azo Dye-Containing Wastewater by Activated Sludge: A Critical Review. *World J. Microbiol. Biotechnol.* **2021**, *37*, 101. [CrossRef] [PubMed]
3. Lade, H.S.; Waghmode, T.R.; Kadam, A.A.; Govindwar, S.P. Enhanced Biodegradation and Detoxification of Disperse Azo Dye Rubine GFL and Textile Industry Effluent by Defined Fungal-Bacterial Consortium. *Int. Biodeterior. Biodegrad.* **2012**, *72*, 94–107. [CrossRef]
4. Bilińska, L.; Gmurek, M.; Ledakowicz, S. Comparison between Industrial and Simulated Textile Wastewater Treatment by AOPs—Biodegradability, Toxicity and Cost Assessment. *Chem. Eng. J.* **2016**, *306*, 550–559. [CrossRef]
5. Paździor, K.; Bilińska, L.; Ledakowicz, S.; Pa, K.; Bili, L.; Paździor, K.; Bilińska, L.; Ledakowicz, S. A Review of the Existing and Emerging Technologies in the Combination of AOPs and Biological Processes in Industrial Textile Wastewater Treatment. *Chem. Eng. J.* **2019**, *376*. [CrossRef]
6. Paździor, K.; Wrębiak, J.; Klepacz-Smółka, A.; Gmurek, M.; Bilińska, L.; Kos, L.; Sójka-Ledakowicz, J.; Ledakowicz, S. Influence of Ozonation and Biodegradation on Toxicity of Industrial Textile Wastewater. *J. Environ. Manag.* **2017**, *195*, 166–173. [CrossRef] [PubMed]
7. Zhu, Y.; Xu, J.; Cao, X.; Cheng, Y.; Zhu, T. Characterization of Functional Microbial Communities Involved in Diazo Dyes Decolorization and Mineralization Stages. *Int. Biodeterior. Biodegrad.* **2018**, *132*, 166–177. [CrossRef]
8. Rodrigues, F.K.; Salau, N.P.G.; Dotto, G.L. New Insights about Reactive Red 141 Adsorption onto Multi-Walled Carbon Nanotubes Using Statistical Physics Coupled with Van Der Waals Equation. *Sep. Purif. Technol.* **2019**, *224*, 290–294. [CrossRef]

9. Manekar, P.; Patkar, G.; Aswale, P.; Mahure, M.; Nandy, T. Detoxifying of High Strength Textile Effluent through Chemical and Bio-Oxidation Processes. *Bioresour. Technol.* **2014**, *157*, 44–51. [CrossRef]
10. Zhang, B.; Xu, X.; Zhu, L. Activated Sludge Bacterial Communities of Typical Wastewater Treatment Plants: Distinct Genera Identification and Metabolic Potential Differential Analysis. *AMB Express* **2018**, *8*, 184. [CrossRef]
11. Li, W.; Mu, B.; Yang, Y. Feasibility of Industrial-Scale Treatment of Dye Wastewater via Bio-Adsorption Technology. *Bioresour. Technol.* **2019**, *277*, 157–170. [CrossRef]
12. Meerbergen, K.; Willems, K.A.; Dewil, R.; Van Impe, J.; Appels, L.; Lievens, B. Isolation and Screening of Bacterial Isolates from Wastewater Treatment Plants to Decolorize Azo Dyes. *J. Biosci. Bioeng.* **2018**, *125*, 448–456. [CrossRef] [PubMed]
13. Ashfaq, M.Y.; Da'na, D.A.; Al-Ghouti, M.A. Application of MALDI-TOF MS for Identification of Environmental Bacteria: A Review. *J. Environ. Manag.* **2022**, *305*, 114359. [CrossRef] [PubMed]
14. Avanzi, I.R.; Gracioso, L.H.; Baltazar, M.d.P.G.; Karolski, B.; Perpetuo, E.A.; Nascimento, C.A.O.D. Rapid Bacteria Identification from Environmental Mining Samples Using MALDI-TOF MS Analysis. *Environ. Sci. Pollut. Res.* **2017**, *24*, 3717–3726. [CrossRef] [PubMed]
15. Abdel Samad, R.; Al Disi, Z.; Mohammad Ashfaq, M.Y.; Wahib, S.M.; Zouari, N. The Use of Principle Component Analysis and MALDI-TOF MS for the Differentiation of Mineral Forming: Virgibacillus and Bacillus Species Isolated from Sabkhas. *RSC Adv.* **2020**, *10*, 14606–14616. [CrossRef] [PubMed]
16. Marklein, G.; Josten, M.; Klanke, U.; Müller, E.; Horré, R.; Maier, T.; Wenzel, T.; Kostrzewa, M.; Bierbaum, G.; Hoerauf, A.; et al. Matrix-Assisted Laser Desorption Ionization-Time of Flight Mass Spectrometry for Fast and Reliable Identification of Clinical Yeast Isolates. *J. Clin. Microbiol.* **2009**, *47*, 2912–2917. [CrossRef]
17. Alves, L.A.C.; Souza, R.C.; da Silva, T.M.C.; Watanabe, A.; Dias, M.; Mendes, M.A.; Ciamponi, A.L. Identification of Microorganisms in Biofluids of Individuals with Periodontitis and Chronic Kidney Disease Using Matrix-Assisted Laser Desorption/Ionization Time-of-Flight Mass Spectrometry. *Rapid Commun. Mass. Spectrom.* **2016**, *30*, 1228–1232. [CrossRef]
18. Han, H.W.; Chang, H.C.; Hunag, A.H.; Chang, T.C. Optimization of the Score Cutoff Value for Routine Identification of Staphylococcus Species by Matrix-Assisted Laser Desorption Ionization-Time-of-Flight Mass Spectrometry. *Diagn. Microbiol. Infect. Dis.* **2015**, *83*, 349–354. [CrossRef]
19. Normand, A.C.; Cassagne, C.; Gautier, M.; Becker, P.; Ranque, S.; Hendrickx, M.; Piarroux, R. Decision Criteria for MALDI-TOF MS-Based Identification of Filamentous Fungi Using Commercial and in-House Reference Databases. *BMC Microbiol.* **2017**, *17*, 1–17. [CrossRef]
20. Sreedharan, V.; Saha, P.; Rao, K.V.B. Dye Degradation Potential of Acinetobacter Baumannii Strain VITVB against Commercial Azo Dyes. *Bioremediation J.* **2021**, *25*, 347–368. [CrossRef]
21. Unnikrishnan, S.; Khan, M.H.; Ramalingam, K. Dye-Tolerant Marine Acinetobacter Baumannii-Mediated Biodegradation of Reactive Red. *Water Sci. Eng.* **2018**, *11*, 265–275. [CrossRef]
22. Ameenudeen, S.; Unnikrishnan, S.; Ramalingam, K. Statistical Optimization for the Efficacious Degradation of Reactive Azo Dyes Using Acinetobacter Baumannii JC359. *J. Environ. Manag.* **2021**, *279*, 111512. [CrossRef] [PubMed]
23. Adebajo, S.; Balogun, S.; Akintokun, A. Decolourization of Vat Dyes by Bacterial Isolates Recovered from Local Textile Mills in Southwest, Nigeria. *Microbiol. Res. J. Int.* **2017**, *18*, 1–8. [CrossRef] [PubMed]
24. Yu, L.; Cao, M.; Wang, P.; Wang, S.; Yue, Y.; Yuan, W.; Qiao, W.; Wang, F.; Song, X. Simultaneous Decolorization and Biohydrogen Production from Xylose by Klebsiella Oxytoca GS-4-08 in the Presence of Azo Dyes with Sulfonate and Carboxyl Groups. *Appl. Environ. Microbiol.* **2017**, *83*, 1–13. [CrossRef] [PubMed]
25. Balraj, B.; Hussain, Z.; King, P. Experimental Study on Non Sporulating Escherichia Coli Bacteria in Removing Methylene Blue. *Int. J. Pharma Bio Sci.* **2016**, *7*, B629–B637.
26. Hilda Josephine, S.; Sekar, A.S.S. A Comparative Study of Biodegradation of Textile Azo Dyes by Escherichia Coli and Pseudomonas Putida. *Nat. Environ. Pollut. Technol.* **2014**, *13*, 417–420.
27. Haque, M.M.; Haque, M.A.; Mosharaf, M.K.; Marcus, P.K. Novel Bacterial Biofilm Consortia That Degrade and Detoxify the Carcinogenic Diazo Dye Congo Red. *Arch. Microbiol.* **2021**, *203*, 643–654. [CrossRef]
28. Kiayi, Z.; Lotfabad, T.B.; Heidarinasab, A.; Shahcheraghi, F. Microbial Degradation of Azo Dye Carmoisine in Aqueous Medium Using Saccharomyces Cerevisiae ATCC 9763. *J. Hazard. Mater.* **2019**, *373*, 608–619. [CrossRef]
29. dos Santos, F.E.; Carvalho, M.S.S.; Silveira, G.L.; Correa, F.F.; Cardoso, M.d.G.; Andrade-Vieira, L.F.; Vilela, L.R. Phytotoxicity and Cytogenotoxicity of Hydroalcoholic Extracts from Solanum Muricatum Ait. and Solanum Betaceum Cav. (Solanaceae) in the Plant Model Lactuca Sativa. *Environ. Sci. Pollut. Res.* **2019**, *26*, 27558–27568. [CrossRef]
30. Peduto, T.A.G.; de Jesus, T.A.; Kohatsu, M.Y. Sensibilidade de Diferentes Sementes Em Ensaio de Fitotoxicidade. *Rev. Bras. De Ciência Tecnol. E Inovação* **2019**, *4*, 200. [CrossRef]
31. Barros, F.F.C.; Simiqueli, A.P.R.; de Andrade, C.J.; Pastore, G.M. Production of Enzymes from Agroindustrial Wastes by Biosurfactant-Producing Strains of Bacillus Subtilis. *Biotechnol. Res. Int.* **2013**, *2013*, 103960. [CrossRef]
32. Kandelbauer, A.; Guebitz, G.M. Bioremediation for the Decolorization of Textile Dyes-A Review. *Environ. Chem. Green. Chem. Pollut. Ecosyst.* **2005**, 269–288. [CrossRef]
33. Haque, M.M.; Haque, M.A.; Mosharaf, M.K.; Rahman, A.; Islam, M.S.; Nahar, K.; Molla, A.H. Enhanced Biofilm-Mediated Degradation of Carcinogenic and Mutagenic Azo Dye by Novel Bacteria Isolated from Tannery Wastewater. *J. Environ. Chem. Eng.* **2023**, *11*. [CrossRef]

34. Haque, M.M.; Hossen, M.N.; Rahman, A.; Roy, J.; Talukder, M.R.; Ahmed, M.; Ahiduzzaman, M.; Haque, M.A. Decolorization, Degradation and Detoxification of Mutagenic Dye Methyl Orange by Novel Biofilm Producing Plant Growth-Promoting Rhizobacteria. *Chemosphere* **2024**, *346*. [CrossRef]
35. Khan, R.; Bhawana, P.; Fulekar, M.H. Microbial Decolorization and Degradation of Synthetic Dyes: A Review. *Rev. Environ. Sci. Biotechnol.* **2013**, *12*, 75–97. [CrossRef]
36. Kapoor, R.T.; Danish, M.; Singh, R.S.; Rafatullah, M.; Abdul, A.K. Exploiting Microbial Biomass in Treating Azo Dyes Contaminated Wastewater: Mechanism of Degradation and Factors Affecting Microbial Efficiency. *J. Water Process Eng.* **2021**, *43*, 102255. [CrossRef]
37. Bal, G.; Thakur, A. Distinct Approaches of Removal of Dyes from Wastewater: A Review. In *Proceedings of the Materials Today: Proceedings*; Elsevier Ltd.: Amsterdam, The Netherlands, 2021; Volume 50, pp. 1575–1579.
38. Saratale, G.; Kalme, S.; Bhosale, S.; Govindwar, S. Biodegradation of Kerosene by Aspergillus Ochraceus NCIM-1146. *J. Basic. Microbiol.* **2007**, *47*, 400–405. [CrossRef] [PubMed]
39. Shah, B.; Jain, K.; Jiyani, H.; Mohan, V.; Madamwar, D. Microaerophilic Symmetric Reductive Cleavage of Reactive Azo Dye—Remazole Brilliant Violet 5R by Consortium VIE6: Community Synergism. *Appl. Biochem. Biotechnol.* **2016**, *180*, 1029–1042. [CrossRef]
40. Mullai, P.; Yogeswari, M.K.; Vishali, S.; Tejas Namboodiri, M.M.; Gebrewold, B.D.; Rene, E.R.; Pakshirajan, K. Aerobic Treatment of Effluents From Textile Industry. In *Current Developments in Biotechnology and Bioengineering*; Elsevier: Amsterdam, The Netherlands, 2017; pp. 3–34, ISBN 9780444636768.
41. Chandanshive, V.V.; Kadam, S.K.; Khandare, R.V.; Kurade, M.B.; Jeon, B.H.; Jadhav, J.P.; Govindwar, S.P. In Situ Phytoremediation of Dyes from Textile Wastewater Using Garden Ornamental Plants, Effect on Soil Quality and Plant Growth. *Chemosphere* **2018**, *210*, 968–976. [CrossRef]
42. Chen, Y.; Feng, L.; Li, H.; Wang, Y.; Chen, G.; Zhang, Q. Biodegradation and Detoxification of Direct Black G Textile Dye by a Newly Isolated Thermophilic Microflora. *Bioresour. Technol.* **2018**, *250*, 650–657. [CrossRef]
43. Zhang, C.; Chen, H.; Xue, G.; Liu, Y.; Chen, S.; Jia, C. A Critical Review of the Aniline Transformation Fate in Azo Dye Wastewater Treatment. *J. Clean. Prod.* **2021**, *321*. [CrossRef]
44. Karim, M.E.; Dhar, K.; Hossain, M.T. Decolorization of Textile Reactive Dyes by Bacterial Monoculture and Consortium Screened from Textile Dyeing Effluent. *J. Genet. Eng. Biotechnol.* **2018**, *16*, 375–380. [CrossRef] [PubMed]
45. Mahmood, S.; Khalid, A.; Arshad, M.; Mahmood, T.; Crowley, D.E. Detoxification of Azo Dyes by Bacterial Oxidoreductase Enzymes. *Crit. Rev. Biotechnol.* **2016**, *36*, 639–651. [CrossRef] [PubMed]
46. Sari, I.P.; Simarani, K. Decolorization of Selected Azo Dye by Lysinibacillus Fusiformis W1B6: Biodegradation Optimization, Isotherm, and Kinetic Study Biosorption Mechanism. *Adsorpt. Sci. Technol.* **2019**, *37*, 492–508. [CrossRef]
47. Pandey, A.; Singh, P.; Iyengar, L. Bacterial Decolorization and Degradation of Azo Dyes. *Int. Biodeterior. Biodegrad.* **2007**, *59*, 73–84. [CrossRef]
48. Misal, S.A.; Gawai, K.R. Azoreductase: A Key Player of Xenobiotic Metabolism. *Bioresour. Bioprocess.* **2018**, *5*. [CrossRef]
49. Khlifi, R.; Belbahri, L.; Woodward, S.; Ellouz, M.; Dhouib, A.; Sayadi, S.; Mechichi, T. Decolourization and Detoxification of Textile Industry Wastewater by the Laccase-Mediator System. *J. Hazard. Mater.* **2010**, *175*, 802–808. [CrossRef]
50. Wong, Y.; Yu, J. Laccase-Catalyzed Decolorization of Synthetic Dyes. *Water Res.* **1999**, *33*, 3512–3520. [CrossRef]
51. Chivukula, M.; Renganathan, V. Phenolic Azo Dye Oxidation by Laccase from Pyricularia Oryzae. *Appl. Environ. Microbiol.* **1995**, *61*, 4374–4377. [CrossRef]
52. Neifar, M.; Sghaier, I.; Guembri, M.; Chouchane, H.; Mosbah, A.; Ouzari, H.I.; Jaouani, A.; Cherif, A. Recent Advances in Textile Wastewater Treatment Using Microbial Consortia. *J. Text. Eng. Fash. Technol.* **2019**, *5*, 134–146. [CrossRef]
53. Shindhal, T.; Rakholiya, P.; Varjani, S.; Pandey, A.; Ngo, H.H.; Guo, W.; Ng, H.Y.; Taherzadeh, M.J. A Critical Review on Advances in the Practices and Perspectives for the Treatment of Dye Industry Wastewater. *Bioengineered* **2021**, *12*, 70–87. [CrossRef]
54. Samuchiwal, S.; Gola, D.; Malik, A. Decolourization of Textile Effluent Using Native Microbial Consortium Enriched from Textile Industry Effluent. *J. Hazard. Mater.* **2021**, *402*, 123835. [CrossRef] [PubMed]
55. Thakur, J.K.; Paul, S.; Dureja, P.; Annapurna, K.; Padaria, J.C.; Gopal, M. Degradation of Sulphonated Azo Dye Red HE7B by Bacillus Sp. and Elucidation of Degradative Pathways. *Curr. Microbiol.* **2014**, *69*, 183–191. [CrossRef] [PubMed]
56. Eslami, H.; Shariatifar, A.; Rafiee, E.; Shiranian, M.; Salehi, F.; Hosseini, S.S.; Eslami, G.; Ghanbari, R.; Ebrahimi, A.A. Decolorization and Biodegradation of Reactive Red 198 Azo Dye by a New Enterococcus Faecalis–Klebsiella Variicola Bacterial Consortium Isolated from Textile Wastewater Sludge. *World J. Microbiol. Biotechnol.* **2019**, *35*. [CrossRef] [PubMed]
57. Al-Tohamy, R.; Ali, S.S.; Xie, R.; Schagerl, M.; Khalil, M.A.; Sun, J. Decolorization of Reactive Azo Dye Using Novel Halotolerant Yeast Consortium HYC and Proposed Degradation Pathway. *Ecotoxicol. Environ. Saf.* **2023**, *263*. [CrossRef] [PubMed]
58. Khandare, S.D.; Teotia, N.; Kumar, M.; Diyora, P.; Chaudhary, D.R. Biodegradation and Decolorization of Trypan Blue Azo Dye by Marine Bacteria Vibrio Sp. JM-17. *Biocatal. Agric. Biotechnol.* **2023**, *51*. [CrossRef]
59. Tizazu, S.; Tesfaye, G.; Wang, A.; Guadie, A.; Andualem, B. Microbial Diversity, Transformation and Toxicity of Azo Dye Biodegradation Using Thermo-Alkaliphilic Microbial Consortia. *Heliyon* **2023**, *9*. [CrossRef] [PubMed]
60. Pearce, C.I.; Lloyd, J.R.; Guthrie, J.T. The Removal of Colour from Textile Wastewater Using Whole Bacterial Cells: A Review. *Dye. Pigment.* **2003**, *58*, 179–196. [CrossRef]

61. Rodrigues, C.S.D.; Madeira, L.M.; Boaventura, R.A.R. Synthetic Textile Dyeing Wastewater Treatment by Integration of Advanced Oxidation and Biological Processes—Performance Analysis with Costs Reduction. *J. Environ. Chem. Eng.* **2014**, *2*, 1027–1039. [CrossRef]
62. Siddiqui, S.I.; Fatima, B.; Tara, N.; Rathi, G.; Chaudhry, S.A. *Recent Advances in Remediation of Synthetic Dyes from Wastewaters Using Sustainable and Low-Cost Adsorbents*; Elsevier Ltd.: Amsterdam, The Netherlands, 2018; ISBN 9780081024911.
63. Foster, J.E.; Adamovsky, G.; Gucker, S.N.; Blankson, I.M. A Comparative Study of the Time-Resolved Decomposition of Methylene Blue Dye under the Action of a Nanosecond Repetitively Pulsed Dbd Plasma Jet Using Liquid Chromatography and Spectrophotometry. *IEEE Trans. Plasma Sci.* **2013**, *41*, 503–512. [CrossRef]
64. Ali, H. Biodegradation of Synthetic Dyes—A Review. *Water Air Soil. Pollut.* **2010**, *213*, 251–273. [CrossRef]
65. Sompark, C.; Singkhonrat, J.; Sakkayawong, N. Biotransformation of Reactive Red 141 by Paenibacillus Terrigena KKW2-005 and Examination of Product Toxicity. *J. Microbiol. Biotechnol.* **2021**, *31*, 967–977. [CrossRef] [PubMed]
66. Kalme, S.; Ghodake, G.; Govindwar, S. Red HE7B Degradation Using Desulfonation by Pseudomonas Desmolyticum NCIM 2112. *Int Biodeterior Biodegrad.* **2007**, *60*, 327–333. [CrossRef]
67. Chung, K.T. Mutagenicity and Carcinogenicity of Aromatic Amines Metabolically Produced from Azo Dyes. *J. Environ. Sci. Health C Environ. Carcinog. Ecotoxicol. Rev.* **2000**, *18*, 51–74. [CrossRef]
68. Brown, M.A.; De Vito, S.C. Predicting Azo Dye Toxicity. *Crit. Rev. Environ. Sci. Technol.* **1993**, *23*, 249–324. [CrossRef]

Disclaimer/Publisher's Note: The statements, opinions and data contained in all publications are solely those of the individual author(s) and contributor(s) and not of MDPI and/or the editor(s). MDPI and/or the editor(s) disclaim responsibility for any injury to people or property resulting from any ideas, methods, instructions or products referred to in the content.

Article

Activity Concentration Index Values for Concrete Multistory Residences in Greece Due to Fly Ash Addition in Cement

Stamatia Gavela [1,*] and Georgios Papadakos [2]

1. Department of Civil Engineering, University of West Attica, Petrou Ralli & Thivon 250, GR-122 44 Athens, Greece
2. Domo+Lysis Lab–Laboratory of Structural Assessment and Protection, Distomou 97, GR-104 43 Athens, Greece; papadng@gmail.com
* Correspondence: sgkavela@uniwa.gr; Tel.: +30-6973579307

Abstract: According to 2013/59/Euratom Directive, the activity concentration index (*ACI*) is required to be estimated for each building material that is of concern from a radiation protection point of view. This index applies to building materials and not to constituents that cannot be used as building materials themselves. Fly ash is a byproduct of coal-fired power plants and is one of the main constituents of cement. The radioactivity in fly ash that is produced by Greek lignite power plants cannot be considered insignificant. For example, in the case of the Megalopolis power plant, the concentration for radioisotopes of the ^{226}Ra chain is found to be about 1 kBq/kg. Since natural radionuclide concentrations, which are harmful to human health in terms of radiation exposure, exist in fly ash, *ACI* should be assessed for building materials containing fly ash. The present study evaluates the *ACI* of concrete containing fly ash cement when used in multistory residential buildings. Results showed that cement produced in Greece by the three main Greek cement production plants, containing lignite fly ash, and used as a material for concrete multistory constructions, should not be considered as "of concern from a radiation protection point of view". Each country that wishes to evaluate the use of fly ash into constructions should repeat the method for the *ACI* uncertainty budget proposed in this study, to assess whether it significantly exceeds the reference value (whether it is of concern from a radiation protection point of view).

Keywords: fly ash addition in cement; concrete multistory residence; activity concentration index; *ACI*

Citation: Gavela, S.; Papadakos, G. Activity Concentration Index Values for Concrete Multistory Residences in Greece Due to Fly Ash Addition in Cement. *Eng* **2023**, *4*, 2926–2940. https://doi.org/10.3390/eng4040164

Academic Editor: F. Pacheco Torgal

Received: 25 September 2023
Revised: 15 November 2023
Accepted: 16 November 2023
Published: 20 November 2023

Copyright: © 2023 by the authors. Licensee MDPI, Basel, Switzerland. This article is an open access article distributed under the terms and conditions of the Creative Commons Attribution (CC BY) license (https://creativecommons.org/licenses/by/4.0/).

1. Introduction

Fly ash is a byproduct of coal-fired power plants. Such plants that are established in Greece operate with lignite as their fuel. Lignite in Greece comes from (a) Ptolemais-wide area, but also from even wider territories within the Prefecture of Kozani, Region of Western Macedonia, (b) territories close to Megalopolis town in Peloponnese and (c) territories close to Florina town, also located in the Region of Western Macedonia. Fly ash is used in the cement production process, to a proportion that has been specified by Greek legislation since 1980 [1–3].

Fly ash is an aluminosilicate material, and its chemical/mineral composition depends on the composition of the coal and the combustion conditions. Because of its fineness and pozzolanic nature, fly ash is widely accepted and specified as mineral admixture both in cement and concrete. In concrete, fly ash substitutes a part of cement. Fly ash was recognized as a pozzolanic constituent for use in concrete in 1914. However, research on the material began in 1937. Since then, extensive research has been conducted throughout the world [4]. Fly ash increases concrete workability, reduces water demand for the same slump, increases strength at later curing ages and improves corrosion resistance. It is highly heterogenous, from its particle size to the chemical composition. It can be seen from the several studies in the past that the properties of the final product are dependent on the properties of fly ash. Therefore, it is difficult to arrive at a definitive conclusion on the effect of fly ash on concrete. However, the results available from the extensive studies can be considered as guidelines for further research [5].

Blended fly ash cement can be produced either by intergrinding fly ash with Portland cement clinker or by blending dry fly ash with Portland cement. European Standard EN 197 determines the requirements for fly ash and the composition (percentage by mass) of cement types containing fly ash. The benefits of fly ash blends are that they are cheaper and reduce the amount of clinker needed. This reduction in needed clinker results in the reduction of needed energy and the reduction of the amount of carbon dioxide released into the atmosphere during the production of clinker. Environmental benefits also include the utilization of a waste material and the use of less raw resources. On the other hand, blended fly ash cement has slower setting time, which means its early strength is lower than cement without fly ash, making it unsuitable for use in the precast industry and potentially increasing construction times. It also requires more curing, while its resistance to carbonation is lower, risking corrosion of steel reinforcement. Moreover, with coal-fired power becoming increasingly unpopular (at least in Europe), because of its environmental impact and competition from alternative power sources, such as renewables, the long-term availability of fly ash is in question.

Naturally occurring radioactivity concentrations in lignite used in Thermal Power Plants (TPP) in Greece and the produced fly ash and bottom ash have already been determined since the 1980s [6–8]. These concentrations are generally high, especially in the case of lignite coming from mines in the Megalopolis Peloponnese area, where the concentration for radioisotopes of the ^{226}Ra chain is found to be about 1 kBq/kg [6–12]. As has been already discussed [8], there have been significant differences in the results of various studies relative to the concentration of naturally occurring radioisotopes in lignite-produced fly ash. This could be attributed to differences between the sampling approaches followed in each study. In the same context, the following parameters also contribute:

- Intrinsic variation by time of the under-study concentrations, according to which was the exact deposit where the finally burned lignite was originating from.
- Variations among technical characteristics of the different burning processes.

Article 75 of the European Directive 2013/59/Euratom ([13], called the Directive for the rest of this study) defines the activity concentration index (*ACI*), abbreviated as I for index. This Directive has been adopted by the Greek legislation through article 75 of the Greek Radiation Protection Regulation (Presidential Decree 101/2018, Governments Gazette 194A/20 November 2018, called the Greek Regulation for the rest of this study). Since then, the estimation of *ACI* is obligatory for each "construction material" which is "identified by the Member State (Greece in this case) as being of concern from a radiation protection point of view". The *ACI* formula shown in Equation (1) is provided in Annex VIII of the Directive and adopted in Annex VIII of the Greek Regulation.

$$I = ACI = \frac{C_{Ra-226}}{300 \text{ Bq/kg}} + \frac{C_{Th-232}}{200 \text{ Bq/kg}} + \frac{C_{K-40}}{2000 \text{ Bq/kg}} \quad (1)$$

where C_x is the specific (by mass) activity for x contributor, namely the ^{226}Ra series, the ^{232}Th series and ^{40}K, expressed in Bq/kg.

Essentially, the exposure to ionizing radiation for a person standing inside a structure is an effect of many contributing factors. A similar index has been proposed in Annex A of the 1993 UNSCEAR Report [14]. It aimed at assessing the level of exposure to the gamma radiation field for a person standing on the ground, due to the presence of Naturally Occurring Radioactive Materials (NORM, Figure 1). The UNSCEAR 1993 proposal included the results of the NORMs measurement in Greece [14,15]. As can be seen in Figure 1, the geometry of the person's exposure, when assuming exposure in a standing position on the ground, is relatively simple. The geometry in question is simpler if the soil material is homogeneous. In the same UNSCEAR report, the same index was introduced for assessing personal exposure in indoor spaces, too. The main assumption for this generalization is that the source of the gamma radiation field is construction materials.

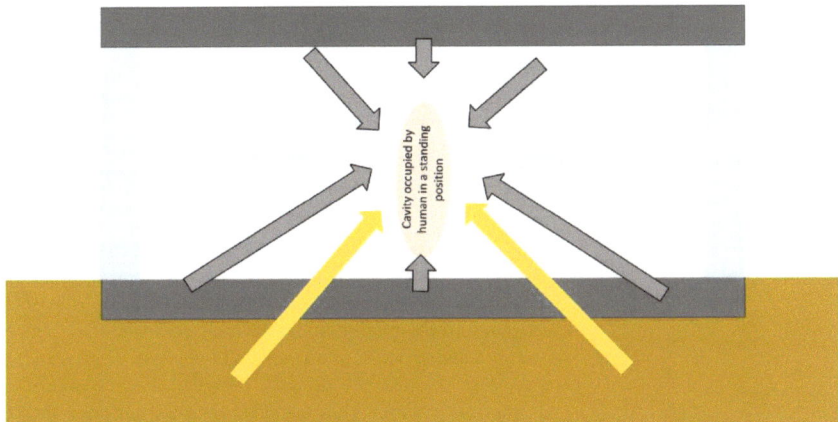

Figure 1. Exposure of a person to gamma radiation field when standing on the ground, due to gamma radiation emitted from soil (yellow arrows) or standing on the floor of a multistory building with concrete as the main structural element, due to gamma radiation emitted by the radioactivity contained in the surrounding materials (dark grey arrows for radiation produced by the floor and the ceiling concrete slabs and light grey arrows for radiation produced by any perpendicular construction materials' masses, e.g. masonry).

Many previous studies estimated *ACI* (or I_γ index, as mentioned in some of them) or other similar indexes such as the ^{226}Ra equivalent activity concentration (Ra_{eq} in Bq/kg) and the external hazard index (H_{ex}, dimensionless as per *ACI*), when gamma radiation is assessed for concrete or even its constituents [16–27]. None of them have proposed a method according to the ISO approach to estimate the uncertainty of the results. Therefore, a producer of construction products/materials or a laboratory contributing to quality control, are not effectively supported to establish a decision rule on whether the product/material is "of concern from a radiation protection point of view".

A certain previous study provides a typology for estimating uncertainty for NORMs measurement in construction materials, according to the ISO methodology [28]. The latter provides a methodology for estimating reliability intervals following measurements on specific specimens that could be either concrete, fresh or hardened, or even samples from concrete constituents. This is a very good approach when establishing a performance indicator, useful for Factory Production Control (FPC). A weakness is that in this way we cannot reveal an overall figure for the entire production or even worse for all production and construction in a country-wide area.

The present study focuses on the case of residences that are part of a multistory concrete construction. Such residential buildings are common in Greece, especially since the 1960s. The exposure geometry of a person standing in the middle of such an indoor space is more complicated than in the case of a person standing on the open field, on plane soil ground. The gamma radiation field is created by concrete, either load bearing or infill, masonry, covering materials such as tiles or plaster, etc., to the extent that they have been used in construction (Figure 1).

In addition, the person is also exposed to the material of the overlying slab, whether it is the floor of the upper floor or the roof of the structure. Also, the person is exposed to the gamma radiation field produced by any vertical structural elements, in a distance that varies significantly by case (Figure 1).

For such complicated exposure geometry, the 1993 UNSCEAR report suggests that *ACI* should be weighed for the mass proportion of the building materials present in the analyzed construction. Equation (2) may be used for estimating the concentration for each of the NORMs that contribute to Equation (1). In this study, where the case is any dwelling in a multistory concrete structure, the activity concentration of NORMs in concrete is the prevailing source external gamma radiation field. Consequently, the result that is assessed in this study is an assessment of the *ACI* only for concrete, so Equation (2) is not used for further incorporation of other building materials (e.g., masonry or tiles).

$$C_x = \frac{\sum_{i=1}^{n} a_i \cdot C_{x,i}}{\sum_{i=1}^{n} a_i} \qquad (2)$$

As known [29], the intensity of the external (to the exposed person) gamma field attenuates according to the inverse square distance to the source. This is not incorporated in Equation (2). Essentially, all construction materials are assumed to be homogeneously distributed in the ground on which the person stands. Consequently, to answer the question whether a construction material is "of concern from a radiation protection point of view", we need to consider the distance between the person and the material. On this basis, it is not practical to apply Equation (2) for construction materials that are located too far away from the person, especially if other materials (e.g., a wall) are in the intermediate space. This means that Equation (2) should only be applied for consequent residential spaces (e.g., one single separate room).

On the other hand, an average person uses multiple residential spaces inside an apartment, according to the use of each room. Generally, different rooms inside the same apartment use different construction materials (e.g., a greater mass of tiles is installed on the internal surfaces of a bathroom). A realistic assessment of the person's exposure to the gamma field emitted by the construction materials in the apartment that he inhabits must consider the distribution of time that the person spends in each room.

An assessment of personal exposure to radioactivity contained in the construction products that are produced in Greece has been performed [30,31] considering, in some cases, even the subsequent exposure to radon concentrations [32–34]. It is recalled that in the case of exposure of a person inside an enclosed space, such as inside a residence in a multistory structure, an important source of internal exposure of the person to radiation comes from radon. Therefore, the estimation of the exposure level of the individual inside a residential building is highly complex. It depends on many factors, not only the concentration of radioactivity in the building materials of the structure, but also on the way the resident uses the apartment, such as the mechanical and natural ventilation.

The exact assessment of the *ACI* for the case of a person inside an apartment in a multistory building requires knowledge in relation to all the construction materials that make up the construction, such as the presence of masonry which is also mentioned as a potentially significant source of gamma radiation [35]. This assessment becomes even more complicated if unusual building materials that are likely to contain a high concentration of NORMs have been used in the construction, as, for example, in the case where certain types of granite are used to cover surfaces or as benches. It is also noted that even in cases where the concentrations of ^{226}Ra, ^{232}Th and ^{40}K in the granites may not be particularly high, the presence of other radioisotopes in this material is also mentioned, increasing the level of the produced gamma radiation field [36]. In these cases, the fact that the *ACI* formula (Equation (1)) cannot incorporate the effect from these radioisotopes should also be considered.

If *ACI* estimation aims at an epidemiological risk assessment due to the individual's exposure to gamma radiation coming from the building materials, there are many factors that must be considered. According to Directive 2013/59/Euratom that has been adopted through the Greek Regulation, and specifically in para. 1 of article 75, "The reference level applying to indoor external exposure to gamma radiation emitted by building materials, in addition to outdoor external exposure, shall be 1 mSv per year". According to Annex

VIII of the same Directive, "The activity concentration index value of 1 can be used as a conservative screening tool for identifying materials that may cause the reference level laid down in Article 75(1) to be exceeded". In this case however, definitional uncertainty, which in paragraph 2.27 of ISO VIM [37] is defined as "component of measurement uncertainty resulting from the finite amount of detail in the definition of a measurand", should be considered. In cases of in situ measurements, not the analysis of samples under strict laboratory conditions, definitional uncertainty can be significant [38]. For reasons analyzed above, the question arises as to whether the *ACI* can fulfill the purpose of assessing the individual's exposure, even if the question is restricted to indoor external exposure to gamma radiation emitted by building materials.

Moreover, due to current legislative status in Greece that is compliant with the European Union's New Legislative Framework, any construction product used in Greece shall both comply with the Construction Products Regulation [39] and Greek Radiation Protection Regulation (adopting the European Directive 2013/59/Euratom). Thus, all construction products that are subject to trade in Greece shall be monitored for their performance on *ACI* values. The above framework becomes particularly interesting when building materials are subject to certification procedures.

This paper aims to assess whether the use of lignite fly ash as an additive in cement produced in Greece should be "of concern from a radiation protection point of view". This assessment is applied to concrete produced in Greece using fly ash cement as a constituent, considering that concrete is the predominant construction material (product) for multistory residential buildings. The evaluation of *ACI* values on an approximately 95% confidence level is presented, based on a corresponding uncertainty budget according to ISO GUM methodology [40]. The *ACI* calculation formula (Equation (1)) was applied for concrete mix designs that are possibly used in concurrent constructions in Greece. The estimates are provided assuming that legislative restrictions on the use of fly ash and the requirements of the Greek Concrete Technology Regulation [41] have been met.

Besides with the above assessment for the Greek case of concrete containing cement with lignite fly ash, this paper proposes for the first time a general simple approach, that could be used by any responsible authority that wants to assess whether a construction product is of concern from a radiation protection point of view. Each country that wishes to evaluate the use of fly ash in constructions, could repeat the method for the *ACI* uncertainty budget proposed in this study and apply the same or equivalent decision rule.

2. Data and Methods

Strictly analyzing the uncertainty of the result when applying Equation (1), the contribution to *ACI* estimation parameters is graphically analyzed through a fishbone diagram in Figure 2. This figure shows that an effective uncertainty budget on the usage of Equation (1) does not rely just on the uncertainty budget on the estimation of activity concentration for each participating NORM. Equation (1) relies on the three major NORMs (^{226}Ra and ^{232}Th series of radioisotopes and ^{40}K) concentration but also on a definition uncertainty caused by a lack of knowledge on the exact placement of the construction materials in the residence.

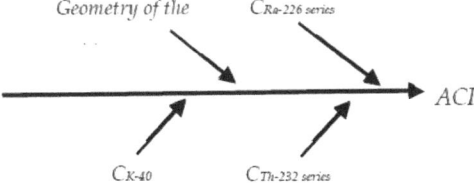

Figure 2. Fishbone diagram of parameters contributing to the estimation of *ACI* when the 2013/59/Euratom Directive formula is used.

As mentioned before, *ACI* estimation should be performed for each individual construction material that is used in the under-assessment structure (Figure 3), but the following analysis focuses only on estimating confidence interval and on applying a decision rule on the estimated *ACI* value only for concrete.

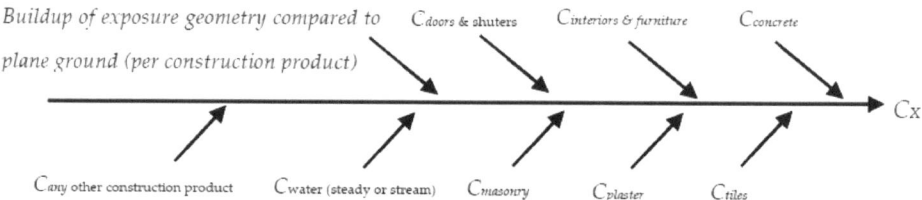

Figure 3. Fishbone diagram showing that the activity concentration that influences the inhabitant of a building is related to many types of construction products (materials).

The presence of fly ash in a concrete construction (i.e., concrete is used as the predominant material) is indirectly affected by the corresponding regulatory context. The status of restriction in the use of fly ash as a cement constituent in Greece could be summarized on a time scale as follows:

Before 1980: There is no existing regulation for the fly ash usage in cement. There is only regulation for natural pozzolana [1].

After 1980: There are two categories of cement containing pozzolana (natural and man-made). In the first category the percentage of pozzolana is determined by the insoluble residue of the cement, which must be 20% maximum. In the second, the insoluble residue of the cement must be 20–40% [2].

After 2002: European Standard EN 197-1 is obligatory, meaning the insoluble residue of the cement must be 35% maximum [3]

Table 1 presents data obtained from the official websites of the three main cement producers in Greece. Those are HERACLES General Cement Company S.A., TITAN Cement Company S.A. and HALYPS Building Materials S.A. In this table, reference is made to the percentage of the presence of cement components, among which is fly ash. The percentage of fly ash present in cement produced in Greece ranges from 0% to 20%. This interval is used as input data for estimating the mass ratio (proportion into cement) in Equation (4) and the corresponding uncertainty budget.

Table 1. Types of cement provided in Greek constructions according to the use of fly ash.

Indicative Types of Cement That the Producer Provides in Year 2023	Cement Producer	Commenting on the Presence of Fly Ash in Cement (As Product)
CEM II/B-M (P-W-L) 32.5N	HERACLES	Natural pozzolana—calcareous fly ash—limestone: 21–35%
CEM II/B-M(W-P-LL) 32.5N	TITAN	Natural pozzolana—calcareous fly ash—limestone: 21–35%
CEM IV/B(P-W) 32.5 (sulfate resistant)	HALYPS	Natural pozzolana—calcareous fly ash—limestone: 36–55%

Table 2 presents results obtained from previous papers on the concentrations of NORMs in samples of fly ash produced by Greek TPPs. This type of fly ash is mainly used by Greek cement producers. The values of mass specific activity of fly ash used in cement that are used in Equation (4) are provided by a review of the values presented in Table 2.

Table 2. Data retrieved from previous works on the potential concentrations of naturally occurring radioisotopes (^{226}Ra, ^{232}Th and ^{40}K) in fly ash produced during lignite combustion in Greek TPPs.

Lignite Origin According to Reference	NORM Concentrations [Bq/kg]		
	^{226}Ra	^{232}Th	^{40}K
Megalopolis (Region of Peloponnese) [7–9,42,43]	293–1058	4–89	308–590
Ptolemais and Kardia (Region of Western Macedonia) [7,8]	204–825	<90	162–299
Aliveri (Evia, Region of Central Greece) [7]	257–357	0.6–2.0	-
Cement producers [31,44]	200–1400	34–84	<650
Greek lignite according to all references	200–1400	<90	<650

Table 3 presents a summary of the results obtained from previous work on the concentrations of radioisotopes of natural origin and in particular the ^{226}Ra and ^{232}Th series and ^{40}K in samples from materials which, according to the current Concrete Technology Regulation, constitute the minimum necessary materials so that the mixture can be called concrete. These values are the input data for Equation (3), specifically for estimating the mass specific activity of concrete constituents other than cement containing fly ash.

Table 3. NORM mass specific concentrations in concrete constituents according to papers related construction materials produced in Greece.

What Was Measured?	NORM Concentrations [Bq/kg]		
	^{226}Ra	^{232}Th	^{40}K
Limestone [31]	5–7	4–10	86–128
Pozzolanic materials [31]	34–40	19–53	298–424
Gypsum [31]	5–9	<LoD *	<LoD
Clinker [31]	12–18	5–23	102–180
Cement not containing fly ash [32]	96	22	200
Total of no fly ash cement constituents	5–96	<53	<424
Cement containing fly ash [32]	215–218	11–26	222–330
Cement (in general) [12]	15–218	10–41	32–457
Sand [31]	7–15	<9.9	<60
Aggregates (in general) [12]	3–46	3–56	19–1048
Aggregates [12,31]	3–46	<56	<1048
Water [42]	<2	<2	<100

* LoD: limit of detection.

It should be noted that all concrete constituents contain NORMs, mainly the aggregates, but even water does so in a significantly lower concentration. Common types of concrete dealt with in this paper mainly contain limestone aggregates, in which the concentration of NORMs is generally low (Table 3). It should be noted that construction materials coming from neighboring countries (e.g., cement imported from Turkey) are used regularly in Greek construction projects. The issue of the NORM presence in them has been studied, providing measurement results like or even higher than those performed for Greek construction products. The intervals reported in Table 3 correspond to the minimum and maximum referenced value. This is assumed to provide intervals at a confidence level of approximately 99.7%. In the same context wherever a reference paper reported a confidence interval based on either the standard deviation of the results or the combined standard uncertainty, Table 3 presents intervals based on these statistics, multiplied by three.

In this paper *ACI* values' reliability limits at an approximately 95% level of confidence are calculated for common types of concrete that may be produced in Greece in year 2023. This proposed calculation could also be repeated for any construction period in the past, considering the corresponding concrete compositions. This paper is restricted to common types of concrete. Consequently, cases of (a) concrete prepared with light weight or heavyweight aggregates and (b) high performance concrete (HPC) were not considered.

Considering that common types of concrete contain cement with fly ash (*fa,cem*), aggregates (*agg*) and water (*w*) in proportions resulting from the corresponding composition study and that any other possible material (e.g., superplasticizer) participates with a non-significant mass ratio, the mass specific activity (concentration) in any case of naturally occurring radioisotopes ^{226}Ra, ^{232}Th and ^{40}K, was calculated according to Equation (3), as also presented graphically in Figure 4.

$$C_{con} = \frac{\alpha_{fa,cem} \cdot C_{fa,cem} + \alpha_{agg} \cdot C_{agg} + \alpha_w \cdot C_w}{\sum \alpha_i} \qquad (3)$$

where a_i are the proportions of the three concrete constituents, expressed in kg/m^2 of fresh concrete, i.e., exactly as determined according to the concrete composition study.

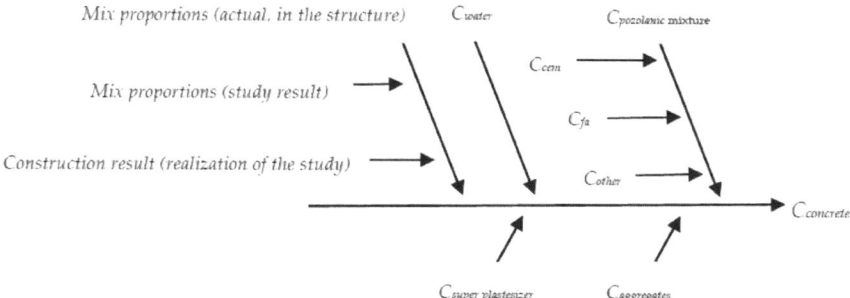

Figure 4. Fishbone diagram showing the parameters that contribute to the concentration of NORMs in a concrete mixture.

In the same equation, no correction has been included according to the possible difference in the moisture content of the fly ash between the values (a) as per the measurement process for determining the activity of NORMs and (b) when mixing with the rest of the cement components. In practice, it was assumed that in these two situations, where the fly ash is expected to be in a dry state, the moisture content is approximately at the same level with no significant difference.

The fresh concrete mix proportions were taken as the ones likely to be used in the Greek construction industry in the year 2022, mainly according to the minimum allowable values resulting from the Greek Concrete Technology Regulation [41] and the maximum possible cement ratio values, considering the economics of residential buildings' construction.

Previous studies were used for estimating the confidence intervals (limits) for mass specific activity values of NORMs in aggregates and water (see Table 2). In the case of the mass specific activity of cement, a calculation was made for each of the three NORMs according to Equation (4):

$$C_{fa,cem} = \alpha_{fa} \cdot C_{fa} + \alpha_{cem} \cdot C_{cem} \qquad (4)$$

where a_{fa}, C_{fa} are the mass ratio (proportion into cement) and mass specific activity of fly ash and a_{cem}, C_{cem} are the mass ratio (proportion into cement) and mass specific activity for the rest of the cement constituents.

Previous studies were also used for estimating confidence intervals (limits) for mass specific concentration values of NORMs in cement and fly ash (Table 1). The a_i proportions of cement and fly ash were obtained according to the permissible limits imposed by the concurrent Greek legislation.

As mentioned above, the result of this work, in the form of a confidence interval of about 95% level for the value of *ACI* was obtained through the application of the ISO GUM methodology [40]. Since the information used in the calculations comes from third-party sources, such as legislation limits or results of previous work and measurements, extensive use was made of type B estimates for the combined standard uncertainty of the result.

3. Results

Table 4 provides the uncertainty budget for the estimation of the mass specific activity of NORMs in cement containing lignite fly ash. The uncertainty budget presented in the table is based on the results of Equation (4). As the ratios of fly ash and other cement constituents are linearly correlated (see Equation (4)), the covariance $cov(\alpha_{fa,cem}, \alpha_{cem})$ was assumed based on a linear correlation coefficient equal to -1.

Table 4. Uncertainty budget for the calculation of the confidence interval for the values of the specific activity of ^{226}Ra, ^{232}Th and ^{40}K in cement to which lignite ash has been added.

Factor	Mean Value	Bounds of Possible Values	Distribution of Possible Values	Standard Uncertainty	Sensitivity Coefficient	Contribution to the Uncertainty of the Result
			^{226}Ra			
$\alpha_{fa,Ra226}$	0.1	0.1	Rectangular	0.06	800 Bq/kg	2133 (Bq/kg)^2
$\alpha_{cem,Ra226}$	0.9	0.1	Rectangular	0.06	50 Bq/kg	9 (Bq/kg)^2
$C_{fa,Ra226}$	800 Bq/kg	600 Bq/kg	Triangular	245 Bq/kg	0.1	600 (Bq/kg)^2
$C_{cem,Ra226}$	50 Bq/kg	46 Bq/kg	Triangular	19 Bq/kg	0.9	292 (Bq/kg)^2
		$Cov(\alpha_{fa,Ra226}, \alpha_{cem,Ra226})$				-288 (Bq/kg)^2
$C_{fa,cem,Ra226}$	125 Bq/kg					2746 (Bq/kg)^2
			^{232}Th			
$\alpha_{fa,Th232}$	0.1	0.1	Rectangular	0.06	45 Bq/kg	7 (Bq/kg)^2
$\alpha_{cem,Th232}$	0.9	0.1	Rectangular	0.06	27 Bq/kg	3 (Bq/kg)^2
$C_{fa,Th232}$	45 Bq/kg	45 Bq/kg	Triangular	18 Bq/kg	0.1	3 (Bq/kg)^2
$C_{cem,Th232}$	27 Bq/kg	27 Bq/kg	Triangular	11 Bq/kg	0.9	98 (Bq/kg)^2
		$Cov(\alpha_{fa,Th232}, \alpha_{cem,Th232})$				-9 (Bq/kg)^2
$C_{fa,cem,Th232}$	29 Bq/kg					102 (Bq/kg)^2
			^{40}K			
$\alpha_{fa,K40}$	0.1	0.1	Rectangular	0.06	325 Bq/kg	380 (Bq/kg)^2
$\alpha_{cem,K40}$	0.9	0.1	Rectangular	0.06	212 Bq/kg	162 (Bq/kg)^2
$C_{fa,K40}$	325 Bq/kg	325 Bq/kg	Triangular	133 Bq/kg	0.1	177 (Bq/kg)^2
$C_{cem,K40}$	212 Bq/kg	212 Bq/kg	Triangular	87 Bq/kg	0.9	6130 (Bq/kg)^2
		$Cov(\alpha_{fa,K40}, \alpha_{cem,K40})$				-496 (Bq/kg)^2
$C_{fa,cem,K40}$	223 Bq/kg					6353 (Bq/kg)^2

Table 4 shows the following:

- The proportion of cement constituents is reported as a dimensionless number, meaning the mass fraction of constituents and the value of the bounds of possible values [40] refer to the half interval on both sides of the mean to produce a confidence interval of about 99.7% level.
- It should be noted that the range of values considered for the proportion of fly ash addition to cement is a worst-case scenario compared to what is known for the actual construction activity in Greece in the last 20 years. The proportion of fly ash added into cement has decreased in recent years due to a decrease in the production of fly ash by Greek TPPs. In the early years of this twenty-year period, the proportion of

fly ash added to cement was approximately up to 15% by mass. In current years, the maximum value of this rate has been reduced to approximately 3%. A range of values for $\alpha_{fa,Ra226}$ = 0–0.2 (i.e., 0% to 20% fly ash percentage, by mass, in the produced cement) was assumed. Such an interval is so wide to includes any effect of a possible difference in the moisture content of the fly ash between samples' moisture content when analyzed in a gamma spectroscopy setup for determining ^{226}Ra, ^{232}Th and ^{40}K activity and the moisture content of all other cement constituents, during their mix.

- The uncertainty budget in Table 4 leads to confidence intervals at an approximately 95% level presented in Table 5 due to applying the "law of error propagation" [40]. In this context, the bounds (a range) of possible values and a corresponding distribution is assumed to produce the standard uncertainty for each variable that contributes to Equation (4). Standard uncertainties are multiplied by the corresponding sensitivity coefficients. These are the partial derivatives of the applied equation as per the variables for which the sensitivity coefficient is estimated. The right-most column of Table 4 shows the contribution of each variable to the finally estimated uncertainty of the result for each NORM as the square value of the result of this multiplication or the value of the covariance between the two mass proportions. The results presented in Table 4 are the sum of these contributions, which is the square of the combined standard uncertainty relative to the result of Equation (4), for each NORM.
- Table 5 presents the result of Equation (4) as an approximately 95% confidence interval. These intervals were estimated as the square root of the results obtained from Table 4, for each NORM, multiplied by a coverage factor k equal to 2.
- The values provided in Table 5 are used as input data in calculations provided in Table 6 in relation to the possible values of the mass specific activity of cement containing fly ash. Table 6 provides the uncertainty budget for the estimation of mass specific activity of fresh concrete constituents: cement, aggregate and water. The uncertainty budget presented in these tables refers to Equation (3). It is noted that the ratios of these three concrete constituents have an expected correlation between them, as any decision to increase the ratio of the one, (e.g., of the cement to seek a greater strength in the hardened concrete), leads to a corresponding readjustment of the ratio of at least one or even both the remaining constituents.

Table 5. Estimated 95% confidence intervals for NORM mass specific concentrations (Bq/kg) in cement produced in Greece during the year 2023.

Estimated for	$C_{fa,cem}$ (Bq/kg)
^{226}Ra	$(1.2 \pm 1.0) \times 10^2$
^{232}Th	29 ± 20
^{40}K	$(2.2 \pm 1.6) \times 10^2$

In Table 6, a weak negative correlation (linear correlation coefficient $r = -0.2$) between the cement ratio and aggregate ratio and between the water and aggregate ratio was assumed. An increased linear correlation coefficient, equal to 0.5, was assumed between the ratio of cement and the ratio of water, as it is known that concrete producers put efforts to maintain a steady value for the cement to water ratio in the mixture.

The uncertainty budget in Table 6 leads to confidence intervals at an approximately 95% level presented in Table 7 because of the application of the "law of error propagation" [40], through the same process that was followed for Tables 4 and 5.

According to the above calculations, and by applying Equation (1), it follows that the value of the ACI for the common types of concrete used in multistory constructions in the Greek area is 0.47 ± 0.20. The decision rule by which the result is compared to the reference value of EURATOM Directive is that the entire 95% confidence interval for the values of ACI should be below 1. Consequently, the maximum value on a 95% confidence interval is

0.67, which means that the entire confidence interval is significantly less than 1 (less than the maximum reference value of EURATOM Directive).

Table 6. Uncertainty budget for the calculation of the confidence interval for the values of the specific activity of ^{226}Ra, ^{232}Th and ^{40}K in fresh concrete.

Factor	Mean Value	Bounds of Possible Values	Distribution of Possible Values	Standard Uncertainty	Sensitivity Coefficient	Contribution to the Uncertainty of the Result
			^{226}Ra			
$\alpha_{fa,cem}$	320 kg/m^2	40	Triangular	16	0.03	0.33 (Bq/kg)^2
α_{agg}	1500 kg/m^2	200	Triangular	82	−0.005	0.16 (Bq/kg)^2
α_w	164 kg/m^2	52	Triangular	21	−0.01	0.1 (Bq/kg)^2
$C_{fa,cem,Ra226}$	125 Bq/kg			42	0.14	35 (Bq/kg)^2
$C_{agg,Ra226}$	24 Bq/kg	22	Triangular	9	0.79	50 (Bq/kg)^2
$C_{w,Ra226}$	2 Bq/kg	2	Triangular	1	0.07	0.003 (Bq/kg)^2
		$Cov(\alpha_{fa,cem}, \alpha_{agg})$ ($r = -0.2$, is assumed)				0.09 (Bq/kg)^2
		$Cov(\alpha_{fa,cem}, \alpha_w)$ ($r = 0.5$, is assumed)				−0.18 (Bq/kg)^2
		$Cov(\alpha_{agg}, \alpha_w)$ ($r = -0.2$, is assumed)				−0.05 (Bq/kg)^2
$C_{con,Ra226}$	35 Bq/kg		-			85 (Bq/kg)^2
			^{232}Th			
$\alpha_{fa,cem}$	320 kg/m^2	40	Triangular	16	0.001	0.0004 (Bq/kg)^2
α_{agg}	1500 kg/m^2	200	Triangular	82	0.0007	0.004 (Bq/kg)^2
α_w	164 kg/m^2	52	Triangular	21	−0.01	0.05 (Bq/kg)^2
$C_{fa,cem,Th232}$	29 Bq/kg			10	0.14	2 (Bq/kg)^2
$C_{agg,Th232}$	28 Bq/kg	28	Triangular	11	0.79	81 (Bq/kg)^2
$C_{w,Th232}$	2 Bq/kg	2	Triangular	0,8	0.07	0.003 (Bq/kg)^2
		$Cov(\alpha_{fa,cem}, \alpha_{agg})$ ($r = -0.2$, is assumed)				−0.0005 (Bq/kg)^2
		$Cov(\alpha_{fa,cem}, \alpha_w)$ ($r = 0.5$, is assumed)				−0.004 (Bq/kg)^2
		$Cov(\alpha_{agg}, \alpha_w)$ ($r = -0.2$, is assumed)				0.006 (Bq/kg)^2
$C_{con,Th232}$	26 Bq/kg		-			83 (Bq/kg)^2
			^{40}K			
$\alpha_{fa,cem}$	320 kg/m^2	40	Triangular	16	−0.10	3 (Bq/kg)^2
α_{agg}	1500 kg/m^2	200	Triangular	82	0.03	7 (Bq/kg)^2
α_w	164 kg/m^2	52	Triangular	21	−0.17	14 (Bq/kg)^2
$C_{fa,cem,K40}$	223 Bq/kg			78	0.14	119 (Bq/kg)^2
$C_{agg,K40}$	524 Bq/kg	524	Triangular	214	0.79	28,422 (Bq/kg)^2
$C_{w,K40}$	50 Bq/kg	50	Triangular	20	0.07	2 (Bq/kg)^2
		$Cov(\alpha_{fa,cem}, \alpha_{agg})$ ($r = -0.2$, is assumed)				2 (Bq/kg)^2
		$Cov(\alpha_{fa,cem}, \alpha_w)$ ($r = 0.5$, is assumed)				6 (Bq/kg)^2
		$Cov(\alpha_{agg}, \alpha_w)$ ($r = -0.2$, is assumed)				4 (Bq/kg)^2
$C_{con,K40}$	448 Bq/kg		-			28,580 (Bq/kg)^2

Table 7. Estimated 95% confidence intervals for NORM mass specific concentrations (Bq/kg) in concrete produced in Greece during the year 2023.

Estimated for	C_{con} (Bq/kg)
^{226}Ra	35 ± 18
^{232}Th	26 ± 18
^{40}K	$(4.5 \pm 3.4) \times 10^2$

4. Discussion

Despite the considered addition of fly ash to cement, none of ^{226}Ra, ^{232}Th and ^{40}K emerged as a prevailing radiological factor as compared to others. It should be noted that ^{40}K concentration in common fresh concrete is unlikely to affect *ACI* values. Covariances between the proportions of the concrete constituents are also unlikely to affect *ACI* values. On the other hand, the sensitivity coefficient related to changes in NORMs concentrations in the aggregates is relatively high. The possible addition of aggregates that do not belong to the—usually used in Greece—limestone category, must be considered.

It is emphasized again that this paper is restricted to the common types of concrete used in the load-bearing structure of a multistory building (slabs, columns and beams), in which case the presence of cement is determined: (a) by the minimum limits for the cement content of the concrete due to the required minimum strength and (b) the economy of construction, which, according to the logic of the construction market, sets the respective upper possible limits of the concrete's cement content.

The ratios shown in the calculations in Table 6 correspond to fresh concrete. If a reduction is attempted in the hardened concrete, the loss of water mass due to evaporation during the hardening period should be considered. As already discussed in the introduction of this paper, it is impractical to attempt to accurately reproduce the exposure geometry of the individual residual in the structure, as this is affected by many significant uncertainty factors. The estimation and uncertainty budget presented in this paper provide an index which makes sense for concrete producers by assisting them in assessing the performance of their product, especially in the context of the regulatory control of trade in the European market.

It should be noted that the above estimate for *ACI* is not a result that corresponds to a specific type of concrete or, even more so, an individual mix proportions' study. It is an estimate of the confidence interval at a level of approximately 95% for all common types of concrete that are expected to have been produced in the Greek area within the last twenty years. The most important feature of this interval is the upper value of *ACI*, equal to 0.67 (0.47 + 0.20). It indicates that the common types of concrete containing fly ash cement are impossible to approach the limit value equal to 1.

ACI can be considered from the perspective of monitoring the performance of construction materials, as required for other performances under the EU Construction Products Regulation [39]. In this context, *ACI* with a reference value equal to 1, could be a performance indicator. A same case that has been regulated in the EU is the ranking of buildings according to their energy performance indicator (EPI), which is obtained as the quotient of the estimated energy consumption in a building to a corresponding reference value [45]. The similarities with the *ACI* are many, as the denominators in Equation (1) provide a clear, and horizontal, reference value.

The methodology presented in this paper may be useful, too, for laboratories performing testing on the NORMs concentration in building materials. According to the International Standard ISO/IEC 17025:2017 [46], the conformity assessment of a product subjected to laboratory testing should be carried out based on a prescribed decision rule. This rule must consider the level of uncertainty of the result of the test(s). If a laboratory is asked to give an opinion or interpretation in relation to the compliance of a construction product, such as concrete, with the *ACI* < 1 criterion, then the laboratory should have established an uncertainty budget, like the one presented in this paper.

5. Conclusions

Fly ash used as an additive in cement for concrete multistory constructions in Greece should not be considered as "of concern from a radiation protection point of view", according to the provisions of 2013/59/Euratom Directive.

Each country that wishes to evaluate the use of fly ash in constructions should repeat the method for the *ACI* uncertainty budget proposed in this study, to assess whether it significantly exceeds the value equal to 1.

When *ACI* is estimated for each construction product but also for the entire construction, given the mass proportions of all the used products in a construction, it could serve as a performance indicator. This is useful in construction product certification procedures, with a special interest for concrete producers.

The above-mentioned use of *ACI*, along with the corresponding uncertainty estimation, could be a useful tool for decision making for all the interested parties, such as construction product producers, building designers and testing laboratories that perform measurements focused on estimating *ACI*.

Author Contributions: Conceptualization, G.P.; Methodology, G.P.; Formal analysis, G.P.; Writing–original draft, S.G. and G.P.; Writing–review & editing, S.G.; Supervision, S.G. All authors have read and agreed to the published version of the manuscript.

Funding: This research received no external funding.

Institutional Review Board Statement: Not applicable.

Informed Consent Statement: Not applicable.

Data Availability Statement: No new data were created.

Conflicts of Interest: The authors declare no conflict of interest.

References

1. Government Gazette of the Hellenic Republic 160A/26.07.1954 "On Regulations for the study and execution of construction projects made of Reinforced Concrete". Available online: https://www.et.gr/api/Download_Small/?fek_pdf=19540100160 (accessed on 15 November 2023).
2. Government Gazette of the Hellenic Republic 69A/28.03.1980 "On the Regulation of Cements for Concrete Projects (Prestressed, Reinforced and Unreinforced". Available online: https://www.et.gr/api/Download_Small/?fek_pdf=19800100069 (accessed on 15 November 2023).
3. Government Gazette of the Hellenic Republic 537B/01.05.2002 "Adaptation of the Concrete Technology Regulation to the Requirements of the Harmonized Standard ELOT EN 197-1 "Cement—Part 1: Composition, Specifications and Conformity Criteria for Common Cements". Available online: https://www.et.gr/api/Download_Small/?fek_pdf=20020200537 (accessed on 15 November 2023).
4. Joshi, R.C.; Lohtia, R.P. *Fly Ash in Concrete: Production, Properties and Uses*, 1st ed.; Gordon and Breach Science Publishers: Amsterdam, The Netherlands, 1997; pp. 4–7.
5. Nayak, D.K.; Abhilash, P.P.; Singh, R.; Kumar, R.; Kumar, V. Fly ash for sustainable construction: A review of fly ash concrete and its beneficial use case studies. *Clean. Mater.* **2022**, *6*, 100143. [CrossRef]
6. Papastefanou, C.; Charalambous, S. On the radioactivity of fly ashes from coal power plants. *Zeit. F. Naturforschung* **1979**, *3402*, 533. [CrossRef]
7. Papastefanou, C.; Charalambous, S. Hazards from Reactivity of Fly Ash of Greek Coal Power Plants. In Proceedings of the 5th International Congress of the IRPA, Jerusalem, Israel, 9–14 March 1980; Volume VII, p. 161.
8. Simopoulos, S.E.; Angelopoulos, M.G. Natural radioactivity releases from lignite power plants in Greece. *J. Environ. Radioact.* **1987**, *5*, 379–389. [CrossRef]
9. Karangelos, D.J.; Rouni, P.K.; Petropoulos, N.P.; Anagnostakis, M.J.; Hinis, E.P.; Simopoulos, S.E. Radioenvironmental survey of the Megalopolis power plants fly-ash deposits. *Radioact. Environ.* **2005**, *7*, 1025–1029.
10. Papaefthymiou, H.; Symeopoulos, B.D.; Soupioni, M. Neutron activation analysis and natural radioactivity measurements of lignite and ashes from Megalopolis basin. *J. Radioanal. Nucl. Chem.* **2007**, *274*, 123–130. [CrossRef]
11. Skodras, G.; Grammelis, P.; Kakaras, E.; Karangelos, D.; Anagnostakis, M.; Hinis, E. Quality characteristics of Greek fly ashes and potential uses. *Fuel Process. Technol.* **2007**, *88*, 77–85. [CrossRef]
12. Trevisi, R.; Leonardia, F.; Risicab, S.; Nuccetelli, C. Updated database on natural radioactivity in building materials in Europe. *J. Environ. Radioact.* **2018**, *187*, 90–105. [CrossRef]

13. Council Directive 2013/59/Euratom of 5 December 2013 Laying down Basic Safety Standards for Protection against the Dangers Arising from Exposure to Ionising Radiation, and Repealing Directives 89/618/Euratom, 90/641/Euratom, 96/29/Euratom, 97/43/Euratom and 2003/122/Euratom. Available online: https://eur-lex.europa.eu/LexUriServ/LexUriServ.do?uri=OJ:L:2014:013:0001:0073:EN:PDF (accessed on 15 November 2023).
14. UNSCEAR 1993 Report, Sources and Effects of Ionizing Radiation, United Nations Scientific Committee on the Effects of Atomic Radiation UNSCEAR 1993 Report to the General Assembly, with Scientific Annexes. Available online: https://www.unscear.org/docs/publications/1993/UNSCEAR_1993_Report.pdf (accessed on 15 November 2023).
15. Simopoulos, S.E. *Natural Radioactivity Analysis Results of Greek Surface Soil Samples*; MPX3 Report; National Technical University: Athens, Greece, 1990.
16. Kumar, V.; Ramachandranb, T.V.; Prasad, R. Natural radioactivity of Indian building materials and by-products. *Appl. Radiat. Isot.* **1999**, *51*, 93–96. [CrossRef]
17. Faheema, M.; Mujahidb, S.A.; Matiullah. Assessment of radiological hazards due to the natural radioactivity in soil and building material samples collected from six districts of the Punjab province-Pakistan. *Radiat. Meas.* **2008**, *43*, 1443–1447. [CrossRef]
18. El-Taher, A.; Makhluf, S.; Nossair, A.; Abdel Halim, A.S. Assessment of natural radioactivity levels and radiation hazards due to cement industry. *Appl. Radiat. Isot.* **2010**, *68*, 169–174. [CrossRef]
19. Al-Sulaiti, H.; Alkhomashi, N.; Al-Dahan, N.; Al-Dosari, M.; Bradley, D.A.; Bukhari, S.; Matthews, M.; Regan, P.H.; Santawamaitre, T. Determination of the natural radioactivity in Qatarian building materials using high-resolution gamma-ray spectrometry. *Nucl. Instrum. Methods Phys. Res. A* **2011**, *652*, 915–919. [CrossRef]
20. Kovler, K. Legislative aspects of radiation hazards from both gamma emitters and radon exhalation of concrete containing coal fly ash. *Constr. Build. Mater.* **2011**, *25*, 3404–3409. [CrossRef]
21. Baykara, O.; Karatepe, S.; Dogru, M. Assessments of natural radioactivity and radiological hazards in construction materials used in Elazig, Turkey. *Radiat. Meas.* **2011**, *46*, 153–158. [CrossRef]
22. Sofilic, T.; Barišic, D.; Sofilic, U.; Đurokovic, M. Radioactivity of some building and raw materials used in Croatia. *Pol. J. Chem. Technol.* **2011**, *13*, 23–27. [CrossRef]
23. Lu, X.; Yang, G.; Ren, C. Natural radioactivity and radiological hazards of building materials in Xianyang, China. *Radiat. Phys. Chem.* **2012**, *81*, 780–784. [CrossRef]
24. Moharram, B.M.; Suliman, M.N.; Zahran, N.F.; Shennawy, S.E.; El Sayed, A.R. External exposure doses due to gamma emitting natural radionuclides in some Egyptian building materials. *Appl. Radiat. Isot.* **2012**, *70*, 241–248. [CrossRef]
25. Sharaf, J.M.; Hamideen, M.S. Measurement of natural radioactivity in Jordanian building materials and their contribution to the public indoor gamma dose rate. *Appl. Radiat. Isot.* **2013**, *80*, 61–66. [CrossRef]
26. Sabbarese, C.; Ambrosino, F.; D'Onofrio, A.; Roca, V. Radiological characterization of natural building materials from the Campania region (Southern Italy). *Constr. Build. Mater.* **2021**, *268*, 121087. [CrossRef]
27. Bulut, H.A.; Sahin, R. Radiological characteristics of Self-Compacting Concretes incorporating fly ash, silica fume, and slag. *J. Build. Eng.* **2022**, *58*, 104987. [CrossRef]
28. de With, G.; Michalik, B.; Hoffman, B.; Dose, M. Special Issue: Use of NORM-containing products in construction. Development of a European harmonized standard to determine the natural radioactivity concentrations in building materials. *Constr. Build. Mater.* **2018**, *171*, 913–918. [CrossRef]
29. Knoll, G. *Radiation Detection and Measurment*, 2nd ed.; John Willey and Sons: Toronto, ON, Canada, 1989; pp. 59–60.
30. Papastefanou, C.; Stoulos, S.; Manolopoulou, M. The radioactivity of building materials. *J. Radioanal. Nucl. Chem.* **2005**, *266*, 367–372. [CrossRef]
31. Papaefthymiou, H.; Gouseti, O. Natural radioactivity and associated radiation hazards in building materials used in Peloponnese. *Radiat. Meas.* **2008**, *43*, 1453–1457. [CrossRef]
32. Pakou, A.A.; Assimakopoulos, P.A.; Prapidis, M. Natural radioactivity and radon emanation factors in building material used in Epirus (north western Greece). *Sci. Total Environ.* **1994**, *144*, 255–260. [CrossRef]
33. Petropoulos, N.P.; Anagnostakis, M.J.; Simopoulos, S.E. Photon attenuation, natural radioactivity content and radon exhalation rate of building materials. *J. Environ. Radioact.* **2002**, *61*, 257–269. [CrossRef]
34. Stoulos, S.; Manolopoulou, M.; Papastefanou, C. Assessment of natural radiation exposure and radon exhalation from building materials in Greece. *J. Environ. Radioact.* **2003**, *69*, 225–240. [CrossRef]
35. Nuccetelli, C.; Risica, S.; D'Alessandro, M.; Trevisi, R. Natural radioactivity in building material in the European Union: Robustness of the activity concentration index I and comparison with a room model. *J. Radiol. Prot.* **2012**, *32*, 349–358. [CrossRef]
36. Pavlidou, S.; Koroneos, A.; Papastefanou, C.; Christofides, G.; Stoulos, S.; Vavelides, M. Natural Radioactivity of Granites Used as Building Materials in Greece. *Bull. Geol. Soc. Greece* **2004**, *36*, 113–120. [CrossRef]
37. JCGM 200:2012. International Vocabulary of Metrology—Basic and General Concepts and Associated Terms (VIM), Joint Committee for Guides in Metrology/Working Group 2 (JCGM/WG 2), 3rd Edition (2008 Version with Minor Corrections). Available online: https://www.bipm.org/documents/20126/2071204/JCGM_200_2012.pdf/f0e1ad45-d337-bbeb-53a6-15fe649d0ff1 (accessed on 15 November 2023).
38. Papadakos, G.N.; Karangelos, D.J.; Petropoulos, N.P.; Chinis, E.P.; Anagnostakis, M.I.; Simopoulos, S.E. Uncertainty introduced into soil radioactivity measurement from sampling definition errors. In Proceedings of the 6th Biennial Panhellenic Metrology Conference "Metrologia 2016", Athens, Greece, 13–14 May 2016.

39. Regulation (EU) No 305/2011 of the European Parliament and of the Council of 9 March 2011 Laying down Harmonised Conditions for the Marketing of Construction Products and Repealing Council Directive 89/106/EEC. Available online: https://eur-lex.europa.eu/LexUriServ/LexUriServ.do?uri=OJ:L:2011:088:0005:0043:EN:PDF (accessed on 15 November 2023).
40. JCGM 100:2008 (ISO GUM with Minor Corrections), Evaluation of Measurement Data—Guide to the Expression of Uncertainty in Measurement, Joint Committee for Guides in Metrology/Working Group 1 (JCGM/WG 1), 1st Edition. Available online: https://www.bipm.org/documents/20126/2071204/JCGM_100_2008_E.pdf/cb0ef43f-baa5-11cf-3f85-4dcd86f77bd6 (accessed on 15 November 2023).
41. Government Gazette of the Hellenic Republic 1561B/02.06.2016 "Concrete Technology Regulation 2016". Available online: https://www.et.gr/api/Download_Small/?fek_pdf=20160201561 (accessed on 15 November 2023).
42. Papadakos, G.N. Stochastic Procedures and Relevant Quantitative and Qualitative Evaluation of Radioenvironmental Consequences in Cohorts Living on Hellenic Ground. Ph.D. Thesis, National Technical University of Athens, Athens, Greece, 2012.
43. Maraziotis, I. Gamma activity of the fly ash from a Greek power plant and properties of fly-ash cement. *Health Phys.* **1985**, *49*, 302–303.
44. Siotis, I.; Wrixon, A.D. Radiological consequences of the use of fly ash in building materials in Greece. *Radiat. Prot. Dosim.* **1984**, *7*, 101–105. [CrossRef]
45. Directive 2010/31/EU of the European Parliament and of the Council of 19 May 2010 on the Energy Performance of Buildings (Recast). Available online: https://eur-lex.europa.eu/LexUriServ/LexUriServ.do?uri=OJ:L:2010:153:0013:0035:en:PDF (accessed on 15 November 2023).
46. *ISO/IEC 17025:2017*; General Requirements for the Competence of Testing and Calibration Laboratories. ISO/CASCO: Geneva, Switzerland, 2017.

Disclaimer/Publisher's Note: The statements, opinions and data contained in all publications are solely those of the individual author(s) and contributor(s) and not of MDPI and/or the editor(s). MDPI and/or the editor(s) disclaim responsibility for any injury to people or property resulting from any ideas, methods, instructions or products referred to in the content.

Article

The Effect of High-Energy Ball Milling of Montmorillonite for Adsorptive Removal of Cesium, Strontium, and Uranium Ions from Aqueous Solution

Iryna Kovalchuk [1,2,*], Oleg Zakutevskyy [1], Volodymyr Sydorchuk [3], Olena Diyuk [3] and Andrey Lakhnik [4]

[1] Department of Sorption and Fine Inorganic Synthesis, Institute for Sorption and Problem of Endoecology, National Academy of Science of Ukraine, 03164 Kyiv, Ukraine
[2] Department of Chemical Technology of Ceramics and Glass, Igor Sikorsky Kyiv Polytechnic Institute, 03056 Kyiv, Ukraine
[3] Department of Oxidative Heterogeneous Catalytic Processes, Institute for Sorption and Problem of Endoecology, National Academy of Science of Ukraine, 03164 Kyiv, Ukraine
[4] Department of Physics of Dispersed Systems, H.V. Kurdyumov Institute of Metal Physics, National Academy of Science of Ukraine, 02000 Kyiv, Ukraine
* Correspondence: kowalchukiryna@gmail.com; Tel.: +380-66-2707-519

Citation: Kovalchuk, I.; Zakutevskyy, O.; Sydorchuk, V.; Diyuk, O.; Lakhnik, A. The Effect of High-Energy Ball Milling of Montmorillonite for Adsorptive Removal of Cesium, Strontium, and Uranium Ions from Aqueous Solution. *Eng* **2023**, *4*, 2812–2825. https://doi.org/10.3390/eng4040158

Academic Editors: Antonio Gil Bravo and Tanay Kundu

Received: 23 August 2023
Revised: 11 October 2023
Accepted: 9 November 2023
Published: 14 November 2023

Copyright: © 2023 by the authors. Licensee MDPI, Basel, Switzerland. This article is an open access article distributed under the terms and conditions of the Creative Commons Attribution (CC BY) license (https:// creativecommons.org/licenses/by/ 4.0/).

Abstract: Clay minerals are widely used to treat groundwater and surface water containing radionuclides. In our study, the method of mechanochemical activation for increasing the sorption capacity of the natural clay mineral montmorillonite was used. By adjusting the grinding time, the increasing sorption parameters of mechanochemically activated montmorillonite were determined. X-ray diffraction method, scanning electron microscopy, and the determination of the specific surface by low-temperature adsorption–desorption of nitrogen to characterize the natural and mechanochemical-activated montmorillonites were used. It was established that the maximal sorption of uranium, strontium, and cesium is found for montmorillonite after mechanochemical treatment for 2 h. It is shown that the filling of the surface of montmorillonite with ions of different natures occurs in various ways during different times of mechanochemical treatment. The appropriateness of the Langmuir and Freundlich models for the sorption parameters of uranium, strontium, and cesium ions on montmorillonite after its mechanochemical activation was established. The effect of natural organic substances—humic acids—on the efficiency of water purification from uranium on mechanoactivated montmorillonite was studied. The obtained sorbents can be effectively used for the removal of trace amounts of radionuclides of different chemical natures (uranium, cesium, and strontium) from polluted surface and ground waters.

Keywords: layered silicate; mechanochemical activation; sorption; radionuclides; surface filling; humic acid

1. Introduction

Contamination of surface water and groundwater by radioactive elements such as uranium, cesium, and strontium is caused by several factors. Among these factors are the obtaining of environmentally hazardous toxicants into groundwater from the mining industry, tailings storage facilities, nuclear power plant accidents, and radioactive waste disposal sites from block waters of radioactive sites of nuclear power plants [1–8]. As a result, the ecological balance is disturbed. The concentration of radioactive elements can vary for different reservoirs depending on the type of drained rocks, the chemical composition of the water, and the hydrogeological regime [9].

In water, uranium forms carbonate and, rarely, sulfate anion complexes, uranyl, and hydroxo-uranyl cations, and can also be in a dispersed state in the composition of detrital minerals. These toxicants can be dissolved in surface water and groundwater or exchange ions in soils or sediments by complex reactions with natural organic matter as pure or mixed mineral phases [3]. In an aqueous medium, cesium exists in a cationic form over a wide pH range. However, for strontium, in addition to the ionic and ion-dispersed form, it is also likely to be found in the form of complex compounds. In surface waters, most of the cesium-137 is transported in the form of positively charged ions, although significant amounts of cesium can be strongly adsorbed on clay particles and transported in the solid phase. The main amounts of strontium-90 in fresh waters are transported in ionic form [10]. There is an interaction of radionuclides with organic components of natural waters, with the formation of sufficiently strong complexes, which can be one of the main channels of transport in the case of a high content of organic substances in waters [11,12].

As a result of migration, radioactive elements enter drinking water sources and are a threat to public health. Because of its radioactivity and toxicity (carcinogenic for humans), uranium is a very dangerous element, and the World Health Organization recommends a limit of 0.015 mg/L for drinking water [13]. Also, U has been added to the US Environmental Protection Agency's National Primary Drinking Water Regulations, with a recommended guideline of 0.030 mg/L [14]. Total beta activity for drinking water consists of $\leq$1.0 Bq/L, specific activity ^{137}Cs consists of $\leq$2.0 Bq/L, and ^{90}Sr consists of $\leq$2.0 Bq/L. Therefore, considerable attention is paid to the study of the features of binding of radioactive elements to the components of natural waters—fulvic acids, and humic acids, as well as possible ways of extracting these compounds from the water environment [15–17].

The problem of preventing the migration of radionuclides and heavy metals from mining and other contaminated areas is huge and requires low-cost technological solutions. The using of the sorption method is one of them. Natural sorbents are usually used to create permeable reactive barriers or as filling materials in storage facilities for the long-term disposal of radioactive waste [18,19]. Clay materials have all the necessary qualities for this, including sufficient resistance to radiation exposure [20,21].

However, these materials in the raw state have only moderate sorption properties. To increase the sorption capacity of clay minerals and selectivity concerning certain toxicants, methods of physical and chemical surface modification are widely used [22]. The simple methods seem to be more promising for the processing of large volumes of clay materials. When grinding solids using energy-intensive equipment, it is possible to achieve a high dispersion of final products, as well as to significantly influence the concentration and nature of surface defects. The latter determines the use of mechanochemical activation (MCA) in the technology of catalysts as well as in sorption processes [23–26].

A large number of works are devoted to studies of the properties of MCA-treated layered materials [27]: bentonite [28], montmorillonite [29–32], vermiculite [33], kaolinite [29,32], talc [34], smectite [35], etc. Mechanochemical activation leads to the decreasing of crystallinity (amorphization) of the solid substance, while the surface energy and surface reactivity of the material and, therefore, chemical activity increase [25,26,31,36,37]. As a result of MCA, the clay structure changes. The following changes are observed: delamination and disorderliness of lays, increasing defects, a reduction in the size of the crystallites, a lowering of the temperature of the dehydroxylation, increasing free surface clay particles, and increasing reaction abilities. The greatest influence on the changing properties of clays in the MCA are time and the intensity of processing [25,26].

Montmorillonite has attracted widespread attention as a potential material for the control of harmful elements in the environment [38–40]. The enhancement of adsorption of organic compounds (gaseous toluene, acetone [41,42]) was investigated on ball milling-treated montmorillonite. The increasing of sorption values by twice and more with the use of mechanochemical treatment were shown on the removal of such elements as lead [33,43], cesium, and strontium [25,43]. The first factor that determines the high selectivity of clay minerals to large cations of alkali metals is structural. At the sorption processes it is caused by the presence of ditrigonal holes on the surface of silica grids proportional to the parameters of large cations. The second feature is an increase in the relative contribution of the side faces to their specific surface and the formation of so-called "frayed" side faces at the fine comminution of air-dried samples of these minerals. It leads to the possible increase in the ^{137}Cs ions' sorption values [25]. As follows, fine comminution with the use of high-energy aggregates [25,44] contributes to increasing the surface activity of clay minerals and, accordingly, increasing their sorption characteristics.

However, the use of mechanoactivated clays, in particular, mechanoactivated montmorillonite, for water purification from heavy metals and radionuclides is not so widely used. Therefore, the use of cheap natural raw materials in combination with high-energy processing will make it possible to obtain effective sorbents for the removal of radionuclides. Our research aimed to obtain montmorillonite-based materials using mechanochemical activation for increasing sorption characteristics. Such sorbents can provide protection to the aqueous environment from cesium, strontium, and uranium.

2. Materials and Methods

2.1. Materials and Chemicals

The layered silicate 2:1-type montmorillonite (Dashukivka, Cherkasy region, Ukraine) with a cation–exchange capacity (CEC) of 0.71 meq/100 g and a structural formula of $(Ca_{0.12}Na_{0.03}K_{0.03})_{0.18}(Al_{1.39}Mg_{0.13}Fe_{0.44})_{1.96}(Si_{3.88}Al_{0.12})_4O_{10}(OH)_2 \cdot nH_2O$ was used. Coarse mineral impurities such as feldspars, quartz, aluminum and iron oxides, carbonates, etc., were withdrawn from montmorillonite rock. Purified montmorillonite, which was used in the next experiments, was obtained by way of sedimentation from an aqueous-clay suspension, centrifugation, drying at 105 °C, and grounding to the fraction < 0.1 mm. The chemical composition of natural montmorillonite, according to SEM/EDS analysis, followed (%, at.): —52.65, Si—27.70, Al—10.02, Fe—8.38, and Mg—1.25.

Hydrochloric acid (HCl), sodium hydroxide (NaOH), sodium chloride (NaCl), salt of uranyl sulfate trihydrate ($UO_2SO_4 \cdot 3H_2O$), cesium chloride (CsCl), and strontium chloride ($SrCl_2 \cdot 6H_2O$) were obtained from Sigma–Aldrich, St. Louis, MO, USA; humic acid was obtained from Fluka. All the chemicals were analytical reagent grade and used without further purification. The solution of the uranium, cesium, and strontium salts were prepared using distilled water. In all experiments, distilled water was used to prepare solutions of different concentrations.

2.2. Synthesis Procedure

Planetary ball mill Pulverisette-6 (Fritsch) was used for the mechanochemical treatment of MMT. Si3N4 balls (d = 20 mm, n = 12) with a total weight of 160 g were used as working bodies. The mass ratio of balls and MMT was 1:10. The milling process continued 0.5–4 h. The rotation speed was equal to 300 rpm. Milling was performed in 10 min cycles; subsequently, a reverse was carried out after each cycle.

2.3. Characterization Methods

X-ray powder diffraction (XRD) patterns of the natural clay mineral and mechanochemical-activated montmorillonite (MCA MMT) were recorded with a DRON-4-07 diffractometer using CuKα (0.154 nm) in the region between 2° and 60° (2 θ). The constant pass energy of U = 30 kV and I = 30 mA were used [45].

Scanning electron microscopy (SEM) method (Jeol JSM-6060, Tokyo, Japan) was used for the investigation of the surface morphology of the natural and mechanochemical-activated MMT. The samples of MMT and MCA MMT were sputter coated with gold for 3 min. In the conditions of analysis, secondary images were used with an acceleration voltage of 30 kV and a magnification of 5000-fold.

The sample of natural montmorillonite was analyzed using a JED-2300 X-ray spectrometer integrated with a JSM 6060 LA scanning electron microscope. Energy-dispersive X-ray (EDS) analysis was performed with an accelerating voltage of 15 kV and a magnification of $\times$500 using standard software.

The low-temperature N_2 adsorption–desorption method (T = −196 °C) (Quantachrome NOVA-2200e Surface Area and Pore Size Analyzer, Boynton Beach, FL, USA) was used for the determination of the porous structure parameters. The ASiQwinTM V 3.0 software was used for the results processed. Specific surface area S (m^2/g) was estimated using the BET equation. The total pore volume V (cm^3/g) was calculated using the maximum adsorbed volume of nitrogen at a relative pressure $p/p_0 \approx 0.99$. The distribution of pore size dV(r) (nm) was determined using BJH and DFT models [46].

Potentiometric titration of the samples was performed using a 781 pH/Ion Meter by the procedure described in [47].

2.4. Adsorption Experiments

The batch-wise method was performed for the adsorption experiment. A series of 0.1 g samples of sorbents to 50 mL aqueous solutions of radionuclides U(VI), Cs(I), and Sr(II) were added. The solutions were prepared from the salts uranyl sulfate (UO$_2$SO$_4$·3H$_2$O), strontium chloride (SrCl$_2$·6H$_2$O), and cesium chloride (CsCl). The ionic strength was created with a solution of 0.01M NaCl. The experiments were conducted in triplicate at pH 6 in a thermostatic cell at 25 °C with continuous shaking for 1 h. After establishing the adsorption equilibrium, the liquid phase was separated by centrifugation (6000 rpm) for 30 min. The effect of humic acid (HA) was studied at a concentration of 100 mg/L for HA and 300 µmol/L for U(VI).

The spectrophotometric method (UNICO 2100UV, United Products and Instruments, Suite E Dayton, NJ, USA) was used for the determination after adsorption of the equilibrium concentration of U(VI); the wavelength was 665 nm, and Arsenazo III was used as a reagent. The atomic adsorption spectroscopy (AA-6300 Shimadzu, Shimadzu Corporation, Tokyo, Japan) was used for the determination of the equilibrium concentrations of Cs(I) and Sr(II) after adsorption.

The kinetic experiments were conducted on samples of natural and mechanical-activated montmorillonite by uranium sulfate, cesium chloride, and strontium chloride solutions with the same initial concentration of 100 µmol/L for different selected times ranging from 5 to 360 min.

The presentation of sorption (PS, %) from U(VI), Cs(I), and Sr(II), was calculated by Equation (1):

$$PS = (C_{in} - C_{eq})/C_{in} \cdot 100 \tag{1}$$

The adsorption capacity (q_{eq}, µmol/g) of radionuclide ions was estimated by Equation (2):

$$q_{eq} = (C_{in} - C_{eq}) \cdot V/m \tag{2}$$

where C_{in} and C_{eq} represent the initial and equilibrium radionuclides concentrations (µmol/L), V is the solution volume (L), and m is the weight of the adsorbent (g).

The distribution of radionuclides between the adsorbent and solution was calculated as the value of the coefficient of distribution K_d, mL/g (Equation (3)):

$$K_d = (C_{in} - C_{eq}) \cdot V/(C_{eq} \cdot m) \cdot 1000 \tag{3}$$

The experimental isotherms of radionuclides adsorption were processed using the Langmuir mathematical model for a homogeneous surface and the Freundlich model for a heterogeneous surface [48].

3. Results and Discussion

3.1. Characteristics of Natural and Mechanical-Activated Montmorillonite

The X-ray diffraction patterns of the MMT and MCA MMT are demonstrated in Figure 1. There were structural changes during the milling process. A gradual destruction of the crystalline phase occurs with the increasing duration of milling. This was evidenced by a progressive decrease in the XRD reflection intensities. Fine grinding leads to a systematic decrease in the intensity of reflections. It is proven to increase the amorphization of the structure of activated MMT during milling. Mechanochemical treatment for 1 h and more leads to the decreasing intensity of the 001 peaks and a shift of the 001 peaks with the concomitant broadening in an asymmetrical fashion. At the same time, the stacking of layers is disturbed and lost, and the initial structural destabilization of the basal plane begins [45,49]. This is probably due to the partial removal of water molecules from the interlayer spaces and may, according to the work [25], be accompanied by the formation of a disordered mixed-layer phase of montmorillonite: partially dehydrated montmorillonite. The broadening of the XRD reflections (Figure 1) indicated that, along with structural changes, a reduction in particle size occurred: from 6.63 nm for MMT to 4.32 nm for 1 h MCA MMT. With prolonged milling (more than 4 h), montmorillonite was transformed into an amorphous solid [25].

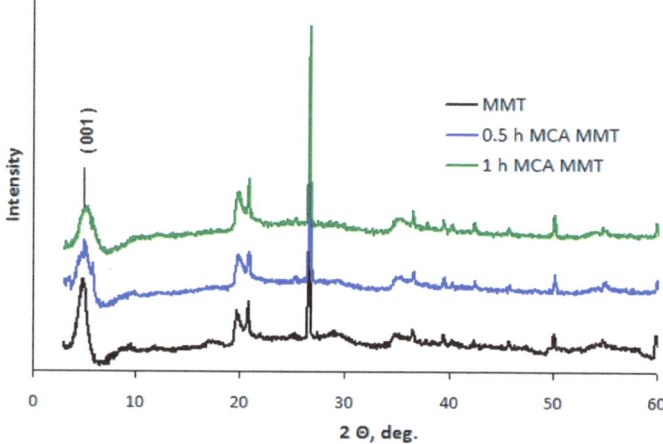

Figure 1. X-ray diffraction patterns of the montmorillonite before and after mechanochemical activation: 1-MMT, 2-0.5 h MCA MMT, 3-1 h MCA MMT.

SEM microphotograph samples of the natural MMT and the most reactive sample (MCA MMT 2 h) are shown in Figure 2. Particles of natural MMT have the shape of plates, which can be seen in Figure 2a. For the series of milled montmorillonite samples, the microstructure is changed with increasing the treatment time. Fine grinding of MMT in a planetary ball mill for 2 h changed the structure of MMT (Figure 2b). Contrary to natural MMT, the morphology of the mechanochemically activated sample (MCA MMT 2 h) is more uniform, with a predominance of spheroidal and (quasi)regular particles. As a result of mechanochemical activation, the flaky-edges particles appeared. So, the mechanochemical treatment of MMT samples leads to the formation of new more reactive surfaces [26,29,32].

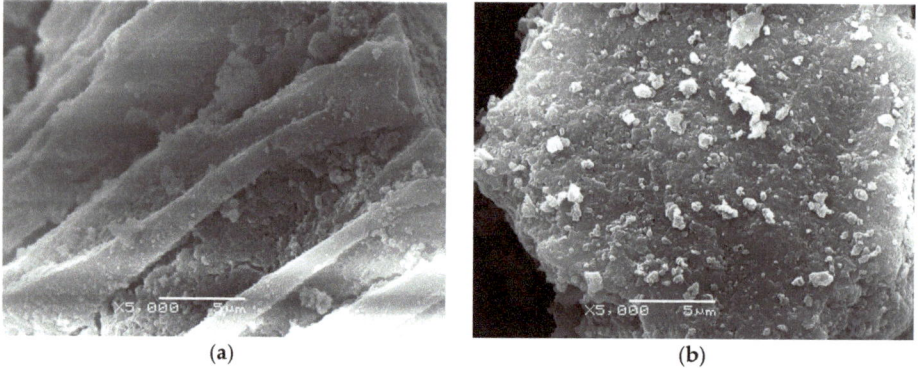

Figure 2. SEM microphotographs of the montmorillonite particles before and after mechanochemical activation: (**a**)—MMT; (**b**)—2 h MCA MMT.

The MCA treatment leads to a decrease in the structural order and an increase in the specific surface area of clays. The data calculated from nitrogen adsorption/desorption isotherms for MMT and MCA MMT samples showed that the dependence of specific surface area, total pore volume, as well as average pore radius r from activation time is non-monotonic, although there is a tendency for the values of these parameters to increase (Table 1).

Table 1. Characteristics of the porous structure for the montmorillonite before and after mechanochemical activation.

Sample	S, m^2/g	V, cm^3/g	r, nm	Distribution of Pore Sizes, nm				
				BJH dV (r)		DFT dV (r)		
				r_1	r_2	r_1	r_2	r_3
MMT	89.1	0.078	1.753	1.98	-	1.41	2.75	-
0.5 h MCA MMT	117.4	0.140	2.377	2.142	-	1.17	2.44	0.64
1 h MCA MMT	93.4	0.131	2.811	2.142	-	1.25	2.64	0.64
2 h MCA MMT	146.8	0.161	2.187	1.92	-	1.13	2.64	0.67
4 h MCA MMT	99.1	0.155	3.128	1.90	15.10	1.21	2.54	0.64

Isotherms of the low-temperature nitrogen sorption–desorption (Figure 3a) for all samples have the waveform. According to the IUPAC classification, the obtained isotherms can be attributed to Type II [50]. The observed hysteresis loops in the range of values $p/p_0 > 0.35$–0.4 can be classified as H3 type, indicating the presence of slit-like pores. Mesopore size distributions were obtained from the low-temperature nitrogen adsorption data according to the currently used Barret–Joyner–Halenda (BJH) method [46]. The maximum in the pore radius r_1 distribution for the natural montmorillonite is centered at 1.98 nm (Figure 3b). This value slightly increases after MCA to 2.14 nm for 0.5 h and 1 h, and again decreases to 1.90 nm for 4 h treatment. In general, the specific surface area of investigated MMT samples increases with the increase in time of mechanochemical activation. The specific surface area increases from 89.1 m^2/g for natural MMT up to 146.8 m^2/g for 2 h MCA MMT. The specific surface area for this sample reaches the maximal value (Table 1). However, dependence on the specific surface area from the time of milling is irregular. This is because several different processes, which occur simultaneously, affect the change in specific surface area during the milling of solids, in general, and minerals, in particular. Thus, the first local maximum on this dependence, which is observed after milling for 0.5 h, is caused by the comminution of montmorillonite particles without significant destruction of crystal structure, as in work [32]. On the contrary, a decrease in specific surface area for samples, milled for 1 h, is associated with aggregation, which often takes place when the milling time is increased [26,33]. Milling for more than 1 h results in significant destruction of the crystal structure of montmorillonite and its amorphization, as mentioned above. This is the reason for the increase in the specific surface area for the sample milled for 2 h. A further increase in the time of mechanochemical activation leads to a decrease in the specific surface of montmorillonite samples, similar to the other clays: kaolinite and ripidolite [32]. This occurs due to the aggregation of amorphized particles. It should be noted that there is a similar irregular dependence of specific surface area from energy supplied by the ball mill for sodium montmorillonite [51].

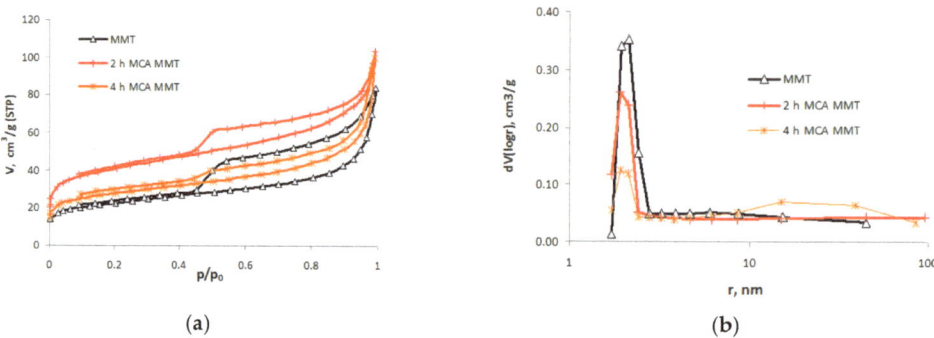

(a) (b)

Figure 3. Nitrogen adsorption–desorption isotherms (**a**) and mesopore size distributions (**b**) of the montmorillonite before and after mechanochemical activation.

So, according to the theory of "mechanochemical activation", the mechanical stress leads to the shifting of atoms from their equilibrium crystal lattice positions, bringing the solid into a high-energy metastable state, and to a further fracture of particles. The particles reach a critical size and solid materials begin to build up crystal defects and develop amorphous phases or other crystalline morphologies [52].

3.2. Adsorption Isotherms Study

Adsorption kinetics of uranium, cesium, and strontium ions from the aqueous solutions onto the natural and 2 h mechanochemical-activated montmorillonite was made up to equilibrium. Adsorption of U(VI), Ce(I), and Sr(II) on the natural and 2 h mechanochemical-activated montmorillonite as a function of the contact time is shown in Figure 4. Kinetic curves of the sorption of uranium, cesium, and strontium ions on all the samples indicate a sufficiently fast attainment of sorption equilibrium, within 15–20 min. In all the next sorption experiments, the duration was established as 1 h. The kinetics of U(VI), Ce(I), and Sr(II) adsorption onto natural and 2 h mechanochemical-activated montmorillonite can be significantly better described by the pseudo-second order rate equation.

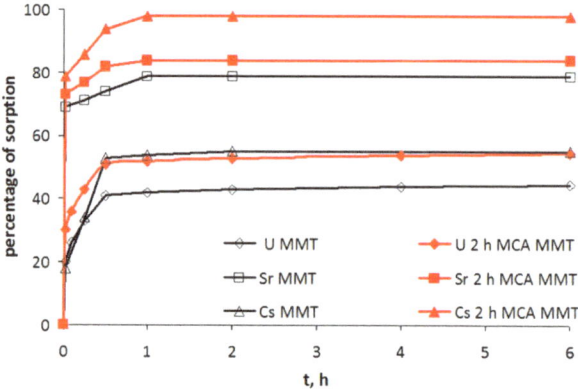

Figure 4. Kinetic study of the sorption of uranium, cesium, and strontium ions on natural montmorillonite and 2 h mechanochemical-activated MMM.

Sorption isotherms of uranium (Figure 5a), strontium (Figure 5b), and cesium (Figure 5c) ions show that the sorption characteristics of the samples for mechanochemically activated MMT are higher compared to those of the natural mineral.

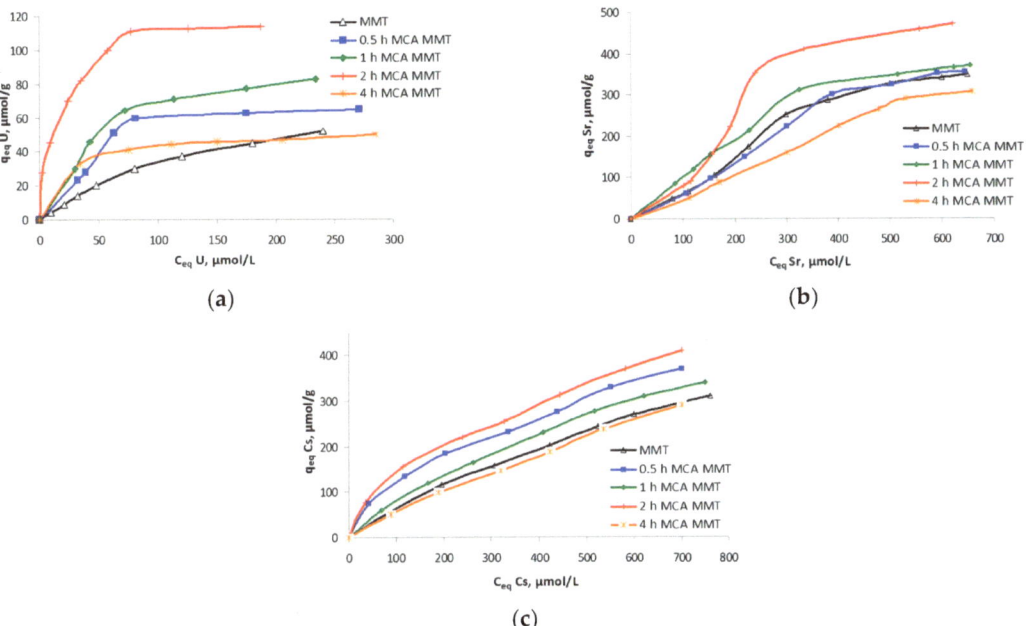

Figure 5. Sorption isotherms of uranium (**a**), strontium (**b**), and cesium (**c**) onto the montmorillonite before and after mechanochemical activation.

With an increase in the time of mechanochemical activation, the values of uranium sorption are 77.5, 105.3, and 123.5 µmol/g for samples activated for 0.5, 1, and 2 h, respectively. However, further activation (4 h) leads to a sharp decrease in the amount of uranium sorption, to 53.5 µmol/g (Table 2). For Sr^{2+} (Figure 5b) and Cs^+ (Figure 5c) ions, the improvement in sorption values correlates with an increase in the duration of MCA in a similar way. The sorption values of the studied metals are correlated with the values of the specific surface area of montmorillonite (Table 1), which increases with increasing the activation time (0.5–2 h) from 89.1 m^2/g to 146.8 m^2/g and decreases with increasing the activation time (4 h) up to 99.1 m^2/g.

The obtained sorption values of the studied metal ions on the natural montmorillonite are related to the peculiarities of its structure. Layered silicates are characterized by the presence of two main types of ion exchange centers, which differ sharply in their properties. These are exchangeable cations associated with non-stoichiometric isomorphic substitutions in the structure and are located on the basal surfaces of minerals, as well as broken silica or aluminum oxide bonds localized on the side faces of the crystals.

U(VI) is characterized by the possibility of existence in surface waters not only as uranyl ions (UO_2^{2+}), but also in the form of positively charged, neutral, or even negatively charged products of hydrolysis and interaction with dissolved CO_2: $[UO_2OH]^+$, $[(UO_2)_3(OH)_5]^+$, $UO_2(OH)_2$, UO_2CO_3, $[(UO_2)_2CO_3(OH)_3]^-$, etc. [10,11]. The dependence of U(VI) sorption values for montmorillonite on pH has an extreme character, with maximum values in a neutral media, which is characterized by a high degree of dissociation of Si(Al)OH groups on the side faces of the mineral particles, which are mainly responsible for the sorption process, and in the presence of positively charged UO_2^{2+} as the dominant form of U(VI) [12].

Table 2. Langmuir and Freundlich parameters for the adsorption of uranium, strontium, and cesium ions onto natural and mechanochemical-activated montmorillonite.

Sample	Radio-Nuclide	Langmuir			Freundlich		
		q_m, µmol/g	K_L, L/µmol	R^2	$1/n$	K_F, L/µmol	R^2
MMT	U(VI)	86.2	6.44	0.995	0.748	9.53	0.980
0.5 h MCA MMT		77.5	21.50	0.967	1.078	19.53	0.765
1 h MCA MMT		105.3	15.83	0.983	0.324	8.49	0.917
2 h MCA MMT		123.5	81.00	0.996	0.317	10.49	0.914
4 h MCA MMT		53.5	46.75	0.999	0.191	6.13	0.953
MMT	Sr(II)	526.3	3.17	0.996	0.984	16.60	0.960
0.5 h MCA MMT		500.0	4.00	0.994	0.984	16.55	0.967
1 h MCA MMT		454.6	7.33	0.999	0.710	15.60	0.956
2 h MCA MMT		588.2	5.67	0.999	0.905	19.09	0.845
4 h MCA MMT		500.0	2.50	0.946	1.055	15.50	0.986
MMT	Cs(I)	714.3	0.93	0.979	0.799	13.52	0.998
0.5 h MCA MMT		384.6	5.20	0.975	0.570	14.20	0.998
1 h MCA MMT		526.3	1.90	0.969	0.729	14.01	0.999
2 h MCA MMT		416.7	6.00	0.967	0.550	14.75	0.998
4 h MCA MMT		588.2	1.06	0.969	0.834	13.32	0.999

q_m—the amount of adsorbate corresponding to complete monolayer coverage; K_L—Langmuir bonding energy coefficient; K_F—Freundlich coefficient of adsorption capacity; n—coefficient of adsorption intensity; R—correlation coefficient.

Sorption of cesium and strontium ions on natural montmorillonite primarily occurs by the ion-exchange mechanism. Taking into account that the ion exchange on layered silicates is determined by the active centers of the basal faces, which account for the majority of the total exchange capacity, this factor is decisive in increasing the selectivity of montmorillonite to cesium ions.

The sorption of cesium and strontium ions on natural montmorillonite (q_m calculated from the Langmuir equation) are 714.3 and 526.3 µmol/g, respectively. In addition, the exchange centers of the basal faces show a high affinity for cesium ions, which is due to the manifestation of a structural factor [25,38]. The ditrigonal holes of silica grids on the surface of clay minerals are proportional to the size of the large alkali metal cations. It is possible to localize large alkali metal ions in pseudo-hexagonal oxygen holes of tetrahedral networks of minerals. Cesium sorption occurs not only on the outer surface of particles. Penetration of cesium ions into the interlayer space occurs, replacing the corresponding ion-exchange positions.

Mechanochemical activation significantly improves the sorption properties of natural MMT by achieving a greater equilibrium sorption capacity in the region of low equilibrium concentrations of uranium ions in the solution. This can be seen visually from the slope of the sorption isotherms (Figure 5a) and can be quantified using K_d values (Figure 6). In the case of the sorption of cesium ions (Figure 6c), a consistent increase in the value of K_d with the time of MCA was observed. For uranium and strontium ions, the character of the dependence of the distribution coefficients has a more complicated form. The maximum values of coefficients of distribution for MMT/2 h MXA MMT are as follows: ml/g: uranium (438/14,000), cesium (641/2191), and strontium (841/1476).

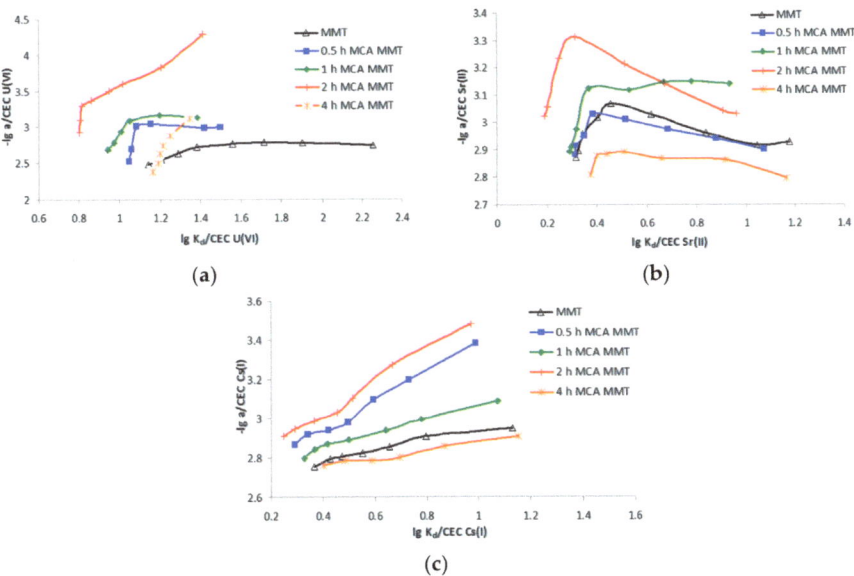

Figure 6. The degree of montmorillonite surface filling before and after mechanochemical activation by uranium (**a**), strontium (**b**), and cesium (**c**) ions.

The observed nature of changes in the sorption properties of minerals after mechanical activation can be explained as follows.

The mechanochemical activation of montmorillonite at the initial stage leads to an increase in the fineness of particle grinding and significant structural changes, which is primarily associated with the disordering of the octahedral mesh and significant distortions in the tetrahedral meshes. At the same time, the main structural motif is preserved and, accordingly, ensures an increase in the number of pseudo-hexagonal silicon–oxygen holes, which are the most important active centers of the sorption of large cations [25,28,32,44].

At the same time, the mechanical activation of air-dried samples leads primarily to the fracture of lamellar particles of clay minerals and, thus, to an increase in the relative contribution of the side faces to the total surface of the samples. A high affinity of the most high-energy sorption centers localized on the so-called cleaved side faces to metal ions is observed. At the same time, it is possible to form strong surface complexes with uranium, cesium, and strontium ions, in the formation of which both surface hydroxyl groups and oxygen atoms of partially destroyed ditrigonal holes participate. Therefore, in the interlayer space of minerals near the side faces, the area easily accessible to metal ions increases [25,38,43].

The presence of various defects on the surface of the mineral determines the significant energy heterogeneity of the sorption centers. These structural defects, and, therefore, the functional groups of different compositions (exchangeable cations, surface hydroxyl groups, and oxygen of the tetrahedral network), play the role of active centers during adsorption and catalytic reactions. Surface $\equiv$Si–OH–, =Al–OH–, etc., groups of the side surface of minerals behave like a mixture of acids of different strengths. During the mechanical activation of samples in air, the growth of the total specific surface area of the particles is largely due to the development of side faces, with partial destruction of the basal faces. It was established that in the process of mechanochemical activation, the number of acid groups per unit mass of the montmorillonite sample increases and amounts to 2.5 meq/g for natural MMT, 0.5 h MCA MMT, and 1 h MCA MMT, while for 2 h MCA MMT is 3.5 meq/g and for 4 h MCA MMT is 4.5 meq/g.

The applicability of the Freundlich equation (Table 2) to the description of sorption data indicates the significant energy heterogeneity of the sorption centers located on the surface. The measure of the energy heterogeneity of the surface is the coefficient $1/n$: the closer that the value of this coefficient is to unity, the more homogeneous the surface is. The obtained results indicate a noticeable increase in the energy heterogeneity of the sorption centers as a result of the mechanical activation of the samples. For cesium ions, the Freundlich correlation coefficient is very high: (R^2 = 0.998–0.999).

For the Langmuir monomolecular sorption equation, which indicates the energy homogeneity of active centers and, accordingly, the proximity of ion sorption energies as the surface is filled, the correlation coefficient is R^2 = 0.946–0.999.

Figure 6 shows the filling of the mineral surface with metal ions. On the abscissa axis, lg (q/CEC) is represented, characterizing the degree of surface filling, where q is the concentration of metal ions in the solid phase (µmol/g) and CEC is the exchange capacity (µmol/g). The ordinate axis shows lg (K_d/CEC), where K_d/CEC is the specific distribution coefficient.

Noticeably different natures of surface filling for uranium, strontium, and cesium ions can be observed in Figure 6. For cesium, lg (K_d/CEC) changes as the montmorillonite surface is filled in the range from 1.13 to 0.37 for the original and from 0.97 to 0.24 for the 2 h mechanochemically activated mineral, which corresponds to the distribution coefficients of 641 and 408 for the original and 2191 and 586 for the mechanically activated for 2 h mineral. For uranium ions, as for strontium ions, there is a gradual filling of the surface with metal ions for the original mineral: lg (K_d/CEC) increases from 2.26 to 1.14 for uranium per MMT with distribution coefficients of 400 and 217, respectively. The much faster filling occurs for mechanochemically activated samples: 2 h (1.41–0.80) and 4 h (1.34–1.16) for uranium, corresponding to distribution coefficients 14,000 and 608 (2 h) and 960 and 176 (4 h).

3.3. Effect of Natural Organic Compounds

It was established that at a uranium concentration of 300 µmol/L, the sorption curves without and in the presence of 100 mg/L humic acid (HA) have a similar appearance (Figure 7). However, the values of uranium sorption in the presence of HA are significantly lower. This is determined by the fact that the presence of natural organic substances in the water environment, in particular, fulvic acids [12,17,53], significantly changes the values of uranium (VI) sorption. Namely, the formation of stable humate complexes with uranium leads to a decrease in sorption indicators for both natural MMT and MCA MMT.

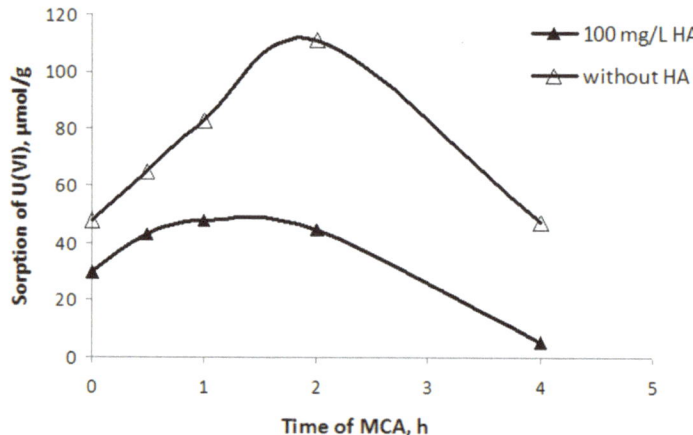

Figure 7. Effect of humic acids on uranium sorption at montmorillonite before and after mechanochemical activation in dependence of MCA time.

4. Conclusions

Mechanochemical activation of montmorillonite during different times (0.5–4 h) leads to certain changes in its structure. First of all, the gradual destruction of the crystalline phase occurred during long-term grinding. Secondly, the morphology of the mechanochemical-activated sample (MCA MMT 2 h) becomes more uniform, with the dominance of spherical and (quasi)regular particles. Thirdly, MCA made it possible to achieve a fairly high degree of increase in the specific surface area, by almost two times (from 80.3 m^2/g to 146.8 m^2/g). A gradual increase in the total pore volume and an increase in the pore radius with increasing activation time were observed.

As a result of structural changes, the sorption properties of natural MMT improve primarily due to the increase in the equilibrium values of sorption and K_d in the region of low equilibrium concentrations of Cs, Sr, and U ions in the solution. The filling of the surface of natural montmorillonite and mechanochemical-activated montmorillonite occurs in various ways for the radionuclides uranium (VI), strontium (II), and cesium (I). The obtained sorbents can be effectively used to remove trace amounts of radionuclides of different chemical natures (uranium, cesium, and strontium) from polluted surface and underground waters. The noted fixation of radionuclides on the surface of mechanoactivated montmorillonite makes it possible to use it as a matrix for burying radioactive waste during heat treatment. The mechanochemically activated clays can be applied to the environmental protection technology in the engineered clay barriers in multibarrier systems for the control of radioactive contamination [18,19].

Author Contributions: Conceptualization, I.K.; Methodology, I.K., O.Z., V.S. and O.D.; Validation, I.K.; Formal analysis, I.K., O.Z. and O.D.; Investigation, I.K., V.S. and A.L.; Writing—original draft, I.K., O.Z. and V.S.; Writing—review & editing, I.K.; Supervision, I.K. All authors have read and agreed to the published version of the manuscript.

Funding: This research received no external funding.

Institutional Review Board Statement: Not applicable.

Informed Consent Statement: Not applicable.

Data Availability Statement: Data are contained within the article.

Acknowledgments: The authors are grateful to the staff of the Institute for Sorption and Problems of Endoecology of the National Academy of Sciences of Ukraine for their support.

Conflicts of Interest: The authors declare no conflict of interest.

References

1. Available online: https://world-nuclear.org/ (accessed on 1 July 2023).
2. Bockris, J.M. *Environmental Chemistry*; Springer: Boston, MA, USA, 1977.
3. Kornilovych, B.Y.; Sorokin, O.G.; Pavlenko, V.M.; Koshyk, Y.I. *Environmental Protection Technologies in the Uranium Mining and Processing Industry*; Norma: Kyiv, Ukraine, 2011. (In Ukrainian)
4. Merkel, B.J.; Hasche-Berger, A. *Uranium, Mining and Hydrogeology*; Springer: Berlin/Heidelberg, Germany, 2008; 955p.
5. Chen, Z.; Wang, S.; Hou, H.; Chen, K.; Gao, P.; Zhang, Z.; Jin, Q.; Pan, D.; Guo, Z.; Wu, W. China's progress in radionuclide migration study over the past decade (2010–2021): Sorption, transport and radioactive colloid. *Chin. Chem. Lett.* **2022**, *33*, 3405–3412. [CrossRef]
6. Taniguchi, K.; Onda, Y.; Smith, H.G.; Blake, W.; Yoshimura, K.; Yamashiki, K.; Takayuki, Y.; Saito, K. Transport and Redistribution of Radiocesium in Fukushima Fallout through Rivers. *Environ. Sci. Technol.* **2019**, *53*, 12339–12347. [CrossRef] [PubMed]
7. Rudenko, L.I.; Khan, V.E.; Panasyuk, N.I. Physicochemical Study of the Mechanism of Radionuclide Migration from the Shelter Object and Its Service Area to Groundwater. *Radiochemistry* **2003**, *45*, 293–297. [CrossRef]
8. Payne, T.E.; Davis, J.A.; Ochs, M.; Olin, M.; Tweed, C.J.; Altmann, S.; Askarieh, M.M. Comparative Evaluation of Surface Complexation Models for Radionuclide Sorption by Diverse Geologic Materials. In *Surface Complexation Modelling*; Lützenkirchen, J., Ed.; Elsevier: Amsterdam, The Netherlands, 2006; Volume 11, pp. 605–633. [CrossRef]
9. Kornilovych, B.; Wireman, M.; Ubaldini, S.; Guglietta, D.; Koshik, Y.; Caruso, B.; Kovalchuk, I. Uranium Removal from Groundwater by Permeable Reactive Barrier with Zero-Valent Iron and Organic Carbon Mixtures: Laboratory and Field Studies. *Metals* **2018**, *8*, 408. [CrossRef]
10. Atwood, D.A. *Radionuclides in the Environment*; Wiley: New York, NY, USA, 2010; 522p.

11. Langmuir, D. *Aqueous Environmental Geochemistry*; Prentice Hall: Upper Saddle River, NJ, USA, 1997.
12. Kornilovich, B.Y.; Pshinko, G.N.; Koval'chuk, I.A. Effect of fulvic acids on sorption of U (VI) on clay minerals of soils. *Radiochemistry* **2001**, *43*, 528–531. [CrossRef]
13. World Health Organization. *Guidelines for Drinking Water Quality*, 4th ed.; World Health Organization: Geneva, Switzerland, 2011; Volume 1.
14. Available online: https://www.epa.gov/ground-water-and-drinking-water/national-primary-drinking-water-regulations# Radionuclides (accessed on 1 July 2023).
15. Chen, L.; Dong, Y. Sorption of Ni(II) to montmorillonite as a function of pH, ionic strength, foreign ions and humic substances. *J. Radioanal. Nucl. Chem.* **2012**, *295*, 2117–2123. [CrossRef]
16. Wen, T.; Chen, Y.; Cai, L. Impact of environmental conditions on the sorption behavior of radiocobalt onto montmorillonite. *J. Radioanal. Nucl. Chem.* **2011**, *290*, 437–446. [CrossRef]
17. Kornilovich, B.Y.; Pshinko, G.N.; Spasenova, L.N.; Kovalchuk, I.A. Influence of humic substances on the sorption interactions between lanthanides and actinides ions and clay minerals. *Adsorpt. Sci. Technol.* **2000**, *18*, 873–880. [CrossRef]
18. Guimarães, V.; Bobos, I. Role of clay barrier systems in the disposal of radioactive waste. In *Sorbents Materials for Controlling Environmental Pollution*; Núñez-Delgado, A., Ed.; Elsevier: Amsterdam, The Netherlands, 2021; Volume 20, pp. 513–541. [CrossRef]
19. Florez, C.; Park, Y.H.; Valles-Rosales, D.; Lara, A.; Rivera, E. Removal of Uranium from Contaminated Water by Clay Ceramics in Flow-Through Columns. *Water* **2017**, *9*, 761. [CrossRef]
20. Yang, G. Sorption and reduction of hexavalent uranium by natural and modified silicate minerals: A review. *Environ. Chem. Lett.* **2023**, *21*, 2441–2470. [CrossRef]
21. Allard, T.; Calas, G. Radiation effects on clay mineral properties. *Appl. Clay Sci.* **2009**, *43*, 143–149. [CrossRef]
22. Bergaya, F.; Theng, B.K.; Lagaly, G. Modified Clays and Clay Minerals. In *Handbook of Clay Science*; Elsevier: London, UK, 2006; 1246p.
23. Samsonenko, M.; Zakutevskyy, O.; Khalameida, S.; Charmas, B.; Skubiszewska-Zięba, J. Influence of mechanochemical and microwave modification on ion-exchange properties of tin dioxide with respect to uranyl ions. *Adsorption* **2019**, *25*, 451–457. [CrossRef]
24. Zakutevskyy, O.; Khalameida, S.; Sydorchuk, V.; Kovtun, M. The Effect of Hydrothermal, Microwave, and Mechanochemical Treatments of Tin Phosphate on Sorption of Some Cations. *Materials* **2023**, *16*, 4788. [CrossRef] [PubMed]
25. Kornilovych, B.Y. *Structure and Surface Properties of Mechanochemically Activated Silicates and Carbonates*; Naukova Dumka: Kyiv, Ukraine, 1994; 127p. (In Russian)
26. Sydorchuk, V.; Vasylechko, V.; Khyzhun, O.; Gryshchouk, G.; Khalameida, S.; Vasylechko, L. Effect of high-energy milling on the structure, some physicochemical and photocatalytic properties of clinoptilolite. *Appl. Catal. A Gen.* **2020**, *610*, 117930. [CrossRef]
27. Tole, I.; Delogu, F.; Qoku, E.; Habermehl-Cwirzen, K.; Cwirzen, A. Enhancement of the pozzolanic activity of natural clays by mechanochemical activation. *Constr. Build Mater.* **2022**, *352*, 128739. [CrossRef]
28. Dellisanti, F.; Valdré, G. Study of structural properties of ion treated and mechanically deformed commercial bentonite. *Appl. Clay Sci.* **2005**, *28*, 233–244. [CrossRef]
29. Baki, V.A.; Ke, X.; Heath, A.; Calabria-Holley, J.; Terzi, C.; Sirin, M. The impact of mechanochemical activation on the phyicochemical properties and pozzolanic reactivity of kaolinite, muscovite and montmorillonite. *Cem. Concr. Res.* **2022**, *162*, 106962. [CrossRef]
30. Tole, I.; Habermehl-Cwirzen, K.; Cwirzen, A. Mechanochemical activation of natural clay minerals: An alternative to produce sustainable cementitious binders—Review. *Mineral. Petrol.* **2019**, *113*, 449–462. [CrossRef]
31. Ramadan, A.R.; Esawi, A.M.K.; Gawad, A.A. Effect of ball milling on the structure of Na$^+$-montmorillonite and organo-montmorillonite (Cloisite 30B). *Appl. Clay Sci.* **2010**, *47*, 196–202. [CrossRef]
32. Vdović, N.; Jurina, I.; Škapin, S.D.; Sondi, I. The surface properties of clay minerals modified by intensive dry milling—Revisited. *Appl. Clay Sci.* **2010**, *48*, 575–580. [CrossRef]
33. Hongo, T.; Yoshino, S.; Yamazaki, A.; Yamasaki, A.; Satokawa, S. Mechanochemical treatment of vermiculite in vibration milling and its effect on lead (II) adsorption ability. *Appl. Clay Sci.* **2012**, *70*, 74–78. [CrossRef]
34. Kim, H.N.; Kim, J.W.; Kim, M.S.; Lee, B.H.; Kim, J.C. Effects of Ball Size on the Grinding Behavior of Talc Using a High-Energy Ball Mill. *Minerals* **2019**, *9*, 668. [CrossRef]
35. Pálková, H.; Barlog, M.; Madejová, J.; Šimon, E.; Zimowska, M. Structural changes in mechanochemically treated smectites investigated by infrared spectroscopy. In *Book of Abstracts of the 8th Workshop of Slovak Clay Group, Clay Minerals and Selected Industrial Minerals in Material Science, Applications and Environmental Technology, Habovka, Slovakia, September 6–8, 2021*; Pálková, H., Šimonová, T., Eds.; 845 36 Slovak Clay Group: Bratislava, Slovenska, 2021; pp. 42–43.
36. Tsuzuki, T. Mechanochemical synthesis of metal oxide nanoparticles. *Commun. Chem.* **2021**, *4*, 143. [CrossRef] [PubMed]
37. Al Bazedi, G.A.; Al-Rawajfeh, A.E.; Abdel-Fatah, M.A.; Alrbaihat, M.R.; AlShamaileh, E. Synthesis of nanomaterials by mechanochemistry. In *Handbook of Greener Synthesis of Nanomaterials and Compounds*; Kharisov, B., Kharissova, O., Eds.; Elsevier: Amsterdam, The Netherlands, 2021; Volume 1, pp. 405–418. [CrossRef]
38. Atun, G.; Bilgin, B.; Mardinli, A. Sorption of cesium on montmorillonite and effects of salt concentration. *J. Radioanal. Nucl. Chem.* **1996**, *211*, 435–442. [CrossRef]

39. Kim, S.J. Sorption mechanism of U(VI) on a reference montmorillonite: Binding to the internal and external surfaces. *J. Radioanal. Nucl. Chem.* **2001**, *250*, 55–62. [CrossRef]
40. Jun, B.-M.; Lee, H.-K.; Park, S.; Kim, T.-J. Purification of uranium-contaminated radioactive water by adsorption: A review on adsorbent materials. *Sep. Purif. Technol.* **2022**, *278*, 119675. [CrossRef]
41. Sun, W.; Li, J.; Li, H.; Jin, B.; Li, Z.; Zhang, T.; Zhu, X. Mechanistic insights into ball milling enhanced montmorillonite modification with tetramethylammonium for adsorption of gaseous toluene. *Chemosphere* **2022**, *296*, 133962. [CrossRef]
42. Sun, W.; Zhang, T.; Li, J.; Zhu, X. Enhanced gaseous acetone adsorption on montmorillonite by ball milling generated Si–OH and interlayer under synergistic modification with H_2O_2 and tetramethylammonium bromide. *Chemosphere* **2023**, *321*, 138114. [CrossRef]
43. Novikau, R.; Lujaniene, G. Adsorption behaviour of pollutants: Heavy metals, radionuclides, organic pollutants, on clays and their minerals (raw, modified and treated): A review. *J. Environ. Manag.* **2022**, *309*, 114685. [CrossRef]
44. Mucsi, G. A review on mechanical activation and mechanicalalloying in stirred media mill. *Chem. Eng. Res. Des.* **2019**, *148*, 460–474. [CrossRef]
45. Brindley, G.W.; Brown, G. *Crystal Structures of Clay Minerals and Their X-ray Identification*; Mineral. Soc.: London, UK, 1980; 496p.
46. Rouquerol, F.; Rouquerol, J.; Sing, K.S.W.; Llewellyn, P.; Maurin, G. *Adsorption by Powders and Porous Solids*; Elsevier: Amsterdam, The Netherlands, 2014.
47. Lützenkirchen, J.; Preočanin, T.; Kovačević, D.; Tomišić, V.; Lövgren, L.; Kallay, N. Potentiometric titrations as a tool for surface charge determination. *Croat. Chem. Acta* **2012**, *85*, 391–417. [CrossRef]
48. Payne, T.E.; Brendler, V.; Comarmond, M.J.; Nebelung, C. Assessment of surface area normalisation for interpreting distribution coefficients (K_d) for uranium sorption. *J. Environ. Radioact.* **2011**, *102*, 888–895. [CrossRef] [PubMed]
49. Xia, M.; Jiang, Y.; Zhao, L.; Li, F.; Xue, B.; Sun, M.; Zhang, X. Wet grinding of montmorillonite and its effect on the properties of mesoporous montmorillonite. *Colloids Surf. A Physicochem. Eng.* **2010**, *356*, 1–9. [CrossRef]
50. Sing, K.S.W.; Everett, D.H.; Haul, R.A.W.; Moscou, L.; Pierotti, R.A.; Rouquerol, J.; Siemieniewska, T. Reporting physisorption data for gas/solid systems with special reference to the determination of surface area and porosity. *Pure Appl. Chem. Res.* **1985**, *57*, 603–619. [CrossRef]
51. Valera-Zaragoza, M.; Agüero-Valdez, D.; Lopez-Medina, M.; Dehesa-Blas, S.; Navarro-Mtz, K.A.; Avalos-Borja, M.; Juarez-Arellano, E.A. Controlled modification of sodium montmorillonite clay by a planetary ball-mill as a versatile tool to tune its properties. *Adv. Powder Technol.* **2021**, *32*, 591–599. [CrossRef]
52. Giovanni, C.; Mohammadtaghi, V. Nonthermal Mechanochemical Destruction of POPs. In *Rersistent Organic Pollutants (POPs)*; Rashed, M., Ed.; IntechOpen: London, UK, 2022; 172p.
53. McBride, M.B. *Environmental Chemistry of Soils*; Oxford University Press: New York, NY, USA, 1994.

Disclaimer/Publisher's Note: The statements, opinions and data contained in all publications are solely those of the individual author(s) and contributor(s) and not of MDPI and/or the editor(s). MDPI and/or the editor(s) disclaim responsibility for any injury to people or property resulting from any ideas, methods, instructions or products referred to in the content.

Article

The Effects of Multistage Fuel-Oxidation Chemistry, Soot Radiation, and Real Gas Properties on the Operation Process of Compression Ignition Engines

Valentin Y. Basevich [1], Sergey M. Frolov [1,2,*], Vladislav S. Ivanov [1,2], Fedor S. Frolov [1,2] and Ilya V. Semenov [2]

[1] Department of Combustion and Explosion, Semenov Federal Research Center for Chemical Physics of the Russian Academy of Sciences, Moscow 119991, Russia; basevichv@yandex.ru (V.Y.B.); ivanov.vls@gmail.com (V.S.I.); f.frolov@chph.ru (F.S.F.)
[2] Department of Computational Mathematics, Federal Science Center "Scientific Research Institute for System Analysis of the Russian Academy of Sciences", Moscow 117218, Russia; ilyasemv@yandex.ru
* Correspondence: smfrol@chph.ras.ru

Citation: Basevich, V.Y.; Frolov, S.M.; Ivanov, V.S.; Frolov, F.S.; Semenov, I.V. The Effects of Multistage Fuel-Oxidation Chemistry, Soot Radiation, and Real Gas Properties on the Operation Process of Compression Ignition Engines. *Eng* 2023, *4*, 2682–2710. https://doi.org/10.3390/eng4040153

Academic Editor: Antonio Gil Bravo

Received: 23 August 2023
Revised: 16 October 2023
Accepted: 19 October 2023
Published: 23 October 2023

Copyright: © 2023 by the authors. Licensee MDPI, Basel, Switzerland. This article is an open access article distributed under the terms and conditions of the Creative Commons Attribution (CC BY) license (https://creativecommons.org/licenses/by/4.0/).

Abstract: The objectives of the study are to reveal the influence of multistage fuel-oxidation chemistry, thermal radiation of soot during the combustion of a small (submillimeter size) fuel droplet, and real gas effects on the operation process of compression ignition engines. The use of the multistage oxidation chemistry of iso-octane in the zero-dimensional approximation reveals the appearance of different combinations of cool, blue, and hot flames at different compression ratios and provides a kinetic interpretation of these phenomena that affect the heat release function. Cool flames are caused by the decomposition of alkyl hydroperoxide, during which a very reactive radical, OH, is formed. Blue flames are caused by the decomposition of H_2O_2 with the formation of OH. Hot flames are caused by the chain branching reaction between atomic hydrogen and molecular oxygen with the formation of OH and O. So-called "double" cool flames correspond to the sequential appearance of a separated cool flame and a low-intensity blue flame rather than two successive cool flames. The use of a one-dimensional model of fuel droplet heating, evaporation, autoignition, and combustion at temperatures and pressures relevant to compression ignition engines shows that the thermal radiation of soot during the combustion of small (submillimeter size) droplets is insignificant and can be neglected. The use of real gas caloric and thermal equations of state of the matter in a three-dimensional simulation of the operation process in a diesel engine demonstrates the significant effect of real gas properties on the engine pressure diagram and on the NO and soot emissions: real gas effects reduce the maximum pressure and mass-averaged temperature in the combustion chamber by about 6 and 9%, respectively, increases the autoignition delay time by a 1.6 crank angle degree, increase the maximum heat release rate by 20%, and reduce the yields of NO and soot by a factor of 2 and 4, respectively.

Keywords: detailed kinetic mechanism; autoignition; cool and blue flames; compression ignition engine; operation process; numerical simulation; real gas equation of state; nitrogen oxides; soot

1. Introduction

In view of increasingly strict regulations on pollutant and greenhouse gas emissions [1], CFD methods are becoming increasingly important for simulating in-cylinder processes and for developing combustion strategies in transportation engines [2,3]. Predictability can be reached with adequate models of all accompanying phenomena, including gas exchange, mixture formation, combustion and pollutant formation, heat transfer to the walls, etc. In this manuscript, we focus on only studying the effects of the detailed chemistry of fuel oxidation, soot radiation during autoignition and combustion of small-size droplets in engine conditions, and real gas thermodynamics on the operation process of the compression ignition engine (CIE). Despite the fact that commercially available CFD codes provide a possibility of applying the state-of-the-art detailed kinetic mechanisms (DKMs) of fuel oxidation, sophisticated soot thermal radiation models, as well as real gas equations of state (EoS), automotive engineers worldwide still apply either semi-empirical models or an overall reaction mechanism for simulating spray autoignition and combustion, zonal methods coupled with the oversimplified models of soot formation for handling radiative heat transfer, and the ideal gas EoS. However, both semi-empirical models and overall reaction mechanisms are not universal and are only applicable within the parametric domains they are developed for. Such models and mechanisms do not usually include low-temperature chemical transformations in the combustible mixture, which can have a noticeable effect on the rate of energy release and the engine operation process as a whole.

This paper has three objectives. The first is to give a kinetic interpretation of the multistage autoignition phenomena observed in CIEs. The second is to reveal the effect of soot thermal radiation on the autoignition and combustion of small (submillimeter size) fuel droplets under the conditions of CIEs. The third is to reveal the real gas effects on the operation process of CIEs and the yields of pollutants (soot and NO) using the accurate thermal and caloric EoS of the main species. These objectives and the obtained results are the novel and distinctive features of this paper. The next three sections in the Introduction provide selective literature reviews on these topics.

1.1. Multistage Autoignition

The concept of the multistage autoignition of a hydrocarbon fuel with separated "cool", "blue", and "hot" flames was introduced in [4] based on studies of the autoignition of iso-octane and other hydrocarbons under engine conditions. Under these conditions, at the parametric plane "mixture composition–compression ratio" (Figure 1), in addition to the domain of mixture autoignition, three more domains of pre-flame luminosity are identified. In the first, a partial reaction of the mixture is found to occur and is accompanied by an increase in pressure and temperature, a decrease in oxygen and fuel concentrations, and the formation of intermediate and final reaction products in certain amounts. The luminous flame of bluish color corresponding to this parametric domain is called a "blue" flame and is caused by the luminosity of excited formyl HCO∗. The second and third domains are located below the domain of blue flames at lower compression ratios. In these domains, a partial reaction is also found to proceed with some increase in the pressure and luminosity of excited formaldehyde H_2CO*. The flames corresponding to these two domains are called a "double" cool flame and a single cool flame, respectively. In [5], based on the review of a large experimental material, a generalization was made and the concept of multistage autoignition with individual cool, blue, and hot flames was introduced. Modeling the kinetics of the autoignition of hydrocarbons under laboratory conditions using a DKM indicates that cool flames are caused by the decomposition of alkyl hydroperoxide, during which a very reactive radical, OH, is formed, and blue flames are caused by the decomposition of H_2O_2 with the formation of OH [6].

Recently, the DKMs of the oxidation of heavy hydrocarbons up to C_{20} containing thousands of species and tens of thousands of elementary reactions have been proposed. For example, the authors of [7] proposed such a DKM composed of 7182 species and 31,721 reactions by merging two mechanisms reported in [8,9]. Later, this mechanism was reduced to contain 2277 reactions and 522 species [10]. Many of the proposed DKMs are capable of describing the low-temperature oxidation of hydrocarbons and predicting the formation of both cool and blue flames. The blue flames are often referred to as "intermediate temperature heat release" [11–15] or "intermediate temperature ignition" [16,17].

Despite the undoubted advantages of the DKMs, their application in multidimensional problems of combustion in engines is difficult because of their complexity. Moreover, consideration of all possible reactions between all possible species, including their isomers, would make such mechanisms much larger. For example, only the inclusion of vibrationally excited molecules with the reactions of their formation and consumption leads to multiple increases in the DKM size [18,19]. In addition, such DKMs involve many uncertainties caused by the lack of reliable data on the reaction rate constants and thermodynamic functions. All these factors greatly affect the accuracy and suitability of the DKMs.

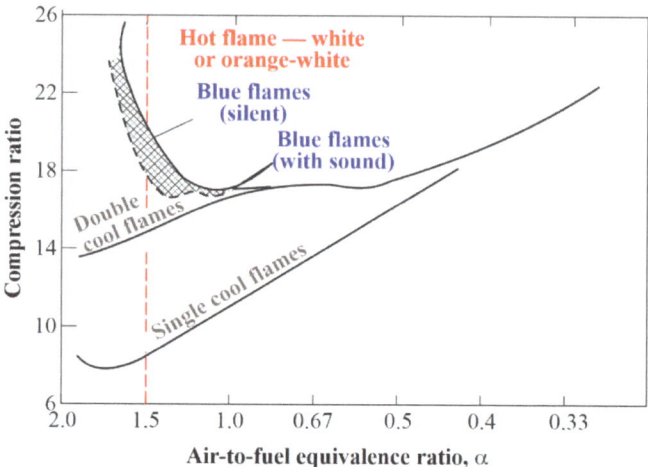

Figure 1. The boundaries of the single and "double" cool flames, blue flame, and hot flame domains during compression induced the autoignition of iso-octane–air mixtures in a compression ignition engine [4] (legends are taken from [5]). Initial temperature $T_0 = 343$ K, initial pressure $P_0 = 1$ bar, engine rotation speed $n = 1500$ rpm. The red dashed vertical line at $\alpha = 1.5$ corresponds to the composition of the fuel–air mixture examined in Section 3.1.

Thus, the known DKMs are generally imperfect and limited to a certain degree. However, in many cases, the optimal and compact DKMs that provide the overall reaction rate and the composition of key intermediate and final products with an acceptable accuracy are required rather than the maximal DKMs, including species and reactions unnecessary for problem solution. Such mechanisms still have the status of non-empirical DKMs, as all included elementary reactions are kinetically substantiated. When simulating the oxidation of hydrocarbon fuels, it is always possible to construct a DKM with a limited number of species and elementary reactions while retaining the main reaction pathways. Such a DKM is proposed for iso-octane [20], which describes the kinetics of multistage autoignition. This DKM has been used to compare the results of calculations with experimental data on the autoignition delay time, laminar flame structure and propagation velocity, and fuel droplet combustion in wide ranges of temperature, pressure, and mixture composition.

1.2. Soot Formation and Radiation

The radiative heat transfer is mainly caused by soot particles formed in fuel-rich locations of the combustion chamber. Contrary to carbon dioxide and water vapor molecules, which radiate in a narrow spectral band, soot particles emit radiation in a wide continuous spectrum, thus producing appreciably more intense radiation than that from the triatomic molecules. Therefore, the key problem in this respect is to evaluate the rate of soot formation and oxidation.

There are many publications in the literature on soot formation, oxidation, and radiation in conditions relevant to CIEs. There exist several noteworthy articles dealing with gaseous diffusion flames [21–25] and hydrocarbon droplet flames [26–28]. These articles discuss the existing approaches to describe the formation and growth of soot particle nuclei, their coagulation, activation and deactivation, and oxidation. As mentioned earlier in this paper, we focus herein on soot radiation in droplet flames under engine conditions. The video frames of large (several millimeters in size) droplet ignitions and combustions in microgravity indicate the appearance of a loose spherical "soot shell" between the flame and the droplet composed of very fine soot particles [29]. This soot shell thickens over time due to the accumulation of soot particles and becomes a kind of thermal screen due to the extraction of a part of the heat flux from the flame to the droplet and thermal radiation to the ambiance. As a result, the rates of droplet evaporation and combustion decrease, and the flame quenches. This phenomenon is called "radiative flame quenching" [30] and has been experimentally [31] and theoretically [32] studied. Interestingly, after flame quenching, the droplet continues unexpectedly and quickly evaporates. Moreover, after some time, a brightly glowing flame spontaneously appears around the droplet and disappears again. Such flame flashes and extinctions can repeatedly occur during the droplet's lifetime. The phenomenology of radiative flame quenching with subsequent flashes of cool and blue flames was computationally reproduced in [33,34] for large n-heptane and n-dodecane droplets. The possibility of the radiative quenching of small (submillimeter-size) droplets is still questionable. Under terrestrial conditions, soot is entrained by the flow of hot gases caused by natural convection, and the flame-quenching ability of the soot shell is less pronounced.

1.3. Real Gas Effects

The ideal gas approximation is only applicable for gases with low density when the interaction between molecules is negligible. Thermodynamic conditions in modern turbocharged diesel engines can go beyond the limits of this approximation due to high gas pressure (up to 200 bar [35] or 140 bar [36]) or density (up to 200 kg/m^3 [37]). Localized regions with low temperatures and high pressure in an engine cylinder can be critical. For example, the maximum pressure in an engine cylinder can reach 200–300 bar, while the temperature in the vicinity of the cooled walls can be quite low (400–550 K). Under such conditions, the real gas effects may manifest themselves. Thus, real gas effects can significantly affect the behavior and structure of the flow in the region of diesel spray [37–40] and change ignition delay times in diesel engines, particularly in the region of the negative temperature coefficient [40,41].

In the literature, multiparameter EoS have usually been used to theoretically approximate the p–ρ–T data with the accuracy of experimental data [42]. In handbooks, the p–ρ–T EoS containing a few tens of parameters (up to 50) have been reported. Such EoS are nothing else but complex interpolations of experimental data. The application of such equations in CFD is accompanied by large computational costs, as the EoS is used at each iteration of each time step in each control volume. Such EoS have only been reported for lower n-alkanes.

There are many approximate real gas EoS, among which the modifications of the Van der Waals EoS are most popular; the Redlich–Kwong equation [43] and Peng–Robinson equation [44].

In modern diesel engines, effects other than real gas effects can also play a certain role. Among them are the two-phase equilibrium effects, the effects of the gas-phase chemical decomposition of fuel molecules, and the effects of dissociation and ionization. In the literature, there are no available approaches to take these effects into account in the real gas EoS. Therefore, these effects are commonly neglected.

The paper is organized as follows. In Section 2, we briefly introduce our own DKM of hydrocarbon fuel oxidation (Section 2.1), our own model of droplet autoignition and combustion (Section 2.1), and our own real gas EoS (Section 2.3) with examples of their validation. Thereafter, we briefly describe the procedures of the numerical solution of the new target problems, namely, the manifestation of multistage autoignition in the CIE (Section 2.4), the manifestation of soot thermal radiation during droplet combustion in CIE (Section 2.5), and the manifestation of real gas effects in CIE (Section 2.6). In Section 3, we show the results of calculations and discuss the corresponding implications. Section 4 summarizes the results of the study and we discuss the directions of future work.

2. Materials and Methods

2.1. Reaction Mechanism

The oxidation of hydrocarbons exhibits great generality [5]. The DKM of iso-octane oxidation and combustion was developed in [20,45]. This DKM includes the main processes that determine the reaction rate and the formation of the main intermediate and final reaction products. Since all included elementary reactions are kinetically substantiated, this DKM has the status of an ab initio mechanism. The DKM is based on the non-extensive approach and the analogy technique [46]. The key point of the non-extensive approach is the generality of the main reaction pathways. The analogy technique is applied for the selection of elementary reactions important for modeling the multistage oxidation. It is assumed, in the DKM, that the first addition of oxygen to the peroxy radical is sufficient, whereas the second addition of oxygen to the isomerized form of the peroxy radical is not essential for the overall reaction progress. The DKM of iso-octane oxidation is obtained by adding nine isomerized derivatives of iso-heptane (2,2-dimethylpentane) and iso-octane (2,2,4-tri-methylpentane) to each of the components of the DKM of oxidation of n-octane C_8H_{18} and isomerized alkanes: iso-butane (2-methylpropane), iso-pentane (2-methylbutane), and iso-hexane (2-methylpentane). Their reactions with each other and other species available in the DKM are also added. The leading role is assumed to be played by deisomerization reactions to form stable intermediate isomerized molecules of 2,2-dimethylpentane, 2-methylpentane, 2-methylbutane, and 2-methylpropane and the corresponding normal-structure hydrocarbons, including methane. All of the other reactions leading to the increase in the "linear" five-membered portion of an iso-octane molecule are assumed to have no effect on the iso-octane oxidation rate. The resulting DKM of iso-octane oxidation is fairly compact: it includes 763 reactions involving normal-structure species and 987 reactions involving isomerized species. The total number of species involved in the DKM is 144.

The enthalpy ΔH°_{f298}, entropy $S^\circ_{298}(T)$, and the specific heat at constant pressure $C_{p0}(T)$ are calculated following known recommendations and additivity rules [47]. The Arrhenius parameters of rate constants for some reactions are calculated based on the rate constants for the reactions with normal-structure species due to the lack of the corresponding experimental data. For this purpose, a two-parameter form for the rate constant of an elementary reaction is used with the preexponential factor A and activation energy E given by [46]:

$$A_{i(i)} = A_{i(n)} \exp\left[\left(\Delta S_{i(i)} - \Delta S_{i(n)}\right)/R\right] \quad (1)$$

$$E_{i(i)} = E_{i(n)} - 0.25\left(\Delta H_{i(i)} - \Delta H_{i(n)}\right) \quad (2)$$

for exothermic reactions and

$$E_{i(i)} = E_{i(n)} + 0.75\left(\Delta H_{i(i)} - \Delta H_{i(n)}\right) \quad (3)$$

for endothermic reactions. Here, ΔS and ΔH are the entropy and enthalpy changes; indices i, (i), and (n) correspond to the reaction number, to the isomerized-structure species, and to the normal-structure species, respectively. Some corrections of the Arrhenius parameters thus obtained were needed for a limited number (less than ten) of reactions. The eminent feature of the DKM is the manifestation of the occurrence of both cool and blue flames at the low-temperature autoignition of iso-octane and other included hydrocarbons.

Figure 2 demonstrates the predictive capabilities of the DKM. It compares the calculated and measured ignition delays for the homogeneous stoichiometric iso-octane–air (Figure 2a) and n-heptane–air (Figure 2b) mixtures in wide ranges of initial conditions in terms of pressure and temperature. A comparison of the calculation results with the experimental data is seen to yield their satisfactory agreement, as the uncertainty in the ignition delay data can exceed 100% [48,49].

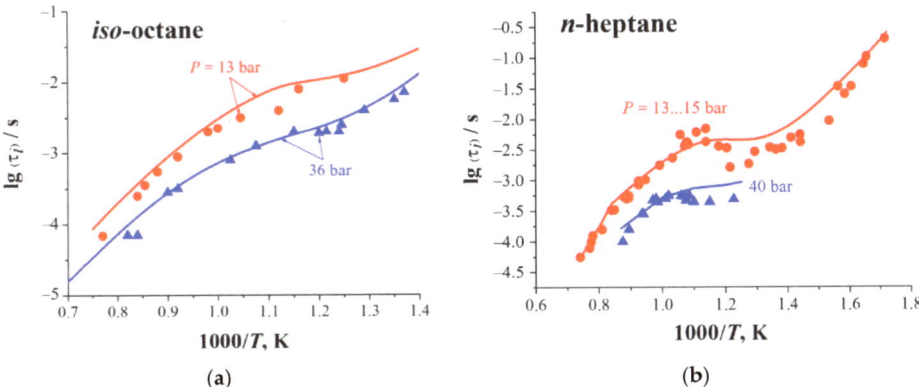

Figure 2. Comparison of predicted (curves) and measured (symbols) ignition delay times for the homogeneous stoichiometric iso-octane–air (**a**) and n-heptane–air (**b**) mixtures in wide ranges of initial conditions in terms of pressure and temperature: (**a**) circles, triangles [48]; (**b**) circles, triangles [49].

2.2. Droplet Autoignition and Combustion

The model of droplet autoignition was reported in [33,34]. The model is built on the concept of multicomponent diffusion in the gas, the DKM of hydrocarbon fuel oxidation, and the constant pressure condition in the gas-droplet system. The eminent feature of the used DKM is that it describes both multistage low-temperature oxidation with cool and blue flames and high-temperature droplet combustion. In the DKM, soot is designated as the C atom and is modeled by an equivalent gas species with a molecular mass of carbon, 12 kg/kmol. Herein, this model is only briefly presented. The set of governing equations includes:

The continuity equation for the liquid ($0 < r < r_m$):

$$\frac{\partial \rho_d}{\partial t} + \frac{1}{r^2}\frac{\partial}{\partial r}\left(r^2 \rho_d u_d\right) = 0 \qquad (4)$$

where r_m is the coordinate of the droplet surface (droplet radius); t is time; ρ is the density; u is the velocity; and index d denotes the liquid parameters;

The energy conservation equation for the liquid ($0 < r < r_m$):

$$c_d \rho_d \frac{\partial T_d}{\partial t} + c_d \rho_d u_d \frac{\partial T_d}{\partial r} = \frac{1}{r^2}\frac{\partial}{\partial r}\left(\lambda_d r^2 \frac{\partial T_d}{\partial r}\right) \qquad (5)$$

$$T_d(0, r) = T_{d0},\ \left.\frac{\partial T_d}{\partial r}\right|_{r=0} = 0,$$

$$T_d(t, r_m) = T_g(t, r_m)$$

where $T_d(t, r)$ is the liquid temperature; $c_d(T_d)$ is the specific heat capacity of the liquid; λ_d is the thermal conductivity of the liquid; index 0 denotes initial values; and index g denotes gas properties;

The equation for the mass fraction of liquid vapor at the droplet surface ($r = r_m$):

$$Y_v = \frac{P_v}{P}\frac{W_v}{\overline{W}} \qquad (6)$$

where P is the pressure; W is the molecular mass; index v refers to the liquid vapor; and the overbar denotes the average value;

The continuity equation for the gas ($r_m < r < R_a$):

$$\frac{\partial \rho_g}{\partial t} + \frac{1}{r^2}\frac{\partial}{\partial r}\left(r^2 \rho_g u_g\right) = 0 \tag{7}$$

$$\rho_d\left(u_d - \frac{\partial r_m}{\partial t}\right)\bigg|_{r=r_m} = \rho_g\left(u_g - \frac{\partial r_m}{\partial t}\right)\bigg|_{r=r_m}$$

where R_a is the radius of the computational domain; and $\partial r_m / \partial t = u_m$ is the instantaneous velocity of the droplet surface due to both thermal expansion and evaporation;

The equation of continuity for gas species ($r_m < r < R_a$):

$$\rho_g \frac{\partial Y_j}{\partial t} + \rho_g u_g \frac{\partial Y_j}{\partial r} = \frac{1}{r^2}\frac{\partial}{\partial r}\left(\rho_g r^2 Y_j V_j\right) + \omega_{gj}$$

$$Y_j(0,r) = Y_{j0}, j = 1, 2, \ldots, N, \tag{8}$$

$$-\rho_d u_d \beta_i|_{r=r_m} = \rho_g Y_j\left(u_g - \frac{\partial r_m}{\partial t}\right) + \rho_g Y_j V_j\bigg|_{r=r_m},$$

$$\frac{\partial \overline{W} Y_j}{\partial r}\bigg|_{r=R} = 0, j = 1, 2, \ldots, N$$

where V_j is the diffusion velocity of the jth species. The rates of chemical reactions ω_{gj} and the coefficients β_i are determined as

$$\omega_{gj} = W_{gj} \sum_{k=1}^{L}\left(v''_{j,k} - v'_{j,k}\right) A_k T_g^{n_k} \exp\left(-\frac{E_k}{RT_g}\right) \prod_{l=1}^{N}\left(\frac{Y_{gl}\rho_g}{W_{gl}}\right)^{v'_{l,k}} \tag{9}$$

$$\beta_i = 1 \text{ at } j = v$$

$$\beta_i = 0 \text{ at } j \neq v$$

where $v'_{j,k}$ and $v''_{j,k}$ are the stoichiometric coefficients for the jth species in the case when it is a reactant and product in the kth reaction, respectively; A_k, n_k, and E_k are the preexponential factor, temperature exponent, and activation energy for the kth reaction;

The equation for the diffusion velocity for the gas ($r_m < r < R_a$):

$$\frac{\partial X_j}{\partial r} = \sum_{k=1}^{N}\left(\frac{X_j X_k}{D_{jk}}\right)(V_k - V_j) \tag{10}$$

where $X_j = Y_j \overline{W} / W_j$ is the mole fraction of the jth component in the mixture;

The equations of conservation of energy for the gas ($r_m < r < R_a$):

$$c_{pg}\rho_g \frac{\partial T_g}{\partial t} + c_{pg}\rho_g u_g \frac{\partial T_g}{\partial r} = \frac{1}{r^2}\frac{\partial}{\partial r}\left(\lambda_g r^2 \frac{\partial T_g}{\partial r}\right) + \Omega - \sigma S_{\text{rad}} Y_s \rho_g T_g^4 \tag{11}$$

$$T_g(0,r) = T_{g0}(r), T_g(t,r_m) = T_d(t,r_m), \frac{\partial T_g}{\partial r}\bigg|_{r=R} = 0$$

where $\rho_g = \rho_g(p, T_g)$, $c_{pg} = c_{pg}(T_g)$, and $\lambda_g = \lambda_g(p, T_g)$ are, respectively, the density, specific heat capacity, and thermal conductivity of the gas mixture. The term Ω in Equation (11) is given by

$$\Omega = \sum_{k=1}^{L} H_k A_k T_g^{n_k} \exp\left(-\frac{E_k}{RT_g}\right) \prod_{j=1}^{N}\left(\frac{Y_{gj}\rho_g}{W_{gj}}\right)^{v'_{l,k}}$$

where H_k is the thermal effect of the kth chemical reaction. The last term in Equation (20) represents the heat loss due to soot radiation, the subject of our primary interest herein. In this term, σ is the Stefan–Boltzmann constant, Y_s is the mass fraction of soot, $S_{\text{rad}} = 6/(d_s\rho_s)$ is the specific surface area of conditional soot particles (here, d_s is the conditional soot particle size, ρ_s is the soot density). If one assumes $d_s \sim 1$ nm and $\rho_s \approx 2000$ kg/m^3, then $S_{\text{rad}} \approx 3 \times 10^6$ m^2/kg [50].

The condition at the droplet surface ($r = r_m$) for determining the droplet surface temperature $T_{d,m}$:

$$\lambda_d \frac{\partial T_d}{\partial r} - \frac{\rho_{d,m} u_{d,m} L_v}{W_v} = \lambda_g \frac{\partial T_g}{\partial r} \tag{12}$$

where L_v is the latent heat of liquid vaporization. Equation (12) is used for matching Equations (5) and (11);

The ideal gas EoS for the gas:

$$\rho_g = \frac{P\overline{W}}{RT_g} \tag{13}$$

The condition of constant pressure:

$$P = \text{const} \tag{14}$$

The set of Equations (4)–(14) allows one to determine the spatial structure of the flow around the droplet and its evolution in time, and to calculate the time dependences of the droplet diameter $d = 2r_m$, droplet surface temperature $T_{d,m}$, the maximum gas temperature $T_m = T_{g,max}$ (flame temperature), etc.

Figures 3–5 demonstrate the predicting capabilities of the model. Figure 3 compares the predicted and measured [51,52] time histories of the normalized squared diameter of n-heptane droplets at vaporization (Figure 3a) and combustion (Figure 3b). Figure 4 compares the predicted and measured [53–59] dependences of the combustion rate constant K on the initial diameter of the n-heptane droplet in normal pressure and temperature conditions. Figure 5 compares the predicted and measured [60,61] ignition delays of single n-heptane droplets at temperatures of 940–1000 K and at different pressures. In general, the agreement between calculations and measurements is encouraging, keeping in mind that the accuracy of optical measurements of the droplet diameter is moderate, while the accuracy of ignition delay measurements can attain 100% [61].

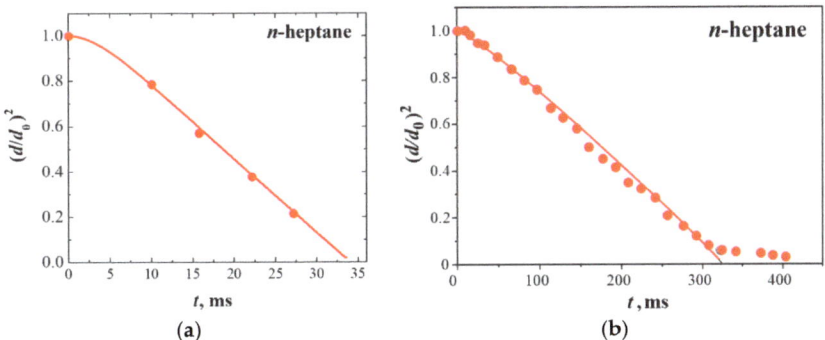

Figure 3. Comparison of predicted (curves) and measured (symbols) dynamics of n-heptane droplet vaporization and combustion in air: (**a**) vaporization, $d_0 = 70$ mm, $T_{d0} = 293$ K, $T_{g0} = 573$ K [51]; (**b**) combustion, $d_0 = 500$ mm, $T_{d0} = 293$ K, $T_{g0} = 293$ K [52].

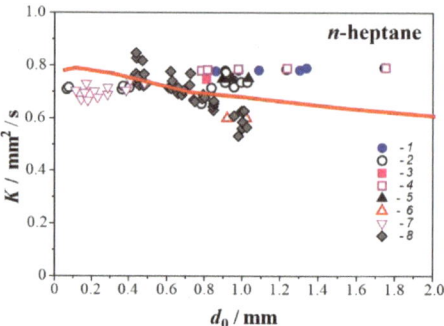

Figure 4. Comparison of predicted (curve) and measured (symbols) dependences of the combustion rate constant K on the initial diameter of the n-heptane droplet. Experimental data: 1—[53], 2—[54], 3—[55], 4, 5—[56], 6—[57], 7—[58], 8—[59].

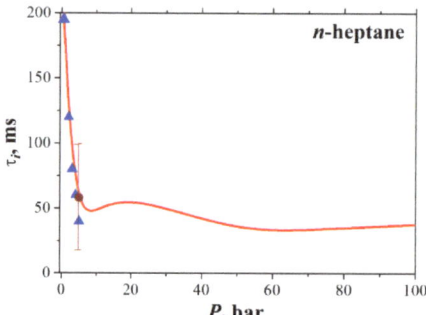

Figure 5. Comparison of predicted (curve) and measured (symbols) dependences of the autoignition delay time of n-heptane droplets on pressure. The curve is obtained for $d_0 = 700$ μm, $T_{d0} = 293$ K, and $T_{g0} = 1000$ K. Triangles correspond to $d_0 = 700$–750 μm [60], circle to $d_0 = 700$ μm, $T_{g0} = 940$ K [61] (the vertical bar shows the scatter of experimental data).

2.3. Real Gas Equation of State

The real gas thermal EoS is commonly written in a "virial" form:

$$\frac{Pv}{RT} = 1 + \frac{B}{v} + \frac{C}{v^2} + \frac{D}{v^3} + \frac{E}{v^4} + \frac{F}{v^5} + \ldots \tag{15}$$

where $B, C, D, E,$ and $F, \ldots$ are the virial coefficients depending on temperature T; P is the pressure; R is the universal gas constant, and v is the molar volume. To describe the operation process in a diesel engine, we apply a relatively accurate real gas EoS proposed in [62]:

$$P = \rho RT \left[1 + B(T)\rho + b^2 \rho^2 + b^3 \rho^3 \right] \tag{16}$$

where $B(T)$ and b are the coefficients of the truncated virial series. To determine the thermodynamic functions of a real gas, the concept of excess thermodynamic functions is used. According to this concept, the excess enthalpy H_{exc} and excess internal energy E_{exc} are given by:

$$H_{exc}(T,\rho) = E_{exc}(T,\rho) + P_{exc}(T,\rho)/\rho \tag{17}$$

$$E_{exc}(T,\rho) = -\int_0^\rho \left[T(\partial p/\partial T)_\rho - P \right] d\rho/\rho^2 \tag{18}$$

where $\rho = 1/v$ is the molar density; and P_{exc} is the excess pressure:

$$P_{\text{exc}} = P - P_0 \tag{19}$$

with index 0 denoting the ideal gas properties.

The enthalpy of a real gas is equal to:

$$H(T,\rho) = H_0(T) + H_{\text{exc}}(T,\rho) \tag{20}$$

where $H_0(T)$ is the ideal gas enthalpy; and $H_{\text{exc}}(T,\rho)$ is the excess enthalpy. The differentiation of Equation (7) with respect to temperature makes it possible to obtain an expression for the specific heat capacity of a real gas at a constant pressure, C_p:

$$C_p(T,\rho) = C_{p0}(T) + C_{p,\text{exc}}(T,\rho) \tag{21}$$

where $C_{p,\text{exc}}$ is the excess heat capacity at constant pressure, and $C_{p0}(T)$ is the specific heat capacity of the ideal gas.

The ideal gas thermal and caloric EoS are written in the standard form:

$$P_0 v = RT \tag{22}$$

$$dE_0 = C_{v0}(T)dT$$

$$dH_0 = C_{p0}(T)dT$$

where $E_0(T)$ is the specific molar internal energy and C_{v0} is the specific heat capacity at constant volume. In accordance with Equation (19), in a real gas, there is a relative excess pressure equal to

$$P_{\text{exc}}/P_0 = (P - P_0)/P_0 = 1 + B(T)\rho + b^2\rho^2 + b^3\rho^3 \tag{23}$$

In [62], the values of coefficients $B(T)$ and b are obtained for n-alkanes from methane to tetradecane, as well as for gases such as O_2, N_2, H_2O, CO, CO_2 and H_2 in wide ranges of pressure (from 0.5 to 200 bar) and temperatures (from 280 to 3000 K). As an example, Table 1 demonstrates the accuracy of Equation (16) for n-hexane and other mentioned gases at some selected isobars and isotherms. It turns out that the error in calculating the pressure according to Equation (16) does not normally exceed tenths of a percent, even in the vicinity of the critical point. Note that Equation (16) is used in the gas-dynamic calculation to determine the density ρ from the known values of pressure P and temperature T, as well as from the known composition of the gas mixture. To solve Equation (16) with respect to density, one has to use numerical methods.

Table 1. Comparison of predicted pressure, P_{calc}, given by the EoS of Equation (16) with measured pressure, P_{exp}, for n-hexane, oxygen, nitrogen, water, carbon monoxide, carbon dioxide, and hydrogen at some selected isobars and isotherms.

	n-Hexane			
T, K	ρ, mol/dm^3	P_{calc}, MPa	P_{exp}, MPa [63]	$\lvert P_{\text{calc}} - P_{\text{exp}} \rvert / P_{\text{exp}}$, %
530	2.526	4.010	4	0.25
550	1.619	4.032	4	0.79
600	1.106	4.008	4	0.19
630	0.9741	4.0025	4	0.06

Table 1. *Cont.*

		Oxygen		
T, K	ρ, kg/m^3	P_{calc}, MPa	P_{exp}, MPa [64]	$\left|P_{calc}-P_{exp}\right|/P_{exp}$, %
500	45.6	6.0008	6	0.013
500	60.51	8.0033	8	0.041
500	75.25	10.007	10	0.07
500	111.31	15.028	15	0.19
500	146.15	20.076	20	0.38
		Nitrogen		
T, K	ρ, mol/dm^3	P_{calc}, MPa	P_{exp}, MPa [65]	$\left|P_{calc}-P_{exp}\right|/P_{exp}$, %
500	0.94635	4.0007	4	0.018
500	1.4070	6.0011	6	0.018
500	1.8590	8.0018	8	0.023
500	2.3020	10.002	10	0.022
500	3.3700	15.004	15	0.027
500	4.3806	20.004	20	0.022
		Water		
T, °C	v, dm^3/g	P_{calc}, MPa	P_{exp}, MPa [66]	$\left|P_{calc}-P_{exp}\right|/P_{exp}$, %
300	5.885	4.0066	4	0.17
300	4.532	5.017	5	0.34
300	3.616	6.033	6	0.55
300	2.976	7.0052	7	0.07
300	2.425	8.090	8	1.1
		Carbon monoxide		
T, K	ρ, mol/dm^3	P_{calc}, MPa	P_{exp}, MPa [67]	$\left|P_{calc}-P_{exp}\right|/P_{exp}$, %
500	0.94818	4.004	4	0.09
500	1.40962	6.002	6	0.04
500	1.86196	7.999	8	0.01
500	2.30518	9.994	10	0.06
500	2.73932	11.990	12	0.08
500	3.16448	13.988	14	0.09
500	3.58073	15.989	16	0.07
500	3.98813	17.994	18	0.03
500	4.38673	20.004	20	0.02
		Carbon dioxide		
T, °C	ρ, g/cm^3	P_{calc}, MPa	P_{exp}, MPa [68]	$\left|P_{calc}-P_{exp}\right|/P_{exp}$, %
300	56.42	6.000	6	0.00
300	75.59	8.000	8	0.01
300	94.89	10.000	10	0.00
300	114.26	12.000	12	0.00
300	133.67	14.001	14	0.01
300	153.09	16.006	16	0.04
300	172.4	18.009	18	0.05
300	191.6	20.014	20	0.07
		Hydrogen		
T, K	ρ, mol/dm^3	P_{calc}, MPa	P_{exp}, MPa [69]	$\left|P_{calc}-P_{exp}\right|/P_{exp}$, %
500	0.94818	3.998	4	0.05
500	1.40962	5.994	6	0.10
500	1.86196	7.990	8	0.13
500	2.30518	9.986	10	0.14
500	2.73932	14.990	15	0.07
500	3.16448	20.03	20	0.15

For mixtures of real gases, the coefficients $B(T)$ and b are calculated according to the following approximate mixing rules [62]:

$$B_{mix} = \sum_i^N B_i x_i;\ b_{mix} = \sum_i^N b_i x_i \qquad (24)$$

where x_i is the volume fraction of substance i; and N is the number of substances in the mixture.

The specific heat capacity at constant pressure is calculated by Equation (21). To determine the specific heat capacity of an ideal gas C_{p0}, a polynomial of the third order is used:

$$C_{p0} = a_1 + a_2 T + a_3 T^2 + a_4 T^3 \qquad (25)$$

where a_1, a_2, a_3, and a_4 are tabulated coefficients [62]. The excess specific heat capacity $C_{p,exc}$ is calculated from analytical relationships that include the logarithmic derivatives of the coefficients $B(T)$ and b with respect to temperature [62].

2.4. Solution Procedure of the Zero-Dimensional Problem

To reveal the possibility of modeling the multistage autoignition phenomena in CIEs with the DKM of iso-octane oxidation, zero-dimensional (0D) calculations of the kinetic processes occurring when a gas volume is compressed by a moving piston are performed using the CHEMKIN computer code [70]. Due to the uncertainty in estimates of the time-varying rate of heat transfer from the fuel charge to the engine wall, the velocity of gases, and turbulence, the process of gas cooling by the engine walls is not taken into account. Due to the short duration of the chemical process, the neglect of heat transfer cannot introduce a noticeable qualitative change in the calculations.

2.5. Solution Procedure of the One-Dimensional Problem

The set of Equations (4)–(11) is numerically integrated using a non-conservative implicit finite difference scheme and a moving adaptive mesh. The thermophysical properties of liquids and gases are taken from [48]. The chemical sources ω_{gj} and Ω are calculated using the DKM of n-heptane oxidation (available in the DKM of iso-octane oxidation) combined with the overall mechanism (OM) of soot formation [71]. The combined mechanism contains 623 reversible and 4 irreversible (Table 2) reactions and 84 species (including soot C). The parameters A_k and E_k in Table 2 are determined using the DKM of soot formation [72]. Contrary to the DKM of [72], in which soot nuclei are formed in processes involving a stable polyaromatic molecule and a radical or two polyaromatic radicals, the role of the soot precursor in the OM of Table 2 is attributed to acetylene C_2H_2. As an example, Figure 6 compares the soot yields predicted by the OM with those predicted by the DKM of [72] for the oxidation of homogeneous fuel-rich n-heptane–air mixture with a fuel-to-air equivalence ratio of $\Phi = 2$. Soot yield refers to the ratio of the mass of carbon contained in soot to the initial mass of carbon contained in the hydrocarbon fuel. As seen, the OM of Table 2 provides satisfactory qualitative and quantitative agreement with the results predicted by the DKM [72].

Table 2. Overall mechanism of soot formation.

Reaction	A_k, (L, mol, s)	E_k/R, K	n_k
$C_2H_2 + C_2H_2 = C + C + C_2H_4$	2×10^{16}	40,000	0
$C + CO_2 = CO + CO$	1×10^{15}	40,000	0
$C + H_2O = H_2 + CO$	1×10^{15}	40,000	0
$C + OH = HCO$	1×10^{12}	0	0

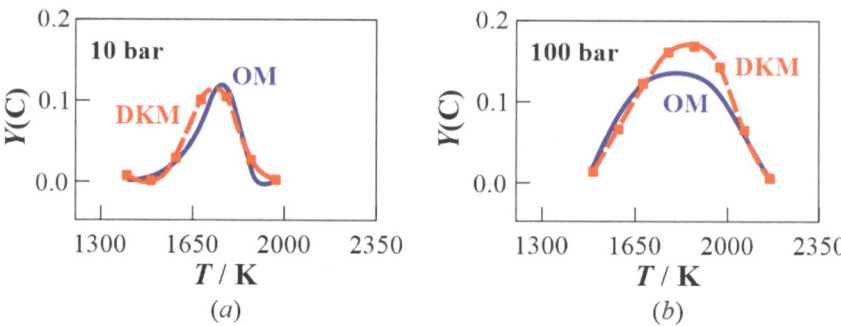

Figure 6. Soot yields $Y(C)$ during the oxidation of fuel-rich n-heptane–air mixture ($\Phi = 2$) as a function of temperature T at $P = 10$ bar (**a**) and 100 bar (**b**) predicted by the DKM [72] and the OM of Table 1.

The calculation is based on the iterative procedure at each time step. The procedure involves the complete linearization of Equation (12) at the droplet surface. The solution accuracy is continuously monitored for compliance with the elemental balances of C and H atoms, as well as with the energy balance. The maximum deviation in the balances is 0.1%.

2.6. Solution Procedure of the Three-Dimensional Problem

To study the influence of real gas effects on the operation process of a diesel engine, the thermal and caloric EoS of [62] is implemented into the AVL FIRE code [73], which is widely used by engine companies worldwide for the design of reciprocating engines. By default, the code applies the ideal gas EoS. The object of the three-dimensional (3D) numerical study is a diesel engine with a semi-separated combustion chamber. Table 3 shows the parameters and conditions of the engine operation mode. This study is interested in the real gas effects on the indicator diagram and the yields of NO and soot.

Table 3. Parameters and conditions of the operation mode of the diesel engine *.

Parameter	Value
Rotation speed, rpm	2000
Cylinder radius, mm	42.5
Compression ratio	16
Start of injection, CAD **	715.78
End of injection, CAD	730.06
Injection angle, deg.	150
Mass of injected fuel, kg	2.8×10^{-5}
Fuel temperature, K	330.15
Mass fraction of exhaust gases	0.233
Equivalence ratio in exhaust gases	0.5606
Flow swirl, 1/min	5800

* The data are provided by AVL LIST GmbH; ** CAD = Crank Angle Degree.

It is assumed that all cylinders of the engine operate in the same way, so only one engine cylinder is considered in the calculations. If we assume that the combustion process is axisymmetric, we can proceed to the consideration of a segment model, which is 1/8 of the combustion chamber in accordance with the number of nozzles in the injector, which greatly simplifies the geometry of the computational domain. However, such a simplification requires the account for the volume of technological recesses in the geometry of the combustion chamber associated with valves, etc. This is usually made by providing the so-called compensation volume (CV) in the geometry of the computational domain. The CV is an additional volume equal to the unaccounted volume of recesses in the piston and in the head cylinder block. In this case, the CV is placed along the cylinder wall. Figure 7 shows a base (Figure 7a) and fine (Figure 7b) computational meshes used in the calculations. The average cell size in the base and fine meshes is 0.5 and 0.25 mm, respectively. The computational mesh is movable.

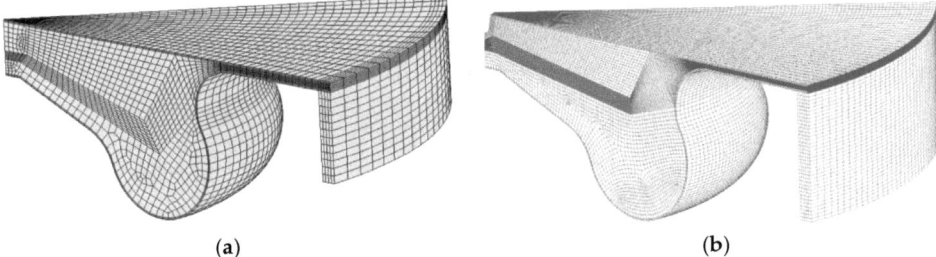

(a) (b)

Figure 7. Base (**a**) and fine (**b**) computational meshes of the diesel segment at a top dead center (TDC); Base mesh contains 30,000 cells, average cell size 0.5 mm; Fine mesh contains 250,000 cells, average cell size 0.25 mm.

Numerical studies of the engine operation process are carried out on the basis of 3D Reynolds-averaged Navier–Stokes (RANS) equations supplemented by the k–ζ–f turbulence model [74], the Lagrangian model of the fuel spray, and the three-zone combustion model ECFM-3Z [75]. The motion, evaporation, and fragmentation of liquid droplets (diesel oil) in the fuel spray model are described by the standard Schiller–Neumann [76], Dukovich [77], and WAVE [78] models, respectively. For simplicity, it is assumed that the chemical surrogate of diesel oil is n-heptane possessing the cetane number ($\approx$56) close to that of diesel oil ($\approx$53) [79]. Thus, the gaseous combustible mixture may contain fuel vapor (C_7H_{16}), O_2, N_2, CO_2, H_2O, CO, and H_2. The formation of nitrogen oxides and soot is described by the Zel'dovich [80] and Kennedy–Hiroyasu–Magnussen [81] standard models. Due to the low concentrations, nitrogen oxides, soot, and active radicals present in the combustion model are not included in the material balance.

Periodic boundary conditions are set on the side surfaces of the segment. The walls of the combustion chamber are considered nonslip, isothermal (piston 475.15 K, head 450.15 K, cylinder walls 375.15 K), impermeable, and noncatalytic. To avoid excessive thickening of the computational mesh near the walls, the formalism of wall functions is used in the calculations. The calculation starts from a crank angle of 566.5 CAD and ends at 860 CAD.

To solve the system of governing mass, momentum, and energy conservation equations, a segregated algorithm of the SIMPLE type (semi-implicit method for pressure-linked equations) is used [82]. The convective transport in the mass conservation law is approximated by the central difference, in the momentum conservation law by the Total Variation Diminishing (TVD) scheme with the MINMOD limiter [83], and for the rest of the conservation equations, the standard first-order UPWIND scheme is used.

Figure 8 demonstrates the predictive capabilities of the model. It compares the measured pressure in the diesel engine under consideration with the pressure histories calculated using the base and fine meshes. In the calculations, the ideal-gas thermal and caloric EoS are used. To fit the measured pressure curve, the empirical constants of the WAVE model were slightly adjusted (within the recommended range) as compared to the default values (C_1 = 0.61 and C_2 = 30), but were the same for both meshes. Despite the use of the simplified engine model, the calculated pressure histories are seen to satisfactorily agree with each other and with the measurements. This comparison justifies the use of the base computational mesh for simulating the engine operation process.

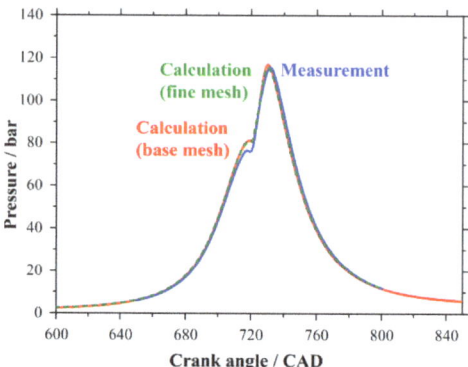

Figure 8. Calculated and measured time histories of the in-cylinder pressure.

3. Results and Discussion

3.1. Multistage Fuel-Oxidation Chemistry

Figure 9 presents the results of the 0D calculations in terms of the time histories of pressure P, temperature T, and volume fractions of OH, iso-octyl hydroperoxide ($CH_3C(CH_3)_2 CH_2CH(CH_3)CH_2O_2H$), and hydrogen peroxide (H_2O_2) in the engine cylinder at an air-to-fuel equivalence ratio $\alpha = 1.5$ ($\Phi = 0.667$) and different compression ratios ε: 12 (Figure 9a); 13 (Figure 9b); 13.5 (Figure 9c); 13.75 (Figure 9d); and 14 (Figure 9e). In all plots, the time is plotted along the X-axis in terms of the crankshaft rotation angle θ (in CAD). The extreme position of the piston (TDC, 0 CAD) is shown by the dashed vertical line. For the sake of convenience, the pressure P (in bar), temperature T (in K), and volume fractions of OH, $CH_3C(CH_3)_2CH_2CH(CH_3)CH_2O_2H$, and H_2O_2 are all plotted along the Y-axis with some scaling factors indicated near the corresponding curves. The mixture composition with $\alpha = 1.5$ ($\Phi = 0.667$) is chosen for calculations because the boundaries of single cool flame, "double" cool flame, blue flame, and hot flame domains for this mixture in the experiments are well separated (see the dashed vertical line in Figure 1). All other calculation conditions are the same as in the experiments.

At $\varepsilon \leq 12$ (see Figure 9a), the reaction is not noticeable in the calculations. The pressure and temperature curves are virtually symmetrical with respect to the dashed vertical line marking the TDC position. The experimental limit for a given mixture with $\alpha = 1.5$ corresponds to $\varepsilon \sim 9$.

At $\varepsilon = 13$ (see Figure 9b), the calculated pressure and temperature curves detect a deviation from the symmetrical shape starting from $\theta \sim 7$ CAD. This deviation corresponds to the heat release in a single cool flame. As a matter of fact, the volume fraction of OH has a peak at $\theta \sim 9$ CAD, where the rate of iso-octyl hydroperoxide decomposition attains the maximum value, while hydrogen peroxide is only accumulated. The experimental range of single cool flames at $\alpha = 1.5$ corresponds to $\varepsilon \sim 9$–15.

At higher compression ratios, the authors of [4] identified the occurrence of so-called "double" cool flames (see Figure 1). As a matter of fact, according to our calculations, the domain of "double" cool flames corresponds to the sequential appearance of a separated cool flame (arising due to the decomposition of iso-octyl hydroperoxide) and a low-intensity blue flame (arising due to the decomposition of H_2O_2), rather than two successive cool flames. In other words, in this parametric domain, the multistage nature of the low-temperature autoignition of hydrocarbons manifests itself. In laboratory conditions, multiple cool flames can only occur when strong cooling is applied to the reactor walls and the reaction progress is insignificant. In this case, almost identical conditions can be reproduced for restarting the cool-flame reactions. In the internal combustion engines, such conditions are not realized.

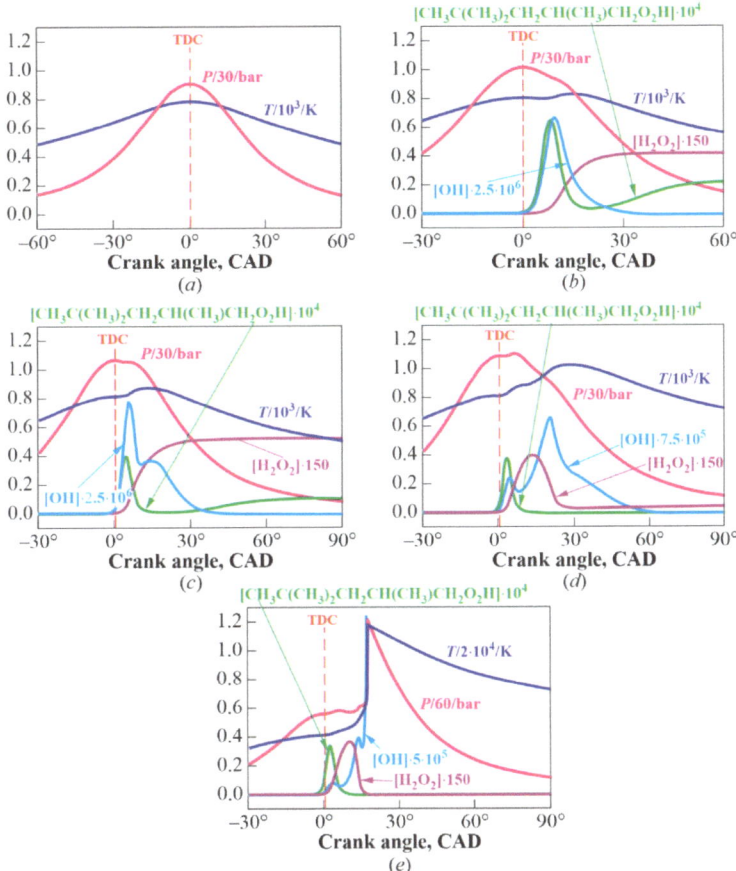

Figure 9. Calculated time histories of pressure P, temperature T, and volume fractions of OH, $CH_3C(CH_3)_2CH_2CH(CH_3)CH_2O_2H$, and H_2O_2 at an air-to-fuel equivalence ratio of 1.5 and different compression ratios ε: (a) 12; (b) 13; (c) 13.5; (d) 13.75; (e) 14.

In the calculations, the "double" cool flames are detected at $\varepsilon = 13.5$ in terms of a double peak of hydroxyl concentration during the expansion of the gas volume after TDC (see Figure 9c). On the pressure and temperature curves, the second peak is virtually not distinguished. The first (large) peak of OH corresponds to the maximum rate of iso-octyl hydroperoxide decomposition at $\theta \sim 6$ CAD. Thereafter, the second (small) peak of OH forms from $\theta \sim 7$–8 to ~ 14 CAD, when iso-octyl hydroperoxide still continues to decompose. Here, H_2O_2 is only accumulated, but, probably, the inverse process of H_2O_2 decomposition is already beginning to noticeably proceed.

The calculated case of clearly separated cool and blue flames corresponds to $\varepsilon = 13.75$ (see Figure 9d). There is also a double peak of OH, but at a significantly increased OH volume fraction (a scaling factor is reduced by a factor of 3). The first peak corresponds to the maximum rate of iso-octyl hydroperoxide decomposition at $\theta = 4.2$ CAD. The second (larger) peak corresponds to the maximum rate of decomposition of the accumulated hydrogen peroxide at $\theta = 20$ CAD. Here, the pressure and temperature are higher than in Figure 9a–c but have the same order of magnitude (the scale factor is retained), and there is no hot autoignition. With an increase in the compression ratio, an intense blue flame arises. The luminosity of the blue flame can clog the weaker luminosity of a cool flame, and then the overall luminosity looks like a true blue flame. Apparently, this is precisely why the authors of [4] marked the lower blurred boundary of the blue flame domain with dashes (see Figure 1e). The experimental range of "double" cool flames for $\alpha = 1.5$ corresponds to $\varepsilon \sim 15$–17, while the range of blue flames corresponds to $\varepsilon \sim 17.5$–19.5.

Finally, at $\varepsilon = 14$ (see Figure 9e), hot autoignition is detected in the calculation. The pressure and temperature peaks are doubled (scale factors are reduced by a factor of 2), and the concentration peak of hydroxyl is orders of magnitude higher (the maximum is beyond the graph). In the experiments with $\alpha = 1.5$, the hot flame corresponds to the values $\varepsilon > 19.5$.

Table 4 additionally compares the results of calculations and experiments at different temperatures at the end of compression, T_c (temperatures at TDC), corresponding to the chosen ε values (for experimental values, we provide an estimate of the compression ratio ε). The comparison of results allows us to make a general conclusion that the calculations are in qualitative agreement with the experimental data.

Table 4. Calculated compression ratios ε and temperatures T_c at the end of compression ($T_0 = 343$ K, $P_0 = 1$ bar, $n = 1500$ rpm, $\alpha = 1.5$).

Reaction Type	ε	T_c, K	Exp. Range T_c, K
No apparent reaction	12.00	<786	<703
Single cool flames	13.00	808	703–838
Double cool flames	13.50	816	838–882
Blue flames	13.75	820	882–914
Hot flames	14.00	>826	>914

3.2. Droplet Autoignition

The self-ignition of droplets is simulated by the instantaneous placement of a droplet in a uniformly heated gas with $T_{g0} = 1000$ K. Below, attention is paid to the autoignition of small (submillimeter) n-heptane droplets with an initial temperature $T_{d0} = 293$ K in air at pressures up to 100 bar.

Figure 10 shows the results of the calculations for droplets of initial diameter $d_0 = 50$, 100, and 200 mm as dependences of the maximum gas temperature T_{max} on the reduced time t/d_0^2. For convenience, all curves are plotted in such a way that the autoignition of droplets occurs at time $t = 0$. Breaks in the curves at $t > 0$ correspond to the complete combustion of droplets. With a decrease in the initial droplet size from 200 to 50 μm, the effect of losses on thermal radiation decreases: during the lifetime of the droplet at 50 μm in initial diameter, the flame temperature decreases by only 300 K, from 2500 to 2200 K, whereas for the droplet 200 μm in initial diameter, the flame temperature decreases by about 550 K.

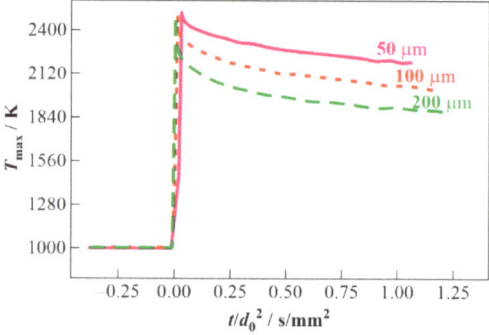

Figure 10. Calculated dependences of the maximum gas temperature in the vicinity of a droplet on the reduced time during the autoignition of droplets of different initial diameters (50 mm; 100 mm; and 200 mm) in air at $T_{g0} = 1000$ K and $P = 100$ bar.

Figure 11 shows the instantaneous spatial distributions of temperature and the mass fraction of soot around a droplet with $d_0 = 50$ µm at $t = 1.5, 2$, and 3 ms. The dimensionless radial distance $2r/d_0$ from the droplet center is plotted along the abscissa. The temperature distribution around the droplet has a dome shape with a pronounced maximum corresponding to the combustion temperature. The distribution of the mass fraction of soot also has a dome shape, and the maximum of this distribution is between the surface of the droplet and the flame, i.e., in the region where fuel vapor is accumulated, and molecular oxygen in the air is absent. The maximum temperature is seen to be attained at a distance up to $2d_0$ from the droplet surface, while the maximum mass fraction of soot is seen to be reached at a distance of $\sim 0.5 d_0$. The characteristic width of the soot "shell" is $\sim d_0$.

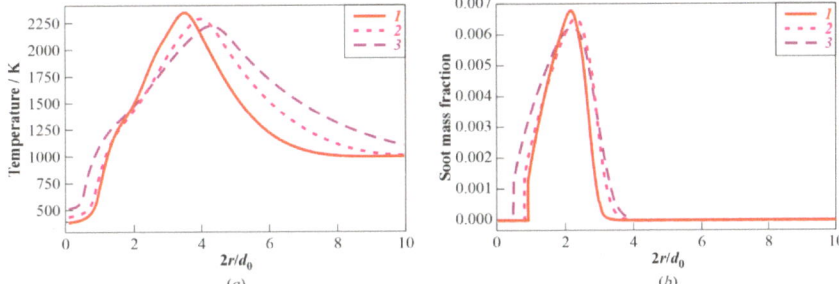

Figure 11. Calculated instantaneous spatial distributions of temperature (**a**) and soot mass fraction (**b**) around a droplet of initial diameter $d_0 = 50$ µm during its autoignition at $T_{g0} = 1000$ K and $P = 100$ bar: 1—$t = 1.5$ ms; 2—2 ms; 3—$t = 3$ ms.

To evaluate the effect of soot thermal radiation on the combustion of small (submillimeter size) droplets, a series of calculations were carried out with and without soot radiation taken into account. Figure 12 shows examples of the calculated dependences of the maximum gas temperature T_{max} around single small droplets with a diameter of 20 (Figure 12a) and 40 µm (Figure 12b) during their two-stage autoignition and subsequent combustion with and without allowance for the thermal radiation of the formed soot. It can be seen that as T_{max} increases to the maximum value, both curves in Figure 12a,b merge, and after reaching the maximum, they slightly diverge. The merging of the curves in the initial section is explained by the negligible intensity of heat losses due to radiation compared to the intensity of chemical energy release during the autoignition of the fuel–air mixture in the vicinity of the droplet. Heat losses due to radiation begin to manifest themselves only after T_{max} reaches its maximum value and the rate of energy release decreases to a level characteristic of the diffusion combustion of the droplet.

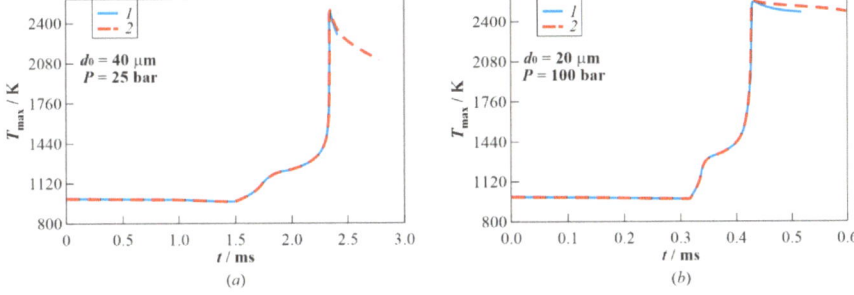

Figure 12. Calculated dependences of the maximum gas temperature T_{max} around the n-heptane droplet during its autoignition and subsequent combustion; $T_{d0} = 293$ K; $T_{g0} = 1000$ K: (**a**) $d_0 = 40$ µm, $P = 25$ bar; (**b**) $d_0 = 20$ µm, $P = 100$ bar; 1—calculation with thermal radiation of soot; 2—calculation without thermal radiation of soot.

The weak effect of thermal radiation on the autoignition and subsequent combustion of small droplets is confirmed by the spatial temperature distributions around the droplets, plotted for time instants closely after autoignition (Figure 13), as well as by the time histories of the squared droplet diameter (Figure 14). As follows from Figures 13 and 14, taking into account losses due to radiation virtually has no effect on the temperature curves or on the value of the combustion rate constant K. Moreover, radiation losses only begin to manifest themselves when the droplets have almost completely evaporated.

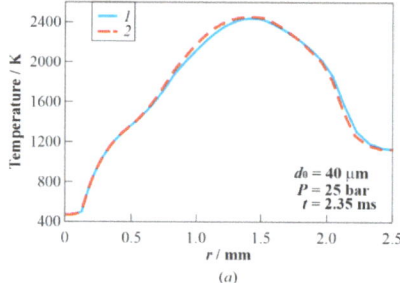 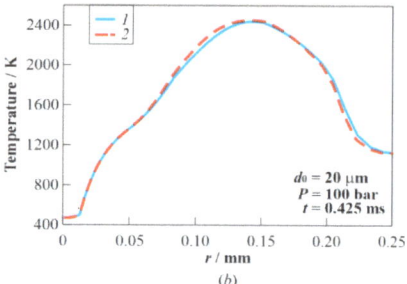

Figure 13. Estimated instantaneous temperature distributions in a droplet and in a gas depending on the distance from the droplet center; T_{d0} = 293 K; T_{g0} = 1000 K: (**a**) d_0 = 40 μm, P = 25 bar, t = 2.35 ms; (**b**) d_0 = 20 μm, P = 100 bar, t = 0.425 ms; 1—calculation with thermal radiation of soot; 2—calculation without thermal radiation of soot.

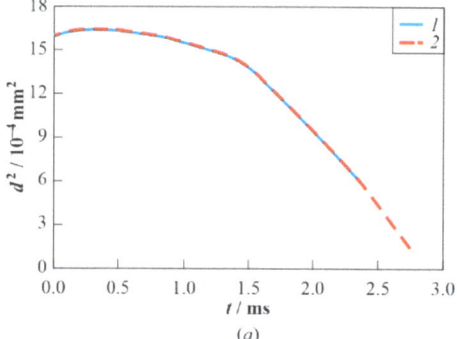 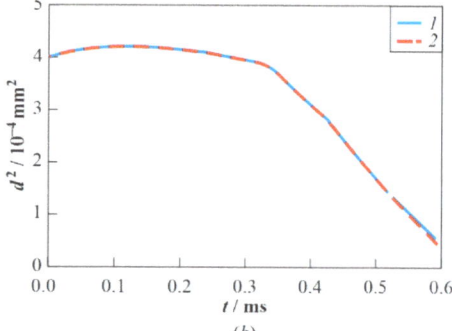

Figure 14. Calculated time histories of the squared droplet diameter during droplet autoignition and subsequent combustion; T_{d0} = 293 K; T_{g0} = 1000 K: (**a**) d_0 = 40 μm, P = 25 bar; (**b**) d_0 = 20 μm, P = 100 bar: 1—calculation with thermal radiation of soot; 2—calculation without thermal radiation of soot.

Thus, the effect of thermal radiation on soot during the autoignition and subsequent combustion of small droplets turns out to be weak: the droplets burn out before the effects of thermal radiation manifest themselves. The latter seems to be important when modeling the operation process in CIEs. The autoignition and combustion of droplets under such conditions can be approximately simulated without taking into account the thermal radiation of soot. In addition, it should be borne in mind that in the presence of directed convective flows in the vicinity of a droplet, the soot shell is deformed and fragmented, which also reduces the effect of soot thermal radiation on droplet combustion. Another implication of this study is that submillimeter-sized droplets of hydrocarbon fuels do not exhibit radiative extinction, contrary to large droplets that are several millimeters in size. The lifetime of such small droplets appears to be shorter than the characteristic time of radiative flame quenching.

3.3. Real Gas Effects

Three test calculations were carried out using the thermal and caloric EoS of real and ideal gases: (I) ideal gas thermal EoS of Equation (22) with the specific heat capacity of Equation (25); (II) real gas thermal EoS of Equation (16) with the specific ideal-gas heat capacity of Equation (25); and (III) real gas thermal EoS of Equation (16) with the specific heat capacity of Equation (21). All calculations were made using the base mesh of Figure 7a under completely identical model settings as well as initial and boundary conditions. Note, that the test calculation II is inherently incorrect and was only carried out to illustrate the important role played by the excess heat capacity in a real gas.

The calculations show the significant role of the real gas effects in the operation process of a diesel engine. Figure 15a,b compares three dependences of the pressure and mass-averaged temperature in the combustion chamber on the crank angle obtained in calculations I, II, and III. It follows from the comparison of curves I and III in Figure 15a,b that the account for the real gas properties decreases the maximum pressure and mass-averaged temperature in the combustion chamber by approximately 7 bar (6%) and 150 K (9%), respectively.

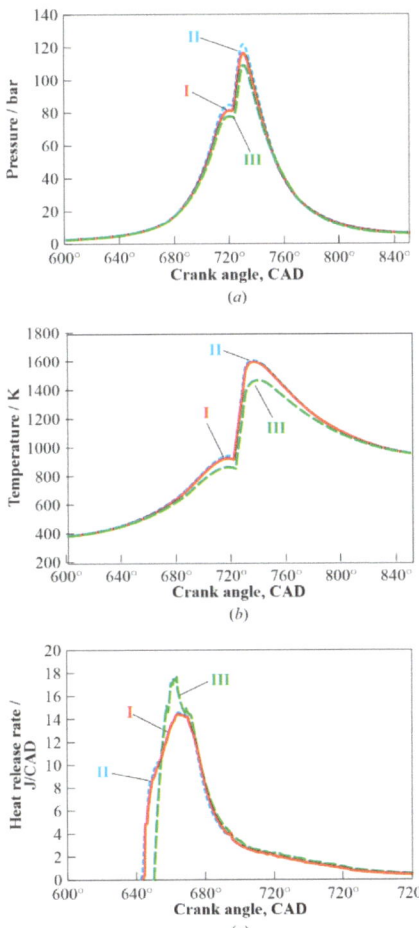

Figure 15. Predicted dependences of pressure (**a**), mass-averaged temperature (**b**), and the total heat release rate (**c**) in the diesel cylinder on the crank angle in test calculations I–III.

The autoignition delay time and the total rate of heat release in the combustion chamber also change. This is clearly seen in Figure 15c, which shows the dependences of the total heat release rate on the crank angle obtained in the three considered test calculations. Compared to calculation I, in calculation III, the autoignition delay increases by 1.6 CAD, and the maximum heat release rate increases by 20%: from 15 to 18 J/CAD.

An increase in the autoignition delay and the total heat release rate is associated with a change in the distributions of the most important flow parameters in the combustion chamber, in particular temperature. Figures 16 and 17 show the instantaneous temperature distributions in the combustion chamber obtained in calculations I and III at different instants of time: from 721 to 728 CAD. It can be seen that, in calculation III, the autoignition of fuel vapors occurs later (~724 CAD) than in calculation I (~722 CAD). In addition, the locations of autoignition centers also differ in these calculations.

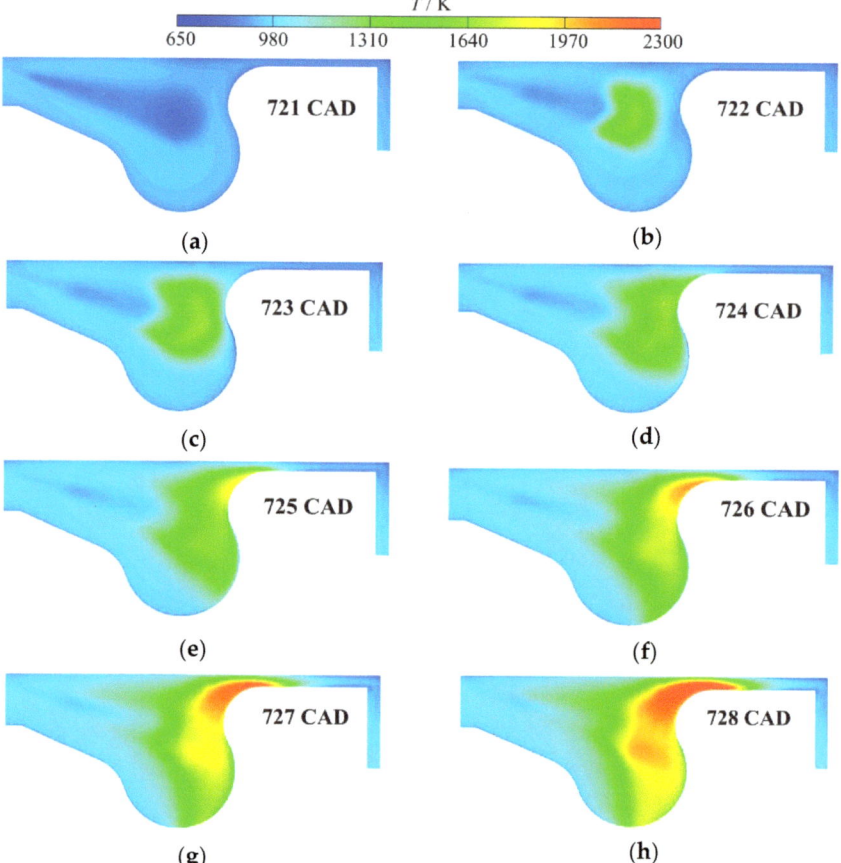

Figure 16. Predicted instantaneous temperature distributions in the combustion chamber at different instants of time obtained in calculation I: (**a**) 721 CAD; (**b**) 722 CAD; (**c**) 723 CAD; (**d**) 724 CAD; (**e**) 725 CAD; (**f**) 726 CAD; (**g**) 727 CAD; and (**h**) 728 CAD.

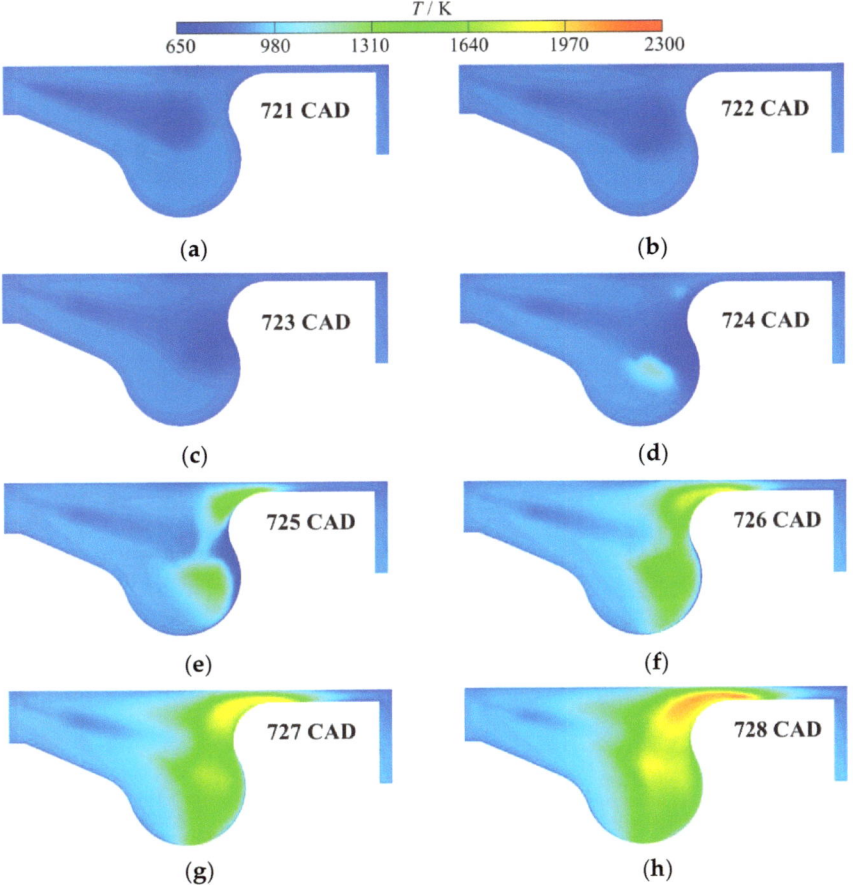

Figure 17. Predicted instantaneous temperature distributions in the diesel combustion chamber at different instants of time obtained in calculation III: (**a**) 721 CAD; (**b**) 722 CAD; (**c**) 723 CAD; (**d**) 724 CAD; (**e**) 725 CAD; (**f**) 726 CAD; (**g**) 727 CAD; and (**h**) 728 CAD.

The changes in the distributions of the flow parameters, autoignition delay time, and the maximum temperature levels in the combustion chamber caused by the account of real gas properties significantly affect the yields of NO and soot (Figure 18). Thus, in calculation III, the final values of NO and soot mass fractions turn out to be lower than in calculation I by a factor of 2 and 4, respectively. Moreover, in comparison with calculation I, an interesting detail is revealed in calculation III. In calculation I, soot forms in the vicinity of the cylinder wall (shown by an arrow in Figure 19), whereas in calculation III, soot does not form in this region. These differences are related to the reduced temperature of the real gas in the vicinity of the wall.

To understand in which zones of the combustion chamber real gas effects are mainly manifested, let us consider Figure 20. Figure 20 shows the spatial distributions of the relative excess heat capacity $C_{p,exc}/C_{p0}$ and the relative excess pressure P_{exc}/P_0 of Equation (6) obtained by calculation III. Prior to autoignition, the relative excess heat capacity of the medium in the combustor is, on average, 1–1.5%, while its maximum value (~2.5%) is achieved near the "cold" walls of the combustion chamber and in the region of the fuel spray with a high volume fraction of fuel vapors and a relatively low temperature; i.e., in those zones of the combustion chamber where the density of the matter is high. In hot combustion products, the value of the relative excess heat capacity is about 0.5%. The average value of the relative excess pressure in the combustion chamber until the moment of autoignition is ~2.5%, whereas its maximum value is also reached near the "cold" walls of the combustion chamber and in the liquid fuel spray (~3%). After the autoignition of the mixture and subsequent increase in pressure, the maximum value of the relative overpressure reaches 4%, and in hot combustion products, it does not exceed 2%.

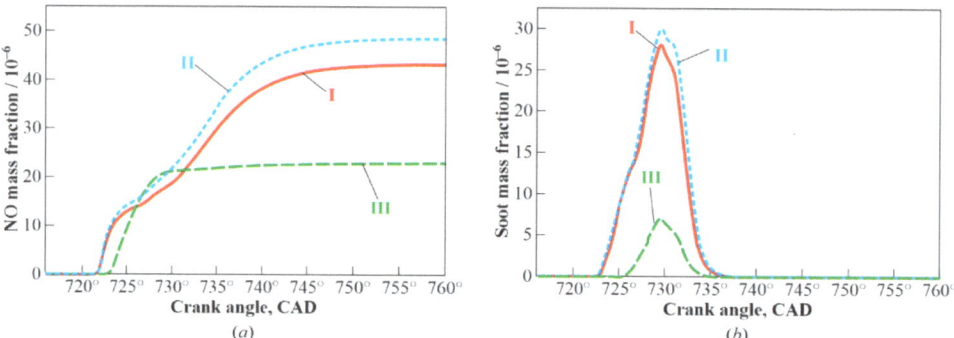

Figure 18. Predicted dependences of the mass fractions of nitrogen oxides (**a**) and soot (**b**) in the combustion chamber on the crank angle obtained in calculations I–III.

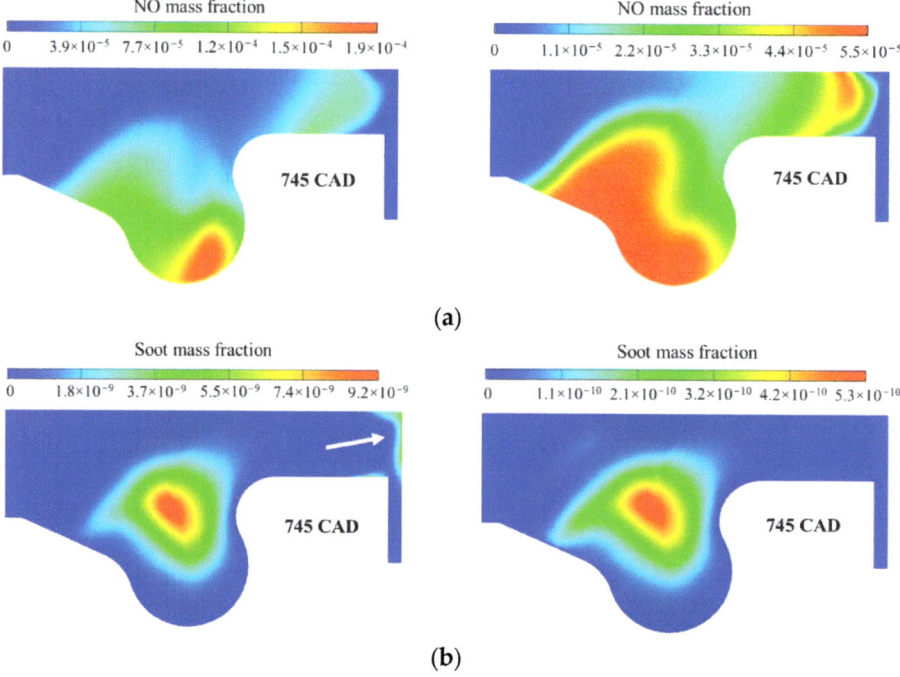

Figure 19. Predicted instantaneous distributions of the mass fractions of NO (**a**) and soot (**b**) in the diesel combustion chamber obtained in calculations I (left column) and III (right column) at 745 CAD. White arrow shows the location in the vicinity of the cylinder wall where soot forms.

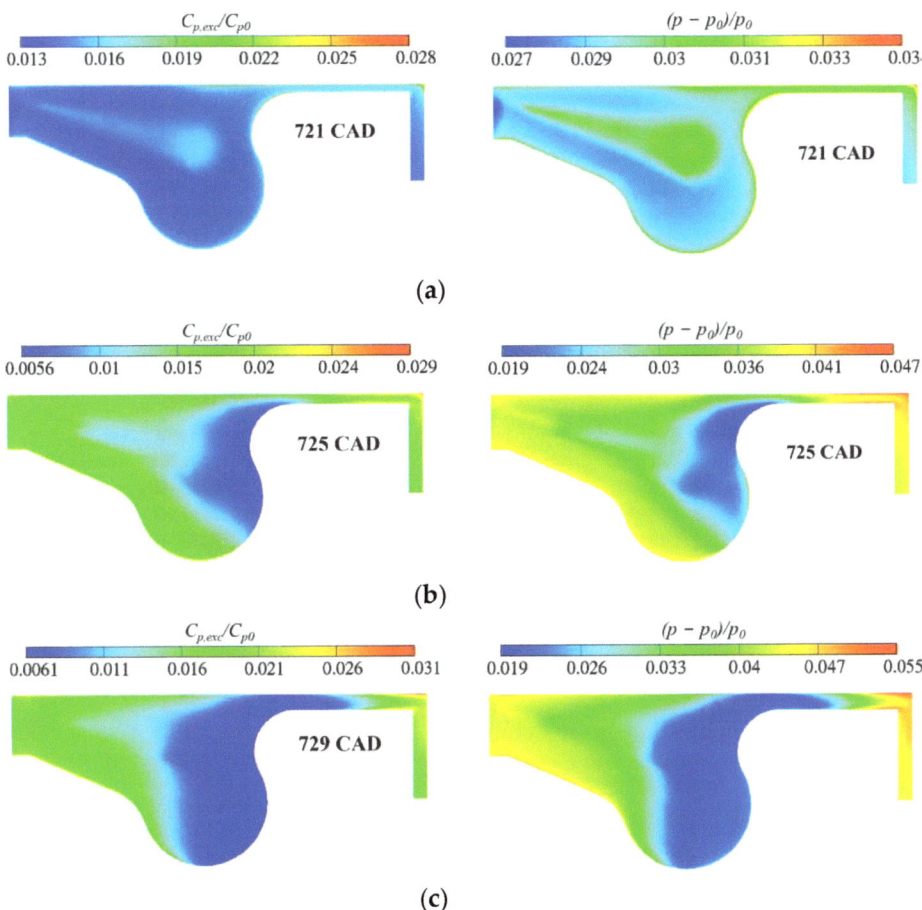

Figure 20. Predicted instantaneous distributions of the relative excess heat capacity $C_{p,exc}/C_{p0}$ (left column) and the relative excess pressure P_{exc}/P_0 (right column) in the combustion chamber at different times obtained in calculation III: (**a**) 721 CAD; (**b**) 725 CAD; and (**c**) 729 CAD.

Interestingly, studies of real gas effects on conventional diesel combustion with n-heptane as a fuel in [36] have shown somewhat different results. First, the Peng–Robinson real gas thermal and caloric EoS overpredicted both the TDC pressure and temperature in a single-cylinder diesel engine as compared to the ideal gas EoS, whereas our Figure 15a,b shows opposite trends. Second, the ignition timing has been found to be slightly advanced for the real gas case, whereas our Figure 15c shows a considerable delay in the ignition timing. Third, the heat release rate was found to be virtually not affected by the account for real gas properties, whereas our Figure 15c shows a considerable increase in the maximum heat release rate. Fourth, the real gas effects on NO and soot emissions have been found to be insignificant, whereas our Figure 18 shows a considerable effect on both NO and soot yields. On the one hand, the indicated qualitative differences could be caused by the chosen different operation modes of diesel engines as well as different computational settings. On the other hand, they could be caused by the poor accuracy of the Peng–Robinson EoS in the vicinity of the critical point of the fuel.

4. Conclusions

The computational studies reported in this paper show the influence of multistage fuel-oxidation chemistry, soot thermal radiation during droplet combustion, and real gas effects on the operation process of compression ignition engines. The use of the detailed reaction mechanism of iso-octane oxidation in the zero-dimensional simulation of the operation process in such an engine reveals the appearance of different combinations of cool, blue, and hot flames at different compression ratios and provides a kinetic interpretation to the phenomena, which essentially affect the heat release function and the engine operation process as a whole.

One-dimensional simulation of fuel droplet autoignition with regard to the detailed chemistry of multistage fuel oxidation and overall chemistry of soot formation shows that the effect of the thermal radiation of soot during autoignition and the subsequent combustion of small droplets under conditions of compression ignition engines turns out to be insignificant; the autoignition and combustion of droplets under such conditions can be approximately simulated without taking into account the thermal radiation of soot. Small (submillimeter size) droplets do not exhibit radiative flame quenching.

Comparative three-dimensional calculations of the operation process in a diesel engine were performed using thermal and caloric real gas and ideal gas equations of state. A significant effect of real gas properties on the engine indicator diagram and the yields of NO and soot was demonstrated. The relative excess heat capacity and relative excess pressure attained their maximum values (~2.5% and ~4%, respectively) near the cold walls of the combustion chamber and in the region of the liquid fuel spray; i.e., in the zones with high gas density. The account for real gas properties:

(1) reduces the maximum pressure and mass-averaged temperature in the combustion chamber by about 7 bar (6%) and 150 K (9%), respectively;
(2) increases the autoignition delay time by a 1.6 crank angle degree;
(3) increases the maximum heat release rate by 20%; and
(4) reduces the yields of NO and soot by a factor of 2 and 4, respectively.

Thus, the thermodynamic conditions in modern turbocharged diesel engines can go beyond the limits of the ideal gas approximation.

Future work will be focused on the three-dimensional simulation of the engine operation process with multistage fuel-oxidation chemistry and real gas effects taken into account. It is expected that the combined consideration of these effects will improve the predictability of calculations in terms of the heat release function and pollutant emissions.

Author Contributions: Conceptualization, V.Y.B. and S.M.F.; methodology, S.M.F.; validation, V.Y.B., S.M.F., V.S.I. and F.S.F.; formal analysis, V.Y.B. and S.M.F.; investigation, V.Y.B., S.M.F., V.S.I., F.S.F. and I.V.S.; resources, S.M.F.; data curation, V.S.I. and F.S.F.; writing—original draft preparation, V.Y.B. and S.M.F.; writing—review and editing, S.M.F.; supervision, S.M.F.; project administration, S.M.F.; funding acquisition, S.M.F. All authors have read and agreed to the published version of the manuscript.

Funding: This study was conducted within the scientific program of the National Center for Physics and Mathematics, section #2 "Mathematical Modeling on Zetta-scale and Exa-scale Supercomputers. Stage 2023–2025" and was partly funded by a subsidy given to the Semenov Federal Research Center for Chemical Physics of the Russian Academy of Sciences to implement the state assignment with registration number 122040500073-4 and by a subsidy given to the Federal State Institution "Scientific Research Institute for System Analysis of the Russian Academy of Sciences" to implement the state assignment on the topic No. FNEF-2022-0005 (Registration No. 1021060708369-1-1.2.1).

Institutional Review Board Statement: Not applicable.

Informed Consent Statement: Not applicable.

Data Availability Statement: Data will be available on request.

Conflicts of Interest: The authors declare no conflict of interest.

Abbreviations

0D	Zero-dimensional
1D	One-dimensional
3D	Three-dimensional
CAD	Crank angle degree
CFD	Computational fluid dynamics
CIE	Compression ignition engines
CV	Compensation volume
DKM	Detailed kinetic mechanism
EoS	Equation of state
OM	Overall mechanism
RANS	Reynolds-averaged Navier–Stokes
SIMPLE	Semi-implicit method for pressure linked equations
TDC	Top dead center
TVD	Total variation diminishing

References

1. Conway, G.; Joshi, A.; Leach, F.; García, A.; Senecal, P.K. A review of current and future powertrain technologies and trends in 2020. *Transport. Eng.* **2021**, *5*, 100080. [CrossRef]
2. Agarwal, A.K.; Kumar, D.; Sharma, N.; Sonawane, U.L. (Eds.) *Engine Modeling and Simulation*; Energy, Environment, and Sustainability Series; Springer Nature Singapore Pte Ltd.: Singapore, 2022. [CrossRef]
3. Frolov, S.M.; Sergeev, S.S.; Basevich, V.Y.; Frolov, F.S.; Basara, B.; Priesching, P. Simulation of multistage autoignition in diesel engine based on the detailed reaction mechanism of fuel oxidation. In *Advances in Engine and Powertrain Research and Technology. Mechanisms and Machine Science*; Parikyan, T., Ed.; Springer: Cham, Switzerland, 2022; Volume 114, pp. 149–165. [CrossRef]
4. Downs, D.; Street, J.C.; Wheeler, R.W. Cool flame formation in a motored engine. *Fuel* **1953**, *32*, 279–295.
5. Sokolik, A.S. *Autoignition, Flame, and Detonation in Gases*; AN SSSR Publs.: Moscow, Russia, 1960.
6. Basevich, V.Y.; Vedeneev, V.I. Kinetic interpretation of self-ignition of paraffin hydrocarbons. *Khim. Fiz.* **1998**, *17*, 73–80.
7. Distaso, E.; Amirante, R.; Calò, G.; De Palma, P.; Tamburrano, P.; Reitz, R.D. Investigation of lubricant oil influence on ignition of gasoline-like fuels by a detailed reaction mechanism. *Energy Procedia* **2018**, *148*, 663–670. [CrossRef]
8. Sarathy, S.M.; Westbrook, C.K.; Mehl, M.; Pitz, W.J.; Togbe, C.; Dagaut, P. Comprehensive chemical kinetic modeling of the oxidation of 2-methylalkanes from C 7 to C 20. *Combust. Flame* **2011**, *158*, 2338–2357. [CrossRef]
9. Wang, H.; Yao, M.; Reitz, R.D. Development of a reduced primary reference fuel mechanism for internal combustion engine combustion simulations. *Energy Fuels* **2013**, *27*, 7843–7853. [CrossRef]
10. Distaso, E.; Amirante, R.; Calò, G.; De Palma, P.; Tamburrano, P.; Reitz, R.D. Predicting lubricant oil induced pre-ignition phenomena in modern gasoline engines: The reduced GasLube reaction mechanism. *Fuel* **2020**, *281*, 118709. [CrossRef]
11. Hwang, W.; Dec, J.; Sjöberg, M. Spectroscopic and chemical-kinetic analysis of the phases of HCCI autoignition and combustion for single- and two-stage ignition fuels. *Combust. Flame* **2008**, *154*, 387–409. [CrossRef]
12. Waqas, M.U.; Hoth, A.; Kolodziej, C.P.; Rockstroh, T.; Gonzalez, J.P.; Johansson, B. Detection of low temperature heat release (LTHR) in the standard cooperative fuel research (CFR) engine in both SI and HCCI combustion modes. *Fuel* **2019**, *256*, 115745. [CrossRef]
13. Singh, E.; Sarathy, S.M. The role of intermediate-temperature heat release in octane sensitivity of fuels with matching research octane number. *Energy Fuels* **2021**, *35*, 4457–4477. [CrossRef]
14. Yoo, K.H.; Voice, A.K.; Boehman, A.L. Influence of intermediate temperature heat release on autoignition reactivity of single-stage ignition fuels with varying octane sensitivity. *Proc. Combust. Inst.* **2021**, *38*, 5529–5538. [CrossRef]
15. Peterson, J.; Mohammed, A.; Gorbatenko, I.; Singh, E.; Sarathy, S.M. The contribution of intermediate-temperature heat release to octane sensitivity. *Fuel* **2023**, *352*, 129077. [CrossRef]
16. Westbrook, C.K. Chemical kinetics of hydrocarbon ignition in practical combustion systems. *Proc. Combust. Inst.* **2000**, *28*, 1563–1577. [CrossRef]
17. Yu, L.; Qiu, Y.; Mao, Y.; Wang, S.; Ruan, C.; Tao, W.; Qian, Y.; Lu, X. A study on the low-to-intermediate temperature ignition delays of long chain branched paraffin: Iso-cetane. *Proc. Combust. Inst.* **2019**, *37*, 631–638. [CrossRef]
18. Lebedev, A.B.; Sekundov, A.N.; Savel'ev, A.M.; Starik, A.M.; Titova, N.S. Numerical analysis of the influence of the operation mode of aviation gas turbine engine on the pollutant emission. In *Nonequilibrium Physicochemical Processes in Gas Flows and New Principles of Combustion Organization*; Starik, A.M., Ed.; Torus Press: Moscow, Russia, 2011; pp. 755–771.
19. Titova, N.S.; Torokhov, S.A.; Starik, A.M. Specific features of ignition and combustion of heavy hydrocarbons and cool-flame phenomena. In *Nonequilibrium Physicochemical Processes in Gas Flows and New Principles of Combustion Organization*; Starik, A.M., Ed.; Torus Press: Moscow, Russia, 2011; pp. 88–110.
20. Basevich, V.Y.; Belyaev, A.A.; Medvedev, S.N.; Posvyanskii, V.S.; Frolov, F.S.; Frolov, S.M. A detailed kinetic mechanism of multistage oxidation and combustion of isooctane. *Rus. J. Phys. Chem. B* **2016**, *10*, 801–809. [CrossRef]

21. Wang, Y.; Chung, S.H. Soot formation in laminar counterflow flames. *Progr. Energy Combust. Sci.* **2019**, *74*, 152–238. [CrossRef]
22. Frenklach, M.; Mebel, A.M. On the mechanism of soot nucleation. *Phys. Chem. Chem. Phys.* **2020**, *22*, 5314–5331. [CrossRef]
23. Lapuerta, M.; Rodríguez–Fernández, J.; Sánchez-Valdepeñas, J. Soot reactivity analysis and implications on diesel filter regeneration. *Progr. Energy Combust. Sci.* **2020**, *78*, 100833. [CrossRef]
24. Gleason, K.; Carbone, F.; Sumner, A.J.; Drollette, B.D.; Plata, D.L.; Gomez, A. Small aromatic hydrocarbons control the onset of soot nucleation. *Combust. Flame* **2021**, *223*, 398–406. [CrossRef]
25. Martin, J.W.; Salamanca, M.; Kraft, M. Soot inception: Carbonaceous nanoparticle formation in flames. *Progr. Energy Combust. Sci.* **2022**, *88*, 100956. [CrossRef]
26. Avedisian, C.T. Recent advances in soot formation from spherical droplet flames at atmospheric pressure. *J. Prop. Power* **2000**, *16*, 628–635. [CrossRef]
27. Abdul Rasid, A.F.; Zhang, Y. Comparison of the burning of a single diesel droplet with volume and surface contamination of soot particles. *Proc. Combust. Inst.* **2021**, *38*, 3159–3166. [CrossRef]
28. Kumar, A.; Chen, H.-W.; Yang, S. Diffusion and its effects on soot production in the combustion of emulsified and nonemulsified fuel droplets. *Energy* **2023**, *267*, 126521. [CrossRef]
29. Shaw, B.; Dryer, F.; Williams, F.; Haggard, J. Sooting and disruption in spherically symmetrical combustion of decane droplets in air. *Acta Astronaut.* **1988**, *17*, 1195–1202. [CrossRef]
30. Nayagam, V.; Haggard, J.B.; Colantonio, R.O.; Marchese, A.J.; Dryer, F.L.; Zhang, B.L.; Williams, F.A. Microgravity n-heptane droplet combustion in oxygen-helium mixtures at atmospheric pressure. *AIAA J.* **1998**, *36*, 1369–1378. [CrossRef]
31. Liu, Y.C.; Xu, Y.; Avedisian, C.T.; Hicks, M.C. The effect of support fibers on micro-convection in droplet combustion experiments. *Proc. Combust. Inst.* **2015**, *35*, 1709–1716. [CrossRef]
32. Nayagam, V.; Dietrich, D.L.; Hicks, M.C.; Williams, F.A. Radiative extinction of large n-alkane droplets in oxygen-inert mixtures in microgravity. *Combust. Flame* **2018**, *194*, 107–114. [CrossRef]
33. Frolov, S.M.; Basevich, V.Y.; Medvedev, S.N. Modeling of low-temperature oxidation and combustion of droplets. *Dokl. Phys. Chem.* **2016**, *470*, 150–153. [CrossRef]
34. Frolov, S.M.; Basevich, V.Y. Simulation of low-temperature oxidation and combustion of n-dodecane droplets under microgravity conditions. *Fire* **2023**, *6*, 70. [CrossRef]
35. Kaario, O.; Nuutinen, M.; Lehto, K.; Larmi, M. Real gas effects in high-pressure engine environment. *SAE Intern. J. Engines* **2010**, *3*, 546–555. Available online: https://www.jstor.org/stable/10.2307/26275498 (accessed on 22 August 2023). [CrossRef]
36. Yue, Z.; Hessel, R.; Reitz, R.D. Investigation of real gas effects on combustion and emissions in internal combustion engines and implications for development of chemical kinetics mechanisms. *Int. J. Engine Res.* **2017**, *19*, 269–281. [CrossRef]
37. Zheng, C.; Coombs, D.M.; Akih-Kumgeh, B. Real gas model parameters for high-density combustion from chemical kinetic model data. *ACS Omega* **2019**, *4*, 3074–3082. [CrossRef] [PubMed]
38. Oefelein, J.; Lacaze, G.; Dahms, R.; Ruiz, A.; Misdariis, A. Effects of real-fluid thermodynamics on high-pressure fuel injection processes. *SAE Int. J. Engines* **2014**, *7*, 1125–1136. [CrossRef]
39. Perini, F.; Busch, S.; Reitz, R. An investigation of real-gas and multiphase effects on multicomponent diesel sprays. *SAE Int. J. Adv. Curr. Prac. Mobil.* **2020**, *2*, 1774–1785. [CrossRef]
40. Ihme, M.; Ma, P.C.; Bravo, L. Large eddy simulations of diesel-fuel injection and auto-ignition at transcritical conditions. *Intern. J. Engine Res.* **2019**, *20*, 58–68. [CrossRef]
41. Kogekar, G.; Karakaya, C.; Liskovich, G.J.; Oehlschlaeger, M.A.; DeCaluwe, S.C.; Kee, R.J. Impact of non-ideal behavior on ignition delay and chemical kinetics in high-pressure shock tube reactors. *Combust. Flame* **2018**, *189*, 1–11. [CrossRef]
42. Span, R. *Multiparameter Equations of State*; Springer: Berlin/Heidelberg, Germany, 2000. [CrossRef]
43. Redlich, O.; Kwong, J.N.S. On the thermodynamics of solutions. V. An equation of state. Fugacities of gaseous solutions. *Chem. Rev.* **1949**, *44*, 233–244. [CrossRef]
44. Peng, D.-Y.; Robinson, D.B. A new two-constant equation of state. *Ind. Eng. Chem. Fundam.* **1976**, *15*, 59–64. [CrossRef]
45. Detailed Kinetic Mechanism of Isooctane. Available online: http://ru.combex.org/lab1313.htm (accessed on 4 August 2023).
46. Basevich, V.Y.; Belyaev, A.A.; Posvyanskii, V.S.; Frolov, S.M. Mechanisms of the oxidation and combustion of normal paraffin hydrocarbons: Transition from C1–C10 to C11–C16. *Russ. J. Phys. Chem. B* **2013**, *7*, 161–169. [CrossRef]
47. Reid, R.C.; Prausnitz, J.M.; Sherwood, T.K. *The Properties of Gases and Liquids*; McGraw-Hill: New York, NY, USA, 1977.
48. Fieweger, K.; Blumenthal, R.; Adomeit, G. Shock-tube investigations on the self-ignition of hydrocarbon-air mixtures at high pressures. *Proc. Combust. Symp.* **1994**, *25*, 1579–1585. [CrossRef]
49. Ciezki, H.K.; Adomeit, G. Shock-tube investigation of self-ignition of n-heptane-air mixtures under engine relevant conditions. *Combust. Flame* **1993**, *93*, 421–433. [CrossRef]
50. Frolov, S.M.; Avdeev, K.A.; Ivanov, V.S.; Vlasov, P.A.; Frolov, F.S.; Semenov, I.V.; Belotserkovskaya, M.S. Evolution of the soot-particle size distribution function in the cylinder and exhaust system of piston engines: Simulation. *Atmosphere* **2023**, *14*, 13. [CrossRef]
51. Massoli, P.; Lazzaro, M.; Beretta, F.; D'Alessio, A. Characterization of hydrocarbon droplets heating in a drop tube furnace. In *Instituto Motori, C.N.R. Report on Research Activities and Facilities*; Di Lorenzo, A., Ed.; Institute for Research on Engines: Napoli, Italy, 1993; pp. 36–37.
52. Avedisian, C.T.; Yang, J.C.; Wang, C.H. On low-gravity droplet combustion. *Proc. R. Soc. A* **1988**, *420*, 183–200. [CrossRef]

53. Okajima, S.; Kumagai, S. Further investigations of combustion of free droplets in a freely falling chamber including moving droplets. *Proc. Combust. Inst.* **1975**, *15*, 401–407. [CrossRef]
54. Hara, H.; Kumagai, S. The effect of initial diameter of free droplet combustion with spherical flame. *Proc. Combust. Inst.* **1994**, *25*, 423–430. [CrossRef]
55. Mikami, M.; Kato, H.; Sato, J.; Kono, M. Interactive combustion of two droplets in microgravity. *Proc. Combust. Inst.* **1994**, *25*, 431–438. [CrossRef]
56. Kumagai, S.; Sakai, T.; Okajima, S. Combustion of free fuel droplets in a freely falling chamber. *Proc. Combust. Inst.* **1971**, *13*, 779–785. [CrossRef]
57. Jackson, G.S.; Avedisian, C.T.; Yang, J.C. Observations of soot during droplet combustion at low gravity: Heptane and heptane/monochloroalkane mixtures. *Int. J. Heat Mass Transf.* **1992**, *35*, 2017–2033. [CrossRef]
58. Monaghan, M.T.; Siddall, R.G.; Thring, M.W. The influence of initial diameter on the combustion of single drops of liquid fuel. *Combust. Flame* **1968**, *12*, 45–53. [CrossRef]
59. Jackson, G.S.; Avedisian, C.T. The effect of initial diameter in spherically symmetric droplet combustion of sooting fuels. *Proc. R. Soc. Lond. A* **1994**, *446*, 255–276.
60. Tanabe, M.; Bolik, T.; Eigenbrod, C.; Rath, H.J.; Sato, J.; Kono, M. Spontaneous ignition of liquid droplets from a view of non-homogeneous mixture formation and transient chemical reactions. *Proc. Combust. Inst.* **1996**, *26*, 1637–1643. [CrossRef]
61. Schnaubelt, S.; Moriue, O.; Coordes, T.; Eigenbrod, C.; Rath Zarm, H.J. Detailed numerical simulations of the multistage self-ignition process of n-heptane, isolated droplets and their verification by comparison with microgravity experiments. *Proc. Combust. Inst.* **2000**, *28*, 953–960. [CrossRef]
62. Frolov, S.M.; Kuznetsov, N.M.; Krueger, C. Real-gas properties of n-alkanes, O_2, N_2, H_2O, CO, CO_2, and H_2 for Diesel engine operation conditions. *Rus. J. Phys. Chemi. B* **2009**, *3*, 1191–1252. [CrossRef]
63. Grigor'ev, B.A.; Rastorguev, Y.L.; Gerasimov, A.A. *Thermodynamic Properties of Normal Hexane*; Standard Publ.: Moscow, Russia, 1990.
64. Sychev, V.V.; Vasserman, A.A.; Kozlov, A.D.; Spiridonov, G.A.; Tsymarnyi, V.A. *Thermodynamic Properties of Oxygen*; Standard Publ.: Moscow, Russia, 1981.
65. Jacobsen, R.T.; Stewart, R.B.; Jahangiri, M. Thermodynamic properties of nitrogen from the freezing line to 2000 K at pressures to 1000 MPa. *J. Phys. Chem. Ref. Data* **1986**, *15*, 735. [CrossRef]
66. Vukalovich, M.P.; Rivkin, S.L.; Alexandrov, A.A. *Tables of Thermodynamic Properties of Water and Water Vapor*; Standard Publ.: Moscow, Russia, 1969.
67. Goodwin, R.D. Carbon monoxide thermophysical properties from 68 to 1000 K at pressures to 100 MPa. *J. Phys. Chem. Ref. Data* **1985**, *14*, 849. [CrossRef]
68. Vukalovich, M.P.; Altunin, V.V. *Thermophysical Properties of Carbon Dioxide*; Atomizdat Publ.: Moscow, Russia, 1965.
69. Vargaftik, N.B. *Handbook on Thermophysical Properties of Gases and Liquids*; Nauka Publ.: Moscow, Russia, 1972.
70. CHEMKIN-PRO Release 15083 Tutorial Manual. Reaction Design, San Diego (CA). 2009. Available online: https://www.ansys.com/products/fluids/ansys-chemkin-pro (accessed on 22 August 2023).
71. Frolov, S.M.; Ivanov, V.S.; Frolov, F.S.; Vlasov, P.A.; Axelbaum, R.; Irace, P.H.; Yablonsky, G.; Waddell, K. Soot formation in spherical diffusion flames. *Mathematics* **2023**, *11*, 261. [CrossRef]
72. Agafonov, G.L.; Bilera, I.V.; Vlasov, P.A.; Kolbanovskii, Y.A.; Smirnov, V.N.; Teresa, A.M. Soot formation at pyrolysis and oxidation of acetylene and ethylene in shock tubes. *Kinet. Catal.* **2015**, *56*, 12–30. [CrossRef]
73. AVL FIRE—Computational Fluid Dynamics for Conventional and Alternative Powertrain Development. Available online: https://www.avl.com/fire (accessed on 4 August 2023).
74. Hanjalic, K.; Popovac, M.; Hadziabdic, M. A robust near-wall elliptic relaxation eddy-viscosity turbulence model for CFD. *Int. J. Heat Fluid Flow* **2004**, *25*, 897–901. [CrossRef]
75. B'eard, P.; Colin, O.; Miche, M. Improved modelling of DI diesel engines using sub-grid descriptions of spray and combustion. In *SAE Paper*; No. 2003-01-0008; SAE International: Warrendale, PA, USA, 2003.
76. Schiller, L.; Naumann, M. A drag coefficient correlation. *VDI Zeitung* **1935**, *77*, 318–320.
77. Dukowicz, J.K. A particle–fluid numerical model for liquid sprays. *J. Comput. Phys.* **1980**, *35*, 229–253. [CrossRef]
78. Liu, A.B.; Reitz, R.D. Modeling the effects of drop drag and breakup on fuel sprays. In *SAE Paper*; No. 930072; SAE International: Warrendale, PA, USA, 1993.
79. Tao, F.; Liu, Y.; RempelEwert, B.A.; Foster, D.E.; Reitz, R.D.; Choi, D.; Miles, P.C. Modeling the effects of EGR and injection pressure on soot formation in a high-speed direct-injection (HSDI) diesel engine using a multi-step phenomenological soot model. In *SAE Paper 2005-01-0121*; SAE International: Warrendale, PA, USA, 2005.
80. Zel'dovich, Y.B.; Sadovnikov, P.Y.; Frank-Kametentskii, D.A. *Oxidation of Nitrogen under Combustion*; Moscow–Leningrad: USSR Acad. Sci. Publ.: Moscow, Russia, 1947.
81. Magnussen, B.F.; Hjertager, B.H. On mathematical modeling of turbulent combustion with special emphasis on soot formation and combustion. *Proc. Combust. Ins.* **1977**, *16*, 719–729. [CrossRef]

82. Ferziger, J.H.; Peric, M. *Computational Methods for Fluid Dynamics*; Springer-Verlag: New York, NY, USA, 1996.
83. Sweby, P.K. High resolution schemes using flux limiters for hyperbolic conservation laws. *SIAMJ. Numer. Anal.* **1984**, *21*, 995. [CrossRef]

Disclaimer/Publisher's Note: The statements, opinions and data contained in all publications are solely those of the individual author(s) and contributor(s) and not of MDPI and/or the editor(s). MDPI and/or the editor(s) disclaim responsibility for any injury to people or property resulting from any ideas, methods, instructions or products referred to in the content.

Review

On the Genesis of a Catalyst: A Brief Review with an Experimental Case Study

Simón Yunes [1,*], Jeffrey Kenvin [1] and Antonio Gil [2,*]

[1] Micromeritics Instrument Corporation, 4356 Communications Drive, Norcross, GA 30093, USA; jeffrey.kenvin@micromeritics.com

[2] INAMAT2, Science Department, Public University of Navarra, 31006 Pamplona, Spain

* Correspondence: simon.yunes@micromeritics.com (S.Y.); andoni@unavarra.es (A.G.)

Abstract: The science of catalysis has a direct impact on the world economy and the energy environment that positively affects the environmental ecosystem of our universe. Any catalyst, before being tested in a reaction, must undergo a specific characterization protocol to simulate its behavior under reaction conditions. In this work, these steps that must be carried out are presented, both generically and with examples, to the support and to the catalyst itself before and after the reaction. The first stage consists of knowing the textural and structural properties of the support used for the preparation of the catalysts. The specific surface area and the pore volume are fundamental properties, measured by N_2 adsorption at $-196\ °C$ when preparing the catalyst, dispersing the active phase, and allowing the diffusion and reaction of the reactants and products on its surface. If knowing the structure of the catalyst is important to control its behavior against a reaction, being able to analyze the catalyst used under the reaction conditions is essential to have knowledge about what has happened inside the catalytic reactor. The most common characterization techniques in heterogeneous catalysis laboratories are those described in this work. As an application example, the catalytic conversion of CO_2 to CH_4 has been selected and summarized in this work. In this case, the synthesis and characterization of Cu and Ni catalysts supported on two Al_2O_3 with different textural properties, 92 and 310 m^2/g, that allow for obtaining various metallic dispersions, between 3.3 and 25.5%, is described. The catalytic behavior of these materials is evaluated from the CO_2 methanation reaction, as well as their stability from the properties they present before and after the reaction.

Keywords: catalyst; catalytic support; metal oxide; textural properties; structural properties; characterization techniques; chemical characterization; physisorption; chemisorption; metal dispersion; catalytic activity; catalytic deactivation; in situ characterization

1. Introduction

The word catalyst is one of the most repeated words in the literature and it is in direct relation to many processes for several applications such as refining, pharmacy, mining, painting, and many others. Catalysts play an important role in converting species into more beneficial products that make a large contribution to the overall world economy, especially processes that convert CO_2 into more beneficial hydrocarbon molecules [1–17]. Catalysis, however, is a process in which a catalyst is involved in a reaction to proceed at a faster rate or under various conditions than otherwise possible, while not being consumed by the reaction.

Several steps are followed to synthesize a heterogeneous catalyst: selection of the support, followed by selection of the oxides that will become the active species, and, finally, synthesis of the catalyst according to one of the well-known methodologies [18,19]. Upon proper characterization of the support, followed by the preparation of the catalyst, testing becomes the critical point to produce the expected products and to have a long-lasting catalyst. Proper tools are required for this study that include in situ characterization of the catalyst, especially after deactivation to elucidate and understand the causes of deactivation and, finally, to produce a catalyst that lasts longer under reaction conditions and produces the best activity and selectivity for any specific reaction.

There are several possible ways to limit the release of greenhouse gases into the environment. The first step is to limit the use of fossil fuels by adopting the practical use of green energy sources, which minimizes gas emissions. Simultaneously, technologies for carbon capture and storage (CCS) and carbon capture and utilization (CCU) should be applied. While CCS processes end with the storage of CO_2, which does not solve the main issue, CCU processes involve capturing CO_2 and converting it into valuable products [20]. This storage and conversion can be performed using a variety of processes. Absorption into liquid is a suitable storage strategy for CO_2, but it is limited by the high energy demands of regenerating the solvent. Adsorption on solids is affected by temperature and pressure and is a current process to generate innovative porous materials [21]. The membrane technique has a variety of advantages over other CCS technologies, such as applicability in isolated areas, simplicity of maintenance, a low-cost installation, and fewer chemical and energy requirements [22,23]. Photocatalytic reduction is initiated through direct sunlight, which is considered a renewable energy source, while electrochemical and plasma technology relies on electricity. Additionally, chemical storage can be implemented on the current line of infrastructure with a high capability [22]. The advanced technology of water electrolysis (power to gas, P2G) has contributed convenient enhancements in carbon dioxide hydrogenation as a source of green fuel [24].

Significant improvements have been achieved in transforming CO_2 into valuable single-carbon materials, such as carbon monoxide, methane, methanol, and formic acid, among others. The reaction of the reverse water gas shift (RWGS) produces carbon monoxide, while heterogeneous catalysis has been utilized to obtain methanol [25]. Thermodynamically, the methanation of CO_2 and CO is quicker than that of other reactions for hydrocarbon production. CO_2 gas is highly stable and molecule separation is costly [26]. This stability has led to low production due to the low adsorption rate of CO_2 in the catalysts. A full understanding of the complexity of the mechanism of CO_2 conversion into hydrogen fuel requires gaining more information, such as from studying the micro mechanisms of the Sabatier reaction [27]. Compounds with small carbonylicity are more reactive during the addition reaction in comparison with the high carbonylicity compounds [28]. CO_2 conversion to methane requires efficient catalysts to be successful. Recently, this has been the aim of much research that has attempted using various catalyst designs [29]. Metal–support catalysts are extensively used for CO_2 fixation reactions with a variety of metals and catalyst supports. Although several transition metals are suitable for the methanation reaction, supported Co, Cu, Fe, and Ni catalysts are extensively used for CO_2 methanation due to their high catalytic performance and cost-effective nature [30], but metal catalysts lose activity quickly during the methanation reaction due to their carbon deposition [29]. The roles of the support, metal loading, the preparation method, and additives have been investigated to enhance the performance of the catalysts. According to the literature, the support might play a role in enhancing metal dispersion and tuning the structure of the catalyst surface, which could improve CO_2 adsorption and affect the reaction mechanism [30]. Alumina, ceria, magnesia, silica, titania, and zirconia are well-studied support materials that are thermally stable and provide significantly high surface areas [31]. They increase the stability of metal-based catalysts by improving their interaction with the active metals [32]. The acidity/basicity of the supports weakens the interaction with CO_2. To improve the interaction, the basicity of the metal catalysts is enhanced with the presence of promoters [33]. Finally, it should not be forgotten that other types of materials such as metal nitrides, hydrides, and carbides could also intervene in the CCS and CCU processes. However, for simplicity, this work has been reduced only to the case of metal oxides and metals supported.

This review work presents the most important steps to produce a catalyst: the properties that determine its performance and what experimental methods available in a large number of academic and business laboratories can be used. Finally, a practical example of the synthesis and characterization of catalysts applied in the methanation of CO_2 is addressed. Special relevance is made to the characterization of the catalyst through techniques that allow its characterization under conditions very similar to those in which the catalytic reaction is carried out. In particular, from the use of the named In Situ Catalyst Characterization System (ICCS).

2. Genesis of a Catalyst

2.1. Selection of the Support

This is the first and most important task to be considered when preparing a catalyst that will later be designated for a certain reaction. In general, they do not have to present specific characteristics for a reaction, their surface properties must allow the dispersion of the active metallic phase. Synthesizing a catalyst is normally a sophisticated and gentle art and is based on the selection of solid support made of refractory oxides (alumina, ceria, silica, titania, zirconia, carbon, and many others) and an active phase, each containing their own physical and chemical properties, which can evolve during preparation. First, the selected support must fulfill certain requirements:

(a) It should withstand high temperature and pressure with minimal sintering;
(b) It should have an adequate texture; that is, sufficient surface area to lodge the necessary oxide particles that will play the role of the active species. It should also have adequately sized pores to facilitate diffusion of the reactants through the pores and the ability for the product to diffuse out of the pore [34,35];
(c) It must have a certain surface acidity/basicity when required by the reaction;
(d) Other interesting characteristics must be sometimes required such as morphology or any structure arrangement that yields to stabilize the active species.

2.2. Selection of the Metal Active Sites

Metal oxides serve as active elements on the catalyst. This second step in this study is also very important and it must be carefully considered in the synthesis of a catalyst. The nature of the active species, also known as active sites, has a direct impact on the surface activity (known as turnover frequency) and the selectivity of the catalyst. This nature is determined by the composition (i.e., type of metal and promoters or bimetallic), size, and shape of the nanoparticles. Reduction of the size of the nanoparticles increases the amount of surface sites per unit weight of metal and generally results in more active catalysts.

The selection of two or more phases to prepare the catalysts will depend on the catalytic process that has to be carried out. For example, for a hydrogenation reaction, the active metal should adsorb and dissociate the molecule of H_2 into atomic hydrogen. The atomic hydrogen can hydrogenate a double-bonded hydrocarbon molecule. For a cracking reaction, a high molecular weight hydrocarbon is converted into more valuable compounds of lower molecular weight. This reaction requires an acid solid, such as zeolite loaded with transition metals, for example.

For the CO_2 methanation processes various noble (Ru, Rh) and non-noble (Ni, Co, Cu) metals have been extensively investigated as active sites. From the data in the literature, the activity of the metals can be presented as: Ru > Rh > Ni > Fe > Co > Os > Pt > Ir > Mo > Pd [13]. It is also indicated in the case of Ru the high CH_4 selectivity and high resistance to oxidizing atmospheres [30]. The main drawback is the cost. Fe is also interesting as a catalyst, mainly due to the reduced cost of the Fe salts. Ni is a metal that combines the advantages of almost all the active metallic phases, with the only drawback being its tendency to oxidize with small amounts of oxygen. The interaction between the support and the metallic active center is also relevant to consider when it comes to catalytic behavior. Many times, metallic oxides are obtained by solid-state reactions that cause the catalytic composting to be modified. Hence, the importance of the characterization of the catalyst and the use of the techniques are described below.

2.3. Preparation of the Catalyst

After the selection of the support and metal oxide specific to the catalysis process, catalyst preparation then proceeds using one of the many well-known preparation methods. Among the most used methods for industrial applications are: precipitation methods, impregnation methods, sol–gel methods, and chemical deposition methods. There are other methods for preparing catalysts, all of which involve the idea of efficiently dispersing the active phase on the surface of a material that acts as an inert support.

(a) Precipitation method: This method is sometimes known as co-precipitation and is one of the most widely used catalyst preparation methods. This method can be used to prepare a single component catalyst or a supported mixed oxide catalyst. The method is based on the precipitation of a single or multi-phase solid by altering the slurry condition, for example, the pH of the solution, applying heat or vaporizing a suitable amount of a precursor, and adsorbing it on a support material. The co-precipitation method slightly differs from the previous one as, in this case, the catalyst is formed by dissolving and mixing the active metal salt and the support to promote nucleation and growth of a combined solid precursor containing both the active element and support.

(b) Impregnation method: In this method, the selected support is immersed in a precursor solution allowing the precursor of the active phase to diffuse into the porous structure of the support. The obtained slurry is slowly dried and later calcined at an appropriate temperature without exceeding the thermal decomposition temperature of the support to prevent its collapse and sinter its texture.

(c) Sol–gel method: This method is very versatile and allows control of the texture, composition, homogeneity, and structural properties of solids, and makes possible production of tailored materials such as dispersed metals, oxidic catalysts, and chemically modified supports. This method involves the formation of a sol from dispersed colloid solutions or from some inorganic precursors as a starting material followed by the formation of a gel. This method yields various configurations such as monoliths, coatings, foams, and fibers without using highly cost processing technologies. The method is based on the hydrolysis and condensation of metal alkoxides such as $SiCl_4$ with alcohol [36–38].

(d) Chemical deposition: This method consists of the formation of thin films on a heated substrate via a chemical reaction of gas-phase precursors, which is known as the Chemical Vapor Deposition (CVD) method. A typical example of a catalyst prepared by this method is a 2D transition metal dichalcogenides deposited on thin polymeric films [39,40].

2.4. Characterization Techniques

Some important characterization techniques widely cited in the literature are listed below. For instance: spectroscopic techniques, adsorption techniques, thermal techniques, etc.

2.4.1. X-ray Diffraction (XRD)

This technique gives information about the crystalline phases present in a sample as well as about the bulk properties of the sample (see Figure 1). The X-ray diffraction pattern can also measure the distance between single planes of atoms in a crystal and also provides a measurement of the layer's height. Due to the difference between cell parameters, symmetry, and space groups that crystalline materials have, characteristic diffraction patterns that work as a fingerprint of the material are produced. To understand what conditions are required for diffraction, Bragg's law (see Equation (1)) provides a simplistic model that also is used to calculate the d-spacing. For parallel planes of atoms (with a space d_{hkl} in between) constructive interference can only occur when this law is satisfied:

$$n\lambda = 2 \cdot d_{hkl} \cdot \sin\theta \tag{1}$$

where d is the interlayer distance between two consecutive layers, n is the diffraction order, θ is Bragg's angle (formed both by the incident X-ray beam and the diffracted by the planes) and λ is the wavelength of the X-rays (0.15418 nm for the copper anode). Additionally, the plane normal (*hkl*) must be parallel to the diffraction vector s (bisects the angle between the incident and the diffracted beam). The crystallite size can also be calculated from the XRD data. It is measured with the Scherrer formulae (see Equation (2)):

$$\tau = \frac{K \cdot \lambda}{\beta_\tau \cdot \cos \vartheta} \quad (2)$$

where K is the shape factor = 0.9; β_τ is the width of the peak at half of the maximum intensity (Full Width at Half Maximum, FWHM) after subtraction of instrumental broadening and θ is the diffraction angle. Normally, the crystallite size is measured in nm which are the units of the wavelength of the X-rays. It must also be taken into account that it is a crystallographic measurement and that it is determined in volume. A particle size can also be determined by other techniques but the material may not need to be crystalline and may be a surface dimension, such as obtained by chemisorption measurements.

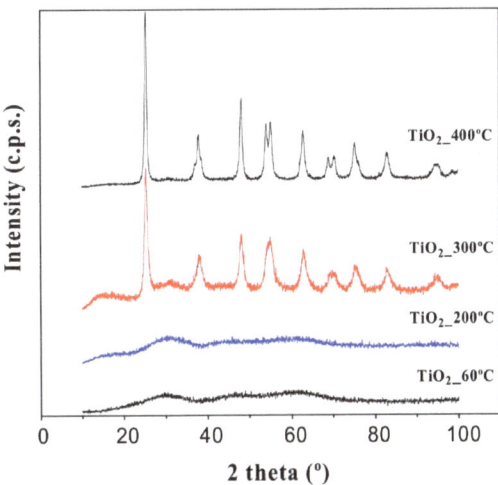

Figure 1. XRD patterns of TiO_2 treated at several temperatures (reprinted with permission from [41]).

In this figure, it can be seen how the treatment temperature allows the crystalline structure of anatase (TiO_2) to develop. At low temperatures, no diffraction peaks are shown because it is an amorphous solid, while the structure is ordered with temperature and allows the characteristic diffraction lines of TiO_2 to appear.

2.4.2. X-ray Photoelectron Spectroscopy (XPS) or Electron Spectroscopy for Chemical Analysis (ESCA)

These surface analysis techniques provide valuable quantitative and chemical state information about the surface of the material and its accessibility for reaction [42]. The techniques yield information at an average depth of 5 nm on the surface of the catalyst. The XPS technique is based on exciting a sample surface with monoenergetic Al kα X-ray causing photoelectrons to be emitted from the solid surface. The emitted electron is then collected at an electron energy analyzer. The binding energy and intensity of the photoelectron peak can be used to determine the elemental identity, chemical state, and quantity of the detected element. An example of the XPS spectrum is presented in Figure 2. The XPS spectrum corresponds to an organic waste (ostrich bones) chemically modified by the reduction of an iron salt with $NaBH_4$ to obtain an adsorbent material [43]. The signals identified in the spectrum provide information about the surface chemical composition. In this case, the presence of oxides of Ca, P, and Fe are observed. In a similar step, in AES the electron emission is used to obtain information about the chemical state of an element. This technique is more sensitive than XPS for lighter elements.

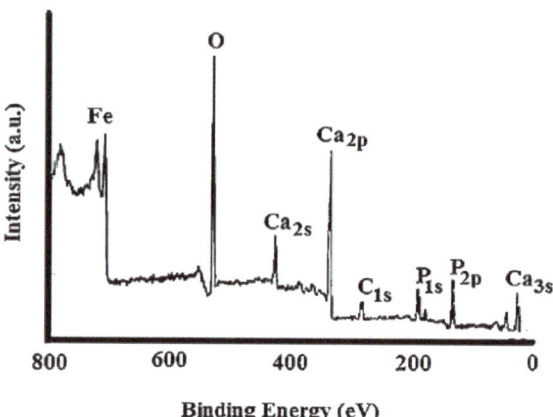

Figure 2. Typical spectrum of X-ray photoelectron spectroscopy (XPS) (reprinted with permission from [43]).

2.4.3. Ion Scattering Spectroscopy (ISS)

This is another surface technique in which a beam of ions is scattered from the surface of the atom and is gathered by a detector, which determines their kinetic energy [44,45]. Peaks are observed at different kinetic energies and related to the mass difference between the ion and the atom. Thus, two atoms on the surface with different masses will scatter ions differently. Because ions are scattered from the surface, the ISS technique is extremely sensitive and therefore analyzes only the topmost atomic layer on the solid. The ISS technique determines the relative coverage of a surface containing a certain atom or element. Adsorption of certain molecules onto a single crystal follows the growth of ultra-thin layers such as in atomic layer deposition. The downside of the technique is that it requires an ultra-high vacuum and is sensitive to contamination. A typical example of this kind of analysis is summarized in Figure 3.

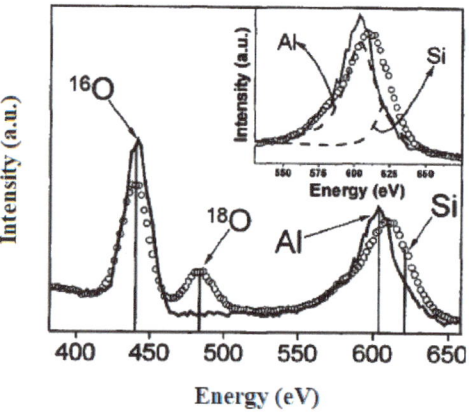

Figure 3. Typical spectrum of ion scattering spectroscopy (ISS) (reprinted with permission from [46]).

2.4.4. Electron Microscopy (EM)

Electron microscopes use signals arising from the interaction of an electron beam with the sample to obtain information about structure, morphology, and composition. Electrons are small particles, like photons in light, that act as waves. A beam of electrons passes through the specimen, and then through a series of lenses that magnify the image. The image results from a scattering of electrons by atoms in the specimen. A heavy atom is more effective in scattering than one of low atomic numbers, and the presence of heavy atoms will increase the image contrast.

Scanning Electron Microscopy (SEM) is a technique that uses an electron beam to produce an image of the sample surface with a resolution down to the nanometer scale. Electrons are emitted from a filament and collimated into a beam in the electron source [47]. The beam is then focused on the sample surface by a set of lenses in the electron column. When high-energy electrons reach the sample, several electrons and X-ray signals are generated. An example is included in Figure 4. Using this technique it is possible to morphologically characterize materials and, in this case, it is possible to observe how a ceramic material based on TiO_2 is deposited on the surface of a metal wire.

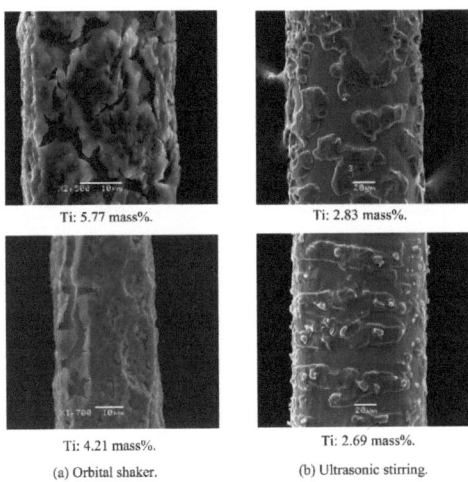

Figure 4. SEM images of TiO_2-coated wires with various stirring methods. The Ti content obtained from EDX analysis is also presented (reprinted with permission from [48]).

These include:
(a) Backscattered electrons (BSE): These are high-energy electrons that are ejected from the solid, losing only a small amount of energy. They originate from deep layers on the surface (a few microns deep). They provide information about the composition of the surface and lower-resolution images. These electrons are reflected after elastic interactions between the beam and the sample.
(b) Secondary Electrons (SE): These electrons originate from a few nanometers into the sample surface, with a lower energy compared to the backscattered electrons. They are very sensitive to surface structure and provide topographic information. Secondary electrons are a result of inelastic interactions between the electron beam and the sample.
(c) X-rays: These characteristic X-rays are produced when electrons hit the sample surface. They give information about the elemental composition of the sample.

BSE comes from deeper regions of the sample, while SE originates from surface regions. Therefore, BSE and SE carry different types of information. BSE images show high sensitivity to differences in atomic number: the higher the atomic number, the brighter the material appears in the image. SE imaging can provide more detailed surface information.

Transmission Electron Microscopy (TEM) is a very powerful tool for material science. A high-energy beam of electrons is shone through a very thin sample, and the interactions between the electrons and the atoms can be used to observe features such as the crystal structure and features in the structure like dislocations and grain boundaries. Chemical analysis can also be performed. TEM can be used to study the growth of layers, their composition, and defects in semiconductors. High resolution can be used to analyze the quality, shape, size, and density of quantum wells, wires, and dots. A typical TEM analysis is included in Figure 5. In this case, these are images of montmorillonites where it is possible to observe the lamellar structure.

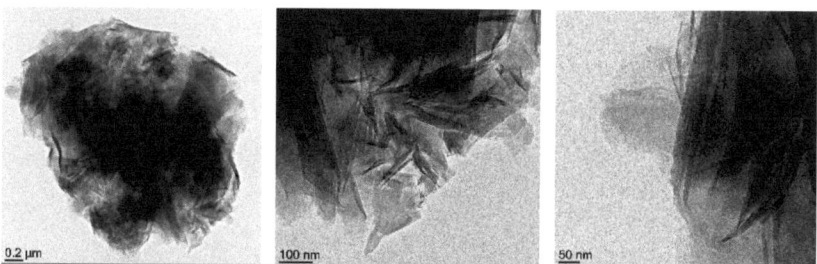

Figure 5. Clay image from transmission electron microscopy (reprinted with permission from [49]).

The TEM operates on the same basic principles as a light microscope but uses electrons instead of light. Because the wavelength of electrons is much smaller than that of light, the optimal resolution attainable for TEM images is many orders of magnitude better than that from a light microscope. Thus, TEMs can reveal the finest details of internal structures in some cases as small as individual atoms.

2.4.5. Infrared Spectroscopy (IR)

IR spectroscopy works on the principle that molecules absorb specific frequencies that are characteristic of their structure [50]. At temperatures above absolute zero, all the atoms in molecules are in continuous vibration with respect to each other. The IR spectrum of a sample is recorded by passing a beam of IR radiation through the sample. When the frequency of a specific vibration is equal to the frequency of the IR radiation directed at the molecule, the molecule absorbs the radiation. The examination of the transmitted light reveals how much energy was absorbed at each frequency (or wavelength) and this is related to the nature of the element. There are basically two types of spectrometers used in IR spectroscopy–Dispersive IR (DIR) spectrometers and Fourier transform IR (FTIR) spectrometers. Dispersive IR consists of radiation from a broadband source passing through the sample and is dispersed by a monochromator into component frequencies. Then, the beams fall on the detector which generates an electrical signal that results in a recorder response. The Fourier transform infrared (FTIR) technique has replaced the dispersive technique for most applications due to its superior speed and sensitivity. Instead of viewing each component frequency sequentially, as in a dispersive IR spectrometer, all frequencies are examined simultaneously in FTIR spectroscopy. A typical IR spectrum, in this case, the spectrum corresponding to activated carbons, is shown in Figure 6. In this case, it is an active carbon whose surface has been activated by treatment with acids. This treatment makes it possible to generate active centers for the adsorption of molecules as well as for catalytic processes, in principle interesting to stabilize the metallic active phase. Using this technique, functional groups generated on the surface can be detected.

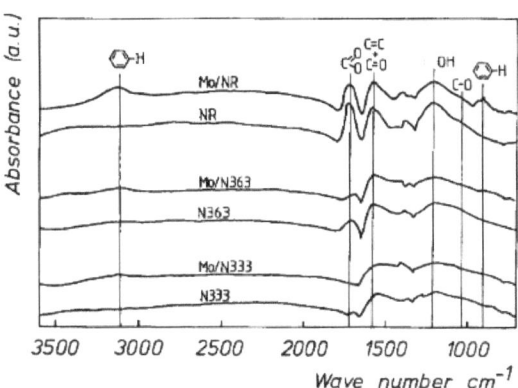

Figure 6. Typical IR spectrum of activated carbons (reprinted with permission from [51]).

2.4.6. Mössbauer Spectroscopy

This technique was discovered by Rudolf Mössbauer in 1958 and is based on the absorption of gamma rays in solids due to the vibrations of the atoms [52]. The unique feature of this technique is that absorption occurs only at the nuclei level and has high-resolution energy sufficient to resolve the hyperfine structures of the nuclear level. It operates on a single gamma ray transition, that is the Mössbauer transition, between the ground state and the exited state of one isotope in the sample. This technique is very sensitive to small changes in the chemical environment of certain nuclei. The number, positions, and intensities of the produced peaks in absorbance yield the characterization of the sample. A typical ^{57}Fe Mössbauer spectrum of clay (Na-Ballarat saponite) is presented in Figure 7.

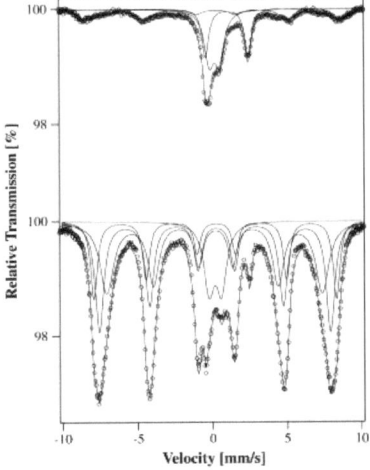

Figure 7. Typical ^{57}Fe Mössbauer spectra at −268.9 °C of Na-Ballarat saponite (**top**) and this clay intercalated with Fe8-tacn polycations (**bottom**) (reprinted with permission from [53]).

2.4.7. Thermal Analysis

(a) Thermogravimetric Analysis (TGA): TGA is a method in which a sample's mass is followed over time as the temperature is increased to yield information about physical and chemical properties of a solid such as transition, absorption–adsorption–desorption, etc. It also gives information about thermal decomposition and solid–gas reactions, for example, oxidation or reduction [54,55]. A typical system consists of a sensitive and precise balance with a sample holder located inside a programmable furnace. As temperature increases, the system monitors the change in mass loss to incur a thermal reaction. This could happen under different atmospheres: ambient, inert gas, oxidation/reducing gases, corrosive gases, vapors or liquids, as well as pressure or vacuum. The result is a collection of mass change versus temperature or time. A typical TGA curve, while the first derivative of this curve determines inflection points useful for further interpretations such as Differential Thermal Analysis known as DTA is included in Figure 8. The weight losses shown in the figure characterize the thermal processes that take place in lamellar double hydroxides (LDH-hydrotalcites). Thus, the first loss in weight at low temperatures can be related to the loss of physisorbed H_2O and the following losses are related to the dehydroxylation of the surface functional groups. Finally, signals appear due to the formation of metal oxides.

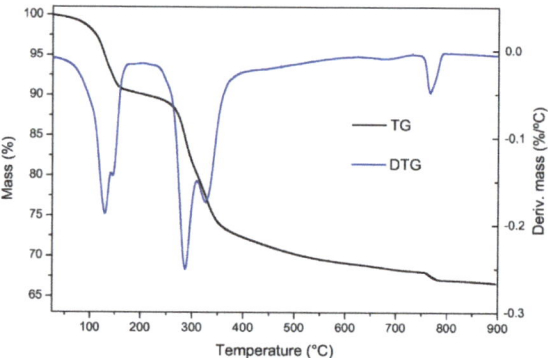

Figure 8. Thermal decomposition of hydrocalumite by TGA (reprinted with permission from [56]).

(b) Differential Scanning Calorimeter (DSC): DSC determines the difference in heat flow between the analysis sample and a reference [57]. The results of this technique measure the amount of heat adsorbed–desorbed during a phase transition and the enthalpies associated phenomena. Therefore, it yields information about melting, crystallization, or even water loss from a hydrated sample as well as some characteristic physical and chemical properties of the solid. A typical analysis of DSC is presented in Figure 9.

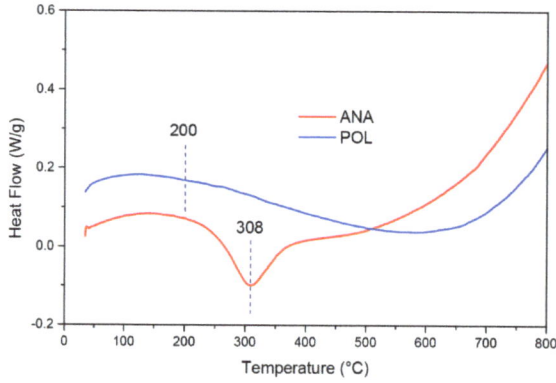

Figure 9. Typical profile of DSC on two zeolites (reprinted with permission from [58]).

2.4.8. Adsorption

Adsorption is the selective transfer of certain components of a fluid phase, called solutes to the surface of an insoluble solid. The adsorbed solute is referred to as an adsorbate, while the solid material is referred to as an adsorbent. When an adsorbent is exposed to a fluid phase, molecules in the fluid phase diffuse to its surface (including its pores if it is a porous adsorbent), where they either chemically bond with the solid surface or are held there physically by weak van der Waals intermolecular forces. When adsorption is caused by van der Waals forces, it is referred to as physical adsorption or physisorption, whereas it is called chemical adsorption or chemisorption if a chemical bond is formed between the adsorbate and the adsorbent. In general, adsorption is a process that involves the accumulation of a substance in molecular species in higher concentrations on the surface of a solid. Adsorption is by nature a surface phenomenon, governed by the unique properties of bulk materials that exist only at the surface due to bonding deficiencies [59–66].

Physical Adsorption Technique: This type of adsorption is due to weak forces between adsorbate and adsorbent. In general, it consists of a solid exposed in a closed space (usually called a sample holder) to a gas or vapors at some definite pressure, and the solid begins to adsorb as soon as the sample holder is immersed in a vessel containing the equivalent liquid of the adsorbate (i.e., N_2 gas at liquid nitrogen, Argon at liquid Argon, etc.). Thus, the pressure starts decreasing to reach equilibrium and becomes constant at a value say P where the sample does not adsorb any more gas. The amount of adsorbed gas can be calculated from the decrease in pressure by application of the gas laws providing a known vessel volume. The adsorption is caused by the forces acting between the solid and the molecules of gas. These forces are the same nature as the van der Wals order (<80 kJ/mol) and are also known as Intrinsic Surface Energy which brings about condensation of the fluid gas to the liquid state. The amount of gas absorbed by the sample (n) is proportional to the mass of the sample and depends also on the temperature of the bath where the sample is immersed, the pressure of the adsorbate over the sample, and the nature of both the solid and the gas. The quantity of gas adsorbed can be expressed as $n = f(P, T, gas, solid)$, and if the temperature is below the critical temperature of the gas, the above form is rather written as $n = f(p/p°)_T$, *gas*, *solid*, with $p°$ being the saturation vapor pressure of the adsorptive. A plot of $n(cm^3/g)$ versus $p/p°$ at a constant temperature yields the adsorption isotherm. The shape of an adsorption isotherm is directly related to the texture of the solid in question. There are six different types of adsorption isotherm, according to BDDT (Brunauer, Deming, Deming y Teller) as shown in Figure 10. Through this classification, it is possible to know if the materials are microporous, mesoporous or microporous solids.

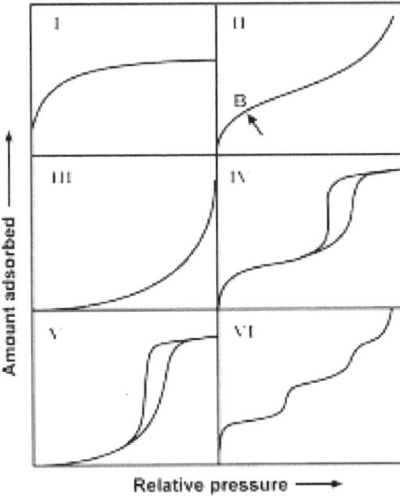

Figure 10. Adsorption isotherms according to BDDT classification [61].

The following important properties can be derived directly from the adsorption–desorption isotherm:

(a) Hysteresis: Hysteresis is defined as the void space between the adsorption and the desorption branch of the isotherms corresponding to the same solid and is associated with the capillary condensation that occurs when the relative pressure over the sample exceeds 0.4, approximately. The hysteresis is related to the shape of a pore in the solid and occurs when a greater pressure change is required to remove an adsorbate from a pore than what was required to adsorb.

(b) The specific surface area: Langmuir assumption [67,68] applies for type I isotherm and corresponds to microporous solids (<2 nm in pore diameter) and BET [69] applies for type II and IV (non-porous and mesoporous, respectively, pore diameter 2 < d < 50 nm).

(c) t-plot analysis: According to Lippens and de Boer [70,71], a comparison of the adsorption isotherm produced by a solid with a standard isotherm (produced by a non-porous material but with a similar nature of the solid in question) yields information about the micropore volume as well as the external surface area of the solid. The external surface area corresponds to the area in the solid that extends above the micropore area. According to Halsey, one of the t-equations, the t value is evaluated from the inverse of the relative pressure according to Equation (3).

$$t = 3.53 \left[\frac{5}{\ln(p^\circ/p)} \right]^{1/3} \quad (3)$$

(d) Pore size and structure: This term is related to mesoporous materials and is directly connected with type IV of adsorption isotherm [71], which shows a hysteresis between the adsorption and the desorption branch. According to Kelvin, there is a direct relationship between the size of the pore and the relative pressure in equilibrium withdrawn directly from the adsorption isotherm (Equation (4)). Finally, the pore size would be the sum of the Kelvin radius and the value of the thickness (t) according to the Pierce method (see Equation (5)) [72–74].

$$r_K = \frac{-2 \cdot V_m \cdot \gamma \cdot \cos\varphi}{RT \cdot \ln(p/p^\circ)} \quad (4)$$

$$r_p = r_K + t \quad (5)$$

where r_K represents the radius of the pore, γ is the surface tension of the adsorbate, V_m is the molar volume, R is the gas constant, and T the analysis temperature.

(e) Total pore volume: The total pore volume of a solid corresponds to the total adsorbed volume at a relative pressure close to unity on the adsorption isotherm, it is normally converted into volume of liquid according to the Gurvitsch rule (see Equation (6)) [75].

$$V_{pT} = V(g) \cdot 0.0015468 \quad (6)$$

Chemical Adsorption Technique: Chemisorption is caused by a reaction on an exposed surface, which creates an electronic bond between the surface and the adsorbate. The heat of chemisorption can exceed 1000 kJ/mole and shows a type I adsorption isotherm. Chemisorption, therefore, consists of one layer of adsorbate on the surface where active sites are located. Hence, the total adsorbed amount of adsorbate determined at saturation corresponds to the number of active sites available and accessible on the surface of the solid providing a known stoichiometry of adsorption between the adsorbate molecule and the active site. Chemisorption encompasses the following important techniques.

Pulse chemisorption technique. This technique serves to titrate the surface-active sites available for the reaction. It is based on dosing a known amount of active gas, normally H_2 or CO, onto a freshly reduced catalyst. The total available surface species are thus quantified providing the stoichiometry factor (SF) of adsorption. SF is normally considered to be 1 for CO and 2 for H_2. Pulse chemisorption evaluates the accessible surface of materials and can quantify the dispersion of the active species on the support. The higher the dispersion the higher the activity of the catalyst. The pulse technique, therefore, provides a rapid comparison of the catalysts and provides a prediction that can be extrapolated to the overall performance of the catalysts. This technique is also widely used to evaluate the total acidity of the catalyst when ammonia or any other basic molecules are used for titration.

2.4.9. Temperature Programmed Reduction (TPR)

This important technique consists of reducing a metal oxide under the effect of a reducible gas, H_2 or CO, and temperature to produce a result known as the TPR profile. Important information can be withdrawn from the TPR profiles and can be enumerated as follows:

(a) The quantity of the total amount of oxides in the solid;
(b) The maximum reduction temperature of the oxides;
(c) The presence of various particle sizes of the oxide that would be indicated by shoulders on the TPR profile;
(d) Shift on the maximum reduction peak indicating the interaction of the active particles with the support. In other words, the role of the support is to stabilize the active species by a certain type of interaction that could be weak or strong; the stronger that interaction is the higher the reduction temperature.

Percentage of reducibility: This indicates the number of oxides that are free to be reduced and become active sites on the solid. The difference between the originally loaded material at preparation and the number of reducible species calculated by TPR indicates the number of species that are capable of being reduced.

Identification of species: TPR is a technique that allows the identification of new species where two or more elements are present in the catalyst. A typical example of when two species are interacting or reacting to produce a completely different species is included in Figure 11. In this case, PtO_2 and RuO_2 are mixed physically and the TPR profiles show a mixture of two profiles that correspond to each individual element. If these two elements are subject to a high temperature, they react to form an alloy. Therefore, the TPR technique demonstrates how different species react to produce new species, in this case, an alloy, and therefore its presence could also be identified in a catalyst formed by these two metal oxides.

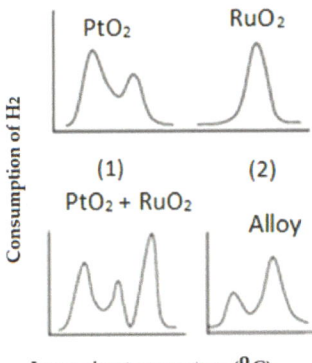

Figure 11. TPR profiles of individual PtO_2 (1) and RuO_2 (2) and the physical mixture (PtO_2 + RuO_2) after thermal treatment to obtain the metal alloy.

2.4.10. Temperature Programmed Oxidation (TPO)

This technique is used to oxidize the freshly reduced elements in a catalyst. As TPR is a bulk reduction technique, which means that all oxide species present in the catalyst area are completely reduced under the effect of a reducible gas and high temperature. TPO evaluates the freshly reduced species to being re-oxidized. Thus, determining the degree of reduction. Cycles of reduction/oxidation, however, can predict the active lifetime of a catalyst after several regeneration cycles. This analysis would produce CO or CO_2 as the result of carbon or graphite oxidation. A mass spectrum of CO and CO_2 is included in Figure 12. Carbon deposits on the active sites of the catalyst lead to deactivation which directly results in lowered yield.

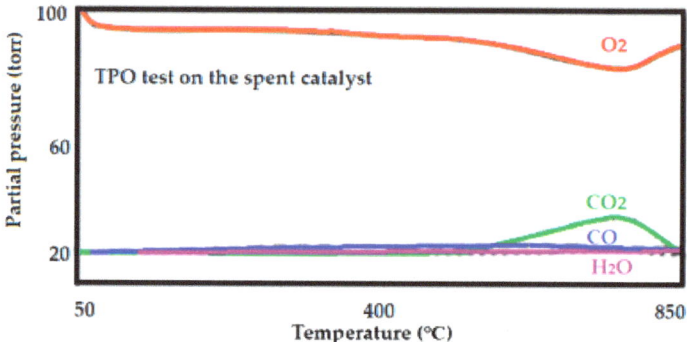

Figure 12. Mass signal of the consumption of O_2 and those corresponding to the production of CO and CO_2.

2.4.11. Temperature Programmed Reaction (TPRe)

This technique shows a simple reaction performed at atmospheric pressure. It helps to simulate and predict a reaction that would produce a larger yield at industrial levels. The following figure shows an interesting reaction in which CO_2 is sequestrated by CaO to produce a more beneficial product such as Calcium carbonate $CaCO_3$. The result shown in Figure 13 indicates the number of cycles that CaO can react with CO_2 before being deactivated. After 4 cycles, the area that was produced on the first cycle is about 4 times the area produced on the 4th cycle. Such that, the number of cycles is related to the capacity of the solid, in this case, the CaO, to react and sequestrate the CO_2. This result provides insight into the reusability of a catalyst.

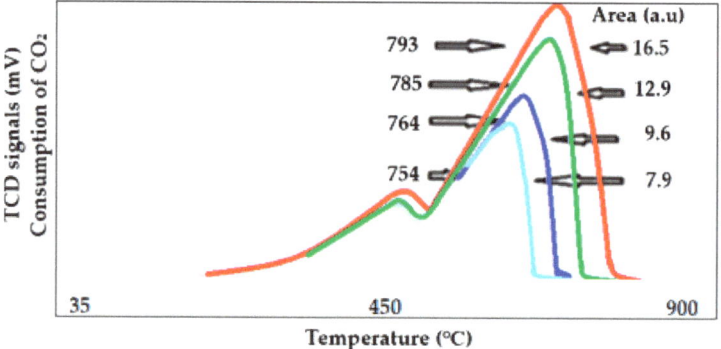

Figure 13. Sequestration cycles of CO_2 by CaO.

2.4.12. Evaluation of Acid Sites in Solids

Some catalytic reactions require acid sites instead of metal active sites. One example is in the cracking reaction. The strength and distribution of acid sites are rather important in catalysis. In this experiment, the sample is saturated with a basic molecule such as ammonia, isopropylamine or any other branched basic molecule. The experiment is carried out at about 50 to 70 °C to minimize the physisorption effect that can occur on the support. Distribution and strength of acid sites, normally Brønsted acid sites (BAS), is then performed by raising the sample temperature which causes desorption of the pre-adsorbed reactants as a function of temperature. If desorption is performed at several temperature ramping rates (2, 5, 10, 15, and 20 °C/min), the heat of desorption can then be determined by applying the Kissinger Equation (7) [76] that is related to the strength of acid sites. An example of the acidity of ZSM5 zeolite evaluated after saturation by ammonia followed by desorption at various ramping rates is presented in Figure 14. This analysis yields very important information about the performance of a catalyst used for cracking. In catalytic cracking, the acid sites are the active sites that break down large hydrocarbon chains into aromatic molecules that are of great interest in the fuel industry. Evaluating the desorption capacity at various heating rates allows obtaining the energy related to the desorption process, that is, the energy necessary for the desorption of the adsorbate.

$$2 ln T_m - ln \beta = \frac{E_d}{R \cdot T_m} + C \qquad (7)$$

$$C = ln\left(\frac{E_d \cdot V_m}{R \cdot K}\right) \qquad (8)$$

where T_m (°C) is the maximum temperature of the temperature-programmed desorption curve, β (°C/min) is the heating rate, E_d (kJ/mol) is the heat of desorption and R is the gas constant.

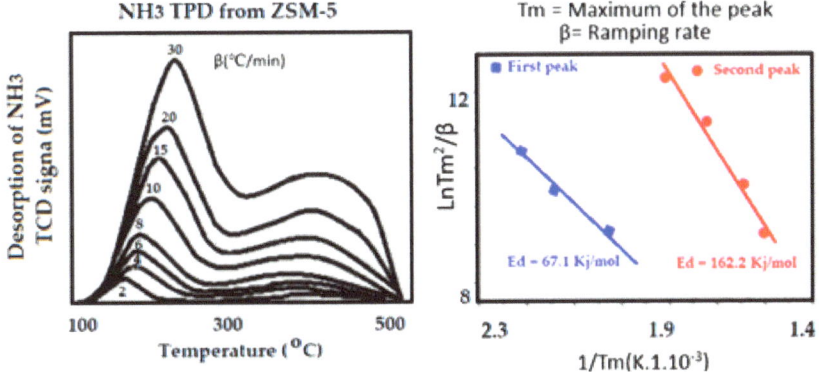

Figure 14. Temperature desorption profiles of NH_3 are used to determine the heat of desorption and the distribution of acid sites according to their strength.

Ammonia as a small probe molecule can penetrate all size pores, micro and mesopores without discrimination. However, in many cases, the very tiny pores will not make any contribution to the overall activity of the catalyst. They, on the contrary, prevent the adsorbate molecules from diffusing inside such a small pore to encounter the active species and react. In this case, larger and more branched base probe molecules are more suitable. I-propylamine as a probe molecule quantifies acidity only in larger pores that make a contribution to the activity of the catalyst. This probe molecule absorbs, reacts, and decomposes over a BAS to produce propylene and other products. The amount of the produced propylene is equivalent to the amount of BAS in the catalyst. The decomposition of the i-propylamine takes place in a similar manner to the Hofmann elimination [77–79] according to the following schematic for the reaction (see Figure 15).

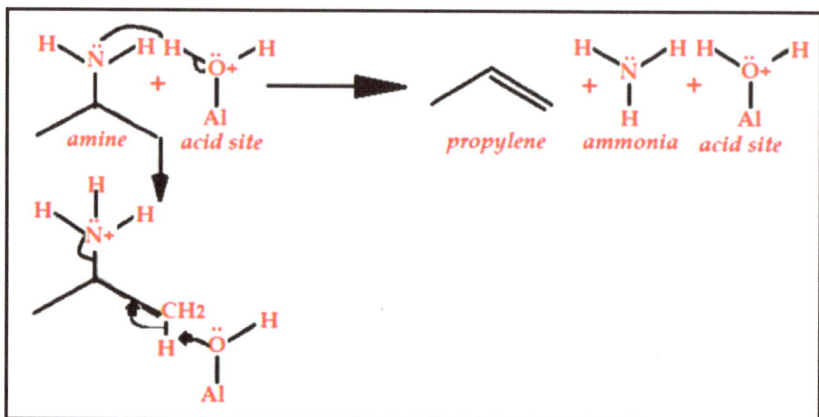

Figure 15. Reaction of the amine with the acid site decomposes to produce propylene and ammonia. The mechanism is similar to the Hofmann elimination.

Methanol is also a suitable molecule to determine active sites on the catalysts [80]. It produces various products according to the active site on which decomposition takes place. A possible mechanism and the products of the reaction are included in Figure 16.

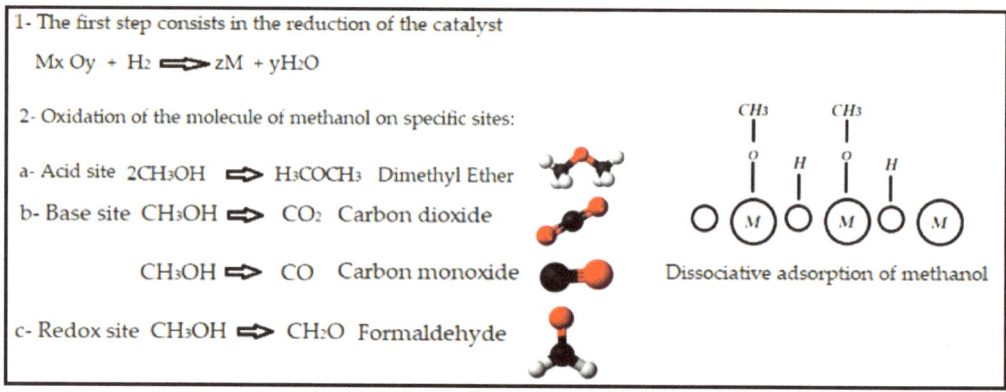

Figure 16. Decomposition of the molecule of methanol over an active site.

2.4.13. Mercury Intrusion

This technique is used to characterize solids that contain pores (0.003–360 µm), where the gas adsorption technique fails to study such large pores. This technique uses mercury to penetrate under pressure the pore. Mercury is a non-wetting liquid; therefore, the instrument applies pressure to provoke mercury to penetrate the pore. The size of the pore is related to the applied pressure according to the Washburn [81] Equation (9).

$$p = \frac{-4 \cdot \gamma \cdot \cos\theta}{d} \tag{9}$$

where p represents the applied pressure, γ is the surface tension of the adsorbate (mercury in this case), θ is the contact angle and d is the pore dimension.

3. Synthesis, Characterization and Catalytic Application

As an example of this work, two series of Ni-supported catalysts were prepared using the impregnation method. The commercial supports used were an alumina trilobe (high surface area material, 310 m^2/g) and an alumina pellet (lower surface area material, 92 m^2/g). Two series of catalysts were prepared on each of the supports, one was with low-loading metal (about 5 wt.%) and the other series was with a higher loading of Ni (about 14 wt.%). Equivalent amounts of salt (Ni(NO$_3$)$_2 \cdot$6 H$_2$O from Aldrich) were taken and dissolved in a volume of DI water corresponding to three times the total pore volume of the support determined from the adsorption isotherm at a relative pressure close to unity. The supports were first degassed at 150 °C for 3 h to completely dry. The samples were then brought to room temperature under an inert atmosphere. The solution was poured over 5 g of the dried support and stirred for 6 h until the solution was completely evaporated. Later, the fresh sample was washed with DI water until the water came out clear. Samples were then placed in the fixed bed reactor where the catalytic tests were carried out and dried at 100 °C by heating the oven at a rate of 1 °C/min and maintained at 100 °C for 12 h. Calcination was carried out by flowing a mixture of 10% O$_2$ in helium and heating the reactor up to 500 °C with a rate of 2 °C/min and maintained for 12 h.

3.1. A Comprehensive Assessment of Characterization Techniques

Both the characterized supports and the freshly prepared catalysts were analyzed by gas adsorption using N$_2$ as the probe molecule at −196 °C (Micromeritics 3flex apparatus, Norcross, GA, USA). The sample was dried at 150 °C under vacuum for 4 h prior to the analysis.

Temperature Programmed Reduction (TPR) (Micromeritics in situ characterization system (ICCS) connected to the Micromeritics FR-100 microreactor): This technique is used to study the role of the support in stabilizing the active species. Ni with various loading on different supports as well as a commercial copper catalyst (13 wt.% Cu-alumina) was used for comparison with the prepared Ni catalysts.

Pulse chemisorption of CO: This technique was used to titrate the surface atoms or active particles present on the surface of the catalyst at a temperature near 35 °C. Carbon monoxide was used as adsorbate and a stoichiometry factor of 1 was considered to correlate the amount of CO chemisorbed to the number of surface-active particles.

N$_2$O was used to determine the dispersion of Cu on the Cu-supported catalyst. The CO was substituted by N$_2$O since Cu does not chemisorb CO or H$_2$. The catalyst was first reduced with H$_2$ by performing a TPR analysis to ensure a complete reduction of the metal. The titration is carried out at 90 °C for a complete surface oxidation of Cu. This analysis produces N$_2$ as the result of Cu oxidation. Quantification of active sites is determined by computing the amount of H$_2$ produced in relation to the stoichiometry factor and the amount of Cu particles on the surface of the catalyst.

Testing the catalysts for the Sabatier reaction was carried out on the Micromeritics FR-100 flowing reactor (see Figure 17). Approximately 0.5 g of catalyst was placed in a fixed bed reactor and reduced with pure H$_2$ at 500 °C for one hour at 30 bar. The reactor temperature was brought to 30 °C prior to flowing the active gases for the reaction. A mixture of 50 cm^3/min of CO$_2$ and 200 cm^3/min of H$_2$ was premixed before passing through the catalyst bed at room temperature. The reactor temperature was then raised to 500 °C at a rate of 2 °C/min. The mass spectrometer was connected to the exhaust of the reactor and served to online monitor the signals for both the flowing mixture as well as for the expected products. In this case, 4 masses (m/z) were selected for this analysis: (2 for H$_2$, 44 for CO$_2$, 28 for CO, and 16 for CH$_4$).

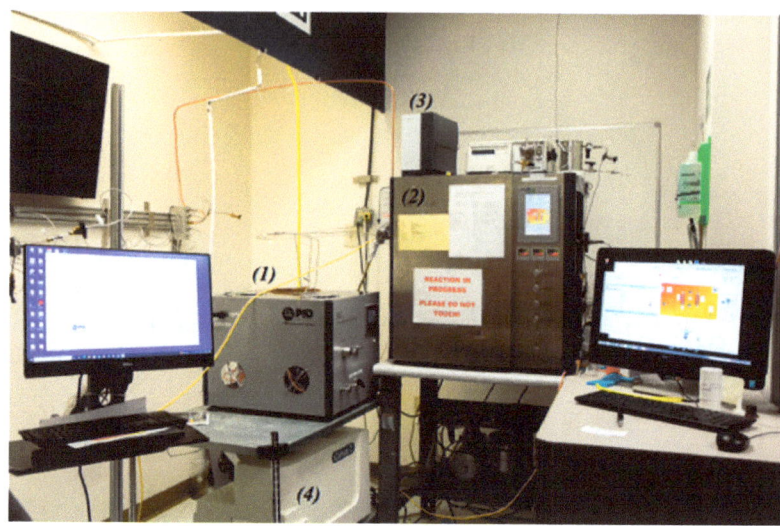

(1) *In-situ* catalyst characterization system (ICCS)
(2) FR-100 Micromeritics micro-reactor
(3) Agilent Micro GC
(4) MKS mass spectrometer

Figure 17. Catalytic equipment.

3.1.1. Physical Adsorption

The isotherms that correspond to both the supports (red) and to the catalysts as well as for the commercial 13 wt.% CuO-supported alumina catalyst (blue) is presented in Figure 18. Both supports showed type IV adsorption isotherms, indicating that both are mesoporous materials. The surface area as well as total pore volume determined from the adsorption isotherms are summarized in Table 1.

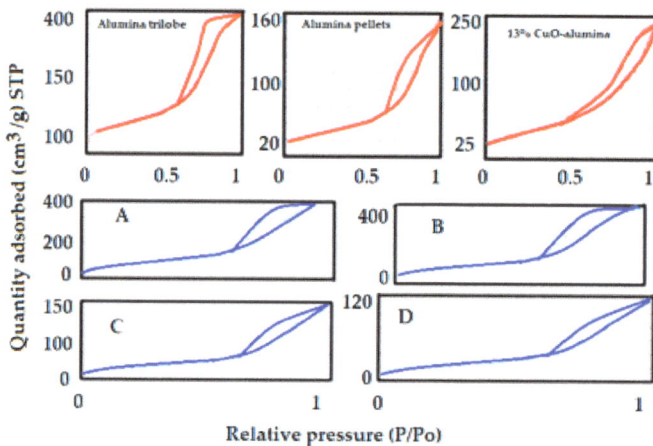

Figure 18. N_2 adsorption isotherms for all supports and all catalysts as well as for the 13 wt.% CuO-alumina catalyst. (**A**,**B**) Ni/Al_2O_3 trilobe, (**C**,**D**) Ni/Al_2O_3 pellet catalysts.

Table 1. BET-specific surface areas, total pore volumes, and pore diameter for all materials.

Sample	S_{BET} (m²/g)	V_{pT} (cm³(STP)/g)	Pore Diameter (nm)
Alumina Trilobe	310	0.67	8.6
Alumina pellet	92	0.26	10.9
13 wt.% CuO-alumina catalyst	165	0.38	9.1

It can be concluded from the adsorption isotherm that the alumina trilobe contains smaller pores than the other support since condensation occurs at a lower relative pressure (0.6 versus 0.7 $p/p°$). The hysteresis shown in the adsorption isotherms for alumina trilobe reflects ink-bottle pores while the alumina pellets and the copper catalyst showed more open slit-shape pores. The specific surface areas as well as total pore volumes of the supported catalysts showed slightly lower values due to the fact that the Ni penetrated the pores, thus decreasing the pore volume and the internal surface area within the pores. This phenomenon is expected and desirable in catalysis as it stabilizes and increases the dispersion of nickel yielding a higher performance of the catalyst when the active element enters the pore. It should be noted that the surface area within the pore is the internal surface area of the solid and has a large contribution to the total surface area, hence a better dispersion of the active elements inside the pores. Results are shown in Table 2.

Table 2. BET specific surface areas, total pore volumes, and pore diameter for all catalysts.

Catalyst	S_{BET} (m²/g)	V_{pT} (cm³(STP)/g)	Pore Diameter (nm)
14.5 wt.% Ni (A)	222	0.51	8.9
4% wt.Ni (B)	280	0.53	8.7
11% wt.Ni (C)	75	0.19	10.0
5.6% wt.Ni (D)	83	0.21	10.4

The surface area was determined by solving the BET equation at relative pressure between 0.05 and 0.3. The pore size was estimated by the Kelvin equation (Equation (4)) and corrected by adding the t value according to de Boer et al. (Equation (3)) to the Kelvin radius. The total pore volume was estimated from the adsorption isotherm at a relative pressure close to 1 and transformed into liquid by using the Gurvitsch rule (Equation (5)).

3.1.2. Chemical Adsorption, Chemisorption or Selective Adsorption Technique

Characterization of the Ni-supported catalysts prepared for this study is used to show examples of the use of some of these techniques.

In general, the low loading in Ni yields a higher dispersion (A: 4 wt.% Ni) of 25.5% while the catalyst (B: 14.5 wt.% Ni) larger loading in Ni yields a much lower dispersion (5%) (see Figure 19A,B). This could happen since large loading could provoke the formation of large agglomeration of Ni, and, hence, lower dispersion. The CO pulse chemisorption profiles for the catalyst supported on alumina pellets are included in Figure 19a,b. As the support in this series of catalysts has a much lower surface area, dispersion also showed much lower values for the high loading (A: 11 wt.% Ni), the dispersion was 3.3%, while the low loading (B: 5.6 wt.% Ni) was 14.5%.

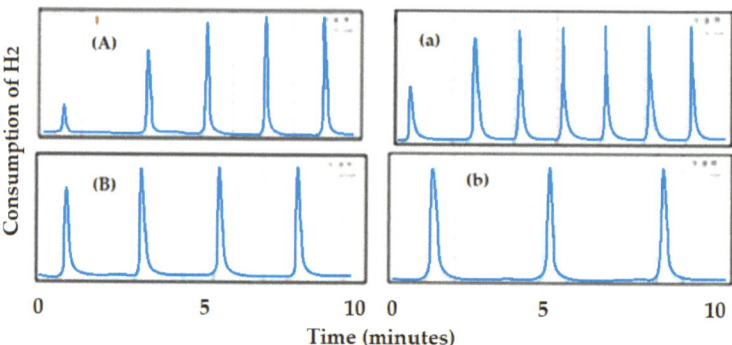

Figure 19. CO pulse chemisorption on Ni catalysts. Ni-alumina trilobe catalysts: (**A**) (25.5% dispersion) and (**B**) (5% dispersion). Ni-alumina pellets catalysts: (**a**) (3.3% dispersion) and (**b**) (14.5% dispersion).

The result of the dispersion on the fresh catalysts would predict a higher conversion for the catalysts-supported alumina trilobe. Thus, a higher surface area support and lower loading catalysts produced higher dispersion, and therefore, they would yield a higher conversion of CO_2 to CH_4.

This analysis was used to characterize the prepared catalysts of Ni supported on alumina, trilobe, and pellet. The role of the support (in this case, supported Ni on high surface area material, alumina trilobe, and other aluminas (pellets) with a lower surface area) and their interaction with the active species is presented in Figure 20. Profile (a) indicates strong interaction between the support and the active species due to the higher reduction temperature, while (b) shows weaker interaction since a lower reduction temperature is shown on the TPR profile. Sometimes the support can strongly interact with the active species, when this occurs, the reaction of the analysis gas on the active species can be disguised up to temperatures of 1000 °C. This strong interaction forms a species called spinels. Spinels are chemically and thermally stable materials and are, generally, not desirable in catalysis except for some biomass reactions. Outside of these biomass reactions, the formed spinels are not as active as the active species and lead to slower product formation.

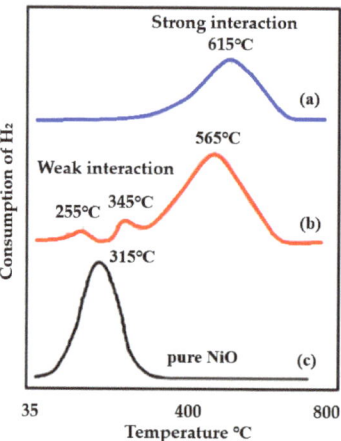

Figure 20. TPR profiles showing the role of the support. (**a**) Ni-trilobe, showing strong interaction with the support compared with unsupported NiO, (**b**) Ni-pellets showing weaker interaction with the support compared with unsupported NiO, and (**c**) pure NiO.

Several TPR profiles of Ni-supported alumina catalysts that have been calcined at various temperatures, between 500 and 1100 °C, under a flow of 10% O_2 in helium are included in Figure 21. It can be observed for these profiles that the calcination temperature can alter the reduction profiles. A higher calcination temperature tends to form new species, this must be avoided to minimize the formation of spinels and to stabilize the active species under the high temperature and pressure reaction conditions.

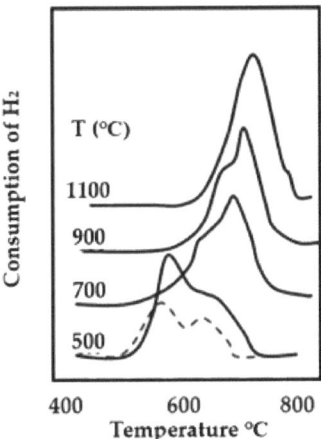

Figure 21. TPR profiles showing the effect of the temperature calcination on the shift of the maximum of the reduction temperature. The TPR profile of the dried catalyst precursor is also included (dried line).

Evaluation of the Cu-alumina commercial catalyst by N_2O. For the copper catalyst, nitrous oxide was used to determine the dispersion of Cu since the metal would not adsorb/react with H_2 nor with CO [82]. In this case, the thermal conductivity detector (TCD) on the ICCS instrument used for the characterization is not capable of separating and identifying the N_2 peak from the peak of N_2O that has not been reacted. To overcome this problem, a Propak separation column from Agilent was installed just before the TCD. This column separates the peak of N_2 as the product of the reaction (see reaction below) from the peak of N_2O as an excess of the pulse that has not been completely consumed by the copper. Quantification of the amount of N_2 produced corresponds to the amount of copper present on the surface of the catalyst providing a stoichiometry factor of 2.

$$N_2O + 2Cu \rightleftarrows Cu-O-Cu + N_2 + (\text{non reactive } N_2O) \qquad (10)$$

The catalyst was first reduced under a flow of pure H_2 at 350 °C and was kept for 1 h at this final temperature to ensure complete reduction of the copper. After completion of the reduction, the catalyst was swept by a current of helium to completely remove the H_2. For the analysis, the sample temperature was reduced to 90 °C, then pulses of N_2O were carried out. The installed column was able to separate the produced N_2 from the N_2O. The registered pulse of N_2O as well as N_2 are summarized in Figure 22. Quantification of N_2O was done upon saturation, where several N_2O peaks produced the same area. The ICCS program monitors the loop temperature and pressure so that the quantification of the loop volume is corrected at each pulse. Having the corrected loop volume at each pulse, the quantification is done by comparing the resultant area peak at each pulse with the peak area at saturation. The freshly reduced copper catalyst produced a dispersion of 29%.

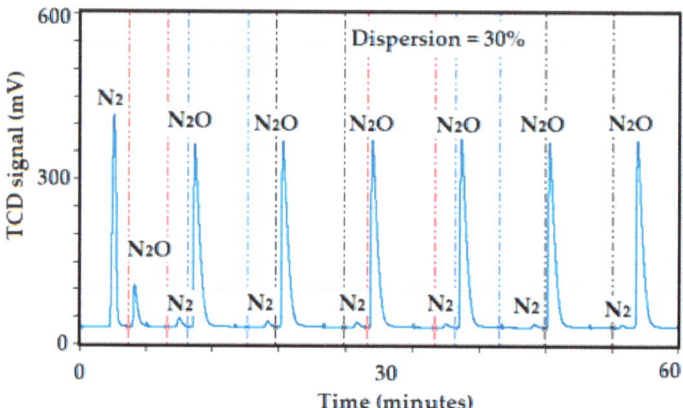

Figure 22. Pulse chemisorption of N_2O on the reduced Cu-catalyst.

Upon characterization of the catalysts, they were then tested for the Sabatier reaction, where CO_2 was reduced by H_2 at elevated temperature and pressure to mainly produce CH_4 with some CO as a co-product and other products. The reaction conditions were reduction of CO_2 by H_2 at 15 bar and heated to 500 °C of temperature with a temperature ramping of 2 °C/minute. Consumption of CO_2 and H_2 as well as production of CO and CH_4 were online recorded by using a mass spectrometer.

Only three catalysts will be shown here as examples: (A) 13 wt.% CuO-alumina, (B) 11 wt.% Ni-alumina pellets and (C) 4 wt.% Ni-alumina trilobe. Before any catalytic test, a blank was performed to ensure that the reactor showed little or no activity for this reaction. The catalyst support was made of quartz wool instead of the normal filter that comes with the reactor, thus, any activity will be attributed to the catalyst.

(A) Catalytic test on the 13 wt.% CuO-alumina catalyst: The reaction conditions were the same as previously described. The analysis was performed using 0.5 g of sample and heating from near room temperature to 550 °C at a rate of 2 °C/min and 15 bar. The produced signals for the products (CH_4 and CO) as well as signals corresponding to the consumption of H_2 and CO_2 are included in Figure 23. The first signal (m/z 28) appeared at about 230 °C and corresponded to the production of CO, while a second signal (m/z 16) appeared at about 360 °C and corresponded to the formation of CH_4. The reaction stabilized at about 525 °C, beyond this temperature the analysis stopped to avoid complete deactivation of the catalyst. This deactivation could be due to excess formation of graphite as the co-product of the reaction, thus, minimizing the contact of the reactant molecules with the active area of the catalyst.

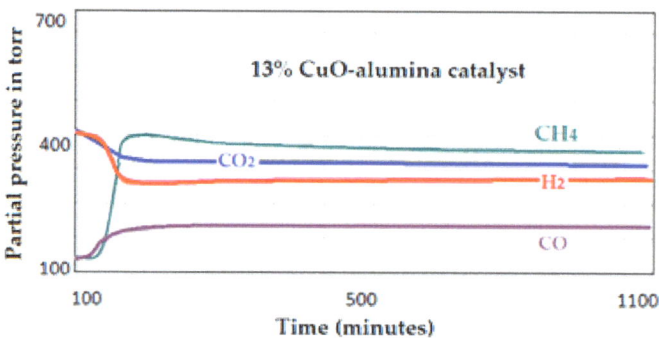

Figure 23. Intensities of the produced signals by the 13 wt.% CuO-alumina catalyst.

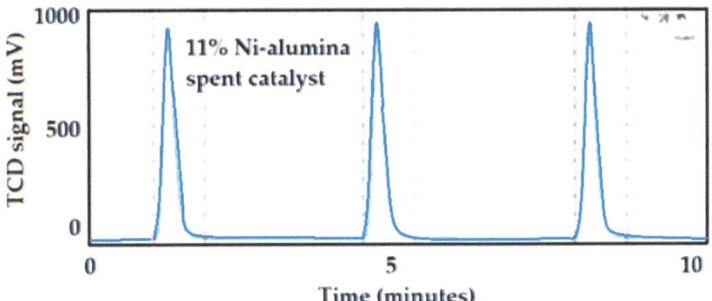

Figure 31. CO pulse chemisorption on the 11 wt.% Ni-supported alumina pellets spent catalyst.

The CO chemisorption on the 4 wt.% Ni-supported alumina trilobe catalyst after 20 h of reaction at 15 bar and 500 °C is included in Figure 32. The catalyst did not seem to be very affected by the reaction conditions as it showed a slightly lower dispersion value (35 versus 43%) for the fresh catalyst.

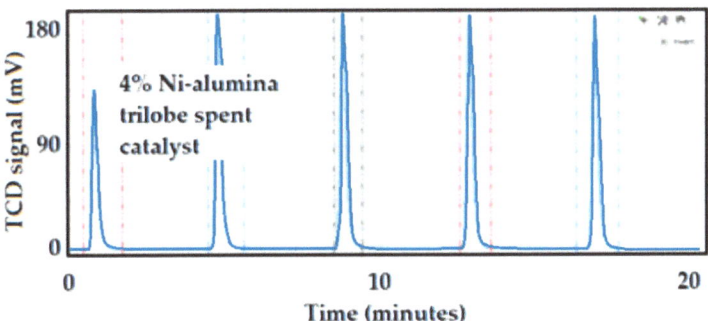

Figure 32. CO pulse chemisorption on the 4 wt.% Ni-supported alumina trilobe spent catalyst.

It can be concluded from the last test that in situ pulse chemisorption was necessary to determine the causes of catalyst deactivation.

Finally, it has been observed that the Ni-supported catalysts showed higher catalytic performance than the Cu-supported alumina commercial catalyst to produce CH_4. It should be noted that the higher the Ni content in the catalyst the higher the activity in producing CH_4 (14.5 wt.% Ni-trilobe < 11 wt.% Ni-alumina pellets < 4 wt.% Ni-trilobe < 13 wt.% CuO-alumina). The 4 wt.% Ni-alumina trilobe catalyst showed the highest dispersion and was not greatly affected by the reaction condition for at least 20 h at 15 bar, while the copper catalyst used for this study did not seem to be an active catalyst for the Sabatier reaction [83] that could be related to the large loading in copper. Possibly a lower loading in copper would yield higher dispersion, and therefore, possibly higher catalytic performance.

4. Summary, Conclusions, and Future Perspectives

Any catalyst, before being tested for a reaction, must undergo a specific protocol of characterization to simulate its behavior under high temperature and pressure reaction conditions. The first task of this study was to understand the importance of the textural and structural properties of the support used for the preparation of catalysts. The gas adsorption analysis completes knowledge of the texture of the support. The surface area provides information about the loading of metal particles that form a surface monolayer and become the active sites for the reaction. The porosity of the support plays an important role in the diffusion of the reactant molecules into the pores to access the active sites for the reaction. The size of the pore will act like a sieve where only the reactant molecules of interest are allowed to diffuse and encounter the active particles that are normally located inside the pores. For an optimal diffusion of the reactant, the size of the pore should have an effective size that is normally five to six times larger than the reactant molecule diameter.

The study of the structure of the freshly prepared catalyst predicts the stability of the active species when the catalyst is subjected to severe reaction conditions, high temperature, and pressure, and the possibility to minimize sintering that is conducive to catalyst deactivation. Properties of the support, also make large contributions to the performance of the catalyst depending on the reaction, as an example, acidity for the hydrocarbon cracking reaction. Basic sites are also widely used in transesterification reactions (biodiesel production by using base catalysts, alkali metal hydroxides (NaOH/KOH), and alkali metals ($NaOCH_3$)) and many other reactions where a special active site is required.

The thermal programmed techniques are not only very attractive but also very accessible to many researchers. Techniques such as TPR are widely used to gain information about the catalyst. For example, the strength of interaction between the support and the active species is highly dependent on the preparation and calcination of the catalyst. It also serves to quantify the total amount of metal oxides available in the catalyst. The TPR profile can also yield information about the active particle size which is revealed by the mechanism of reduction.

Pulse chemisorption is also an important technique for the characterization of the catalyst. For example, the pulse technique is used as a titration method to quantify the number of active species available on the surface for the reaction for the reaction. Only the surface-active particles are accessible for the reaction and yield the activity of the catalyst. Hence, the importance of this technique, it can predict the activity and selectivity of the catalyst before the catalytic operation. This technique is also used to quantify the acidity of the catalyst as well as the strength of the acid sites at the reaction conditions. The total acidity of the catalyst is important; however, the distribution and strength of acid sites are more critical in catalysis.

After the characterization of the catalyst's property, testing becomes the second step in the life of the catalyst. The condition for the testing will depend on the reaction itself. For example, selecting the reaction conditions as well as the reactant molecules to produce the final desired product. In this study, a simple catalytic test was provided as an example, and a reduction of carbon dioxide was selected for the Sabatier reaction to complete the work for this review. Although GC is widely mentioned as a technique in the literature as the preferred analytical method, a mass spectrometer was instead selected for this study, as it allows a researcher to better follow the reaction steps and thus elucidate the formation of different products and the actual temperature at which reaction starts for each product.

Catalyst deactivation was also studied. Usually, there are two main causes of deactivation. During reactions at high temperatures and pressure, cracking can occur. Cracking is the production of carbon atoms as the result of the breakage of hydrocarbon molecules, or the formation of carbon species such as graphite for example as in the case of the Sabatier reaction. Carbon covers the active species preventing direct contact between the active particles and the reactant molecules. Hence, minimizing accessible catalytic sites or even completely deactivating the catalyst. Sintering of the active particles can also lead to deactivation. This occurs mainly at high temperatures when the support interaction with the active species is rather weak. The active particles, however, will start moving on the surface to collapse together forming larger particles and minimizing the active surface area where the reaction takes place. Characterization before the catalytic test is important, however, characterization after deactivation is crucial as well. A special condition for this study is that characterization must be done in situ without having to remove the catalyst from the reactor and expose it to an atmosphere where oxygen from the air can change the surface. One of the important issues mentioned in this study is the use of the new Micromeritics instrument, the ICCS, as was used in situ to characterize the catalyst before and after the reaction. TPO was used to identify the cause of deactivation as the deposition of carbon onto the surface of the catalyst. Through TPO, the catalysts were reoxidized, producing both carbon monoxide and carbon dioxide. These two elements were monitored by the mass spectrometer that was connected to the exhaust of the reactor. The thermal conductivity detector (TCD) fails to differentiate between two elements, in this case, between CO and CO_2. All catalysts used in this study produced carbon during the reaction which contributes to the formation of carbon deposits, especially graphite, as one of the reaction products. The use of the Micromeritics FR-100 reactor for this study has the capability to remove the water produced by the Sabatier reaction at the reaction conditions (500 °C and 15 bar) from the primary product, CH_4 in this case.

The ICCS instrument allowed for titration of CO for the analysis of the Ni catalysts and the use of N_2O for the Cu catalyst. It is to be noted that the ICCS includes a special column that can separate the N_2 from the excess of N_2O without the need to use an external cryogenic trapping device. It has been concluded that the Ni-supported catalysts were not highly affected by the reaction conditions, especially for the low-loading catalyst, as dispersion did not reflect large changes as the Cu-supported catalyst showed. This effect could be because the Cu catalyst in this study contains a large amount of Cu (13 wt.%), which could result in agglomeration of the active particles. It could also be due to the low surface area of the catalyst support. While the Ni catalysts supported on the alumina trilobe enjoy a large surface area of the support and relatively low loading, that could lead to the high dispersion of the active particles, hence, higher activity for the Sabatier reaction.

An important conclusion from this study is that the use of a lower copper loading and larger surface area support could yield a more productive catalyst for the Sabatier reaction. Although it is widely mentioned in the literature that the use of copper as an active element for the Sabatier reaction is capable of producing not only CH_4 but also larger hydrocarbon molecules and some alcohol products.

Finally, for the CO_2 methanation Ni based catalysts results to be very effective and therefore are the most efficient and active catalytic system together with alumina. Even when Ru catalysts, or other precious metals, present better performances, Ru is about 120 times more expensive than Ni. Nickel catalysts have a short lifetime, because of carbon deposition which blocks pores and consequently deactivates the catalyst. A rational design of Ni-based methanation catalysts with high catalytic performance at low temperatures, good redox properties, and better stability at reaction temperatures could lead to a better option for industrial applications of CO_2 hydrogenation to methane. Knowing and understanding the stages that control the synthesis of a catalyst is crucial for its good design. This is the idea that has been tried to convey in this review work and that will be necessary to continue investigating in greater depth.

Author Contributions: All authors contributed to researching data for the article and writing and reviewing/editing the manuscript before submission. All authors have read and agreed to the published version of the manuscript.

Funding: A.G. is grateful for financial support from the Spanish Ministry of Science and Innovation (MCIN/AEI/10.13039/501100011033) through project PID2020-112656RB-C21.

Institutional Review Board Statement: Not applicable.

Informed Consent Statement: Not applicable.

Data Availability Statement: Not applicable.

Acknowledgments: A.G. is grateful for financial support from the Spanish Ministry of Science and Innovation (MCIN/AEI/10.13039/501100011033) through project PID2020-112656RB-C21. The authors are grateful to Micromeritics Instrument Corporation for the support.

Conflicts of Interest: The authors declare no competing interests.

References

1. Bradford, M.C.J.; Vannice, M.A. CO_2 reforming of CH_4. *Catal. Rev. Sci. Eng.* **1999**, *41*, 1–42. [CrossRef]
2. Darensbourg, D.J. Making plastics from carbon dioxide: Salen metal complexes as catalysts for the production of polycarbonates from epoxides and CO_2. *Chem. Rev.* **2007**, *107*, 2388–2410. [CrossRef] [PubMed]
3. Sakakura, T.; Choi, J.-C.; Yasuda, H. Transformation of carbon dioxide. *Chem. Rev.* **2007**, *107*, 2365–2387. [CrossRef] [PubMed]
4. Wang, W.-H.; Himeda, Y.; Muckerman, J.T.; Manbeck, G.F.; Fujita, E. CO_2 hydrogentaion to formate and methanol as an alternative to photo- and electrochemical CO_2 reduction. *Chem. Rev.* **2013**, *115*, 12936–12973. [CrossRef] [PubMed]
5. Aresta, M.; Dibenedetto, A.; Angelini, A. Catalysis for the valorization of exhaust carbon: From CO_2 to chemicals, materials, and fuels. Technological use of CO_2. *Chem. Rev.* **2014**, *114*, 1709–1742. [CrossRef]
6. Jalama, K. Carbon dioxide hydrogenation over nickel-, ruthenium-, and copper-based catalysts: Review of kinetics and mechanism. *Catal. Rev. Sci. Eng.* **2017**, *59*, 95–164. [CrossRef]
7. Bulushev, D.A.; Ross, J.R.H. Heterogeneous catalysts for hydrogenation of CO_2 and bicarbonates to formic acid and formates. *Catal. Rev. Sci. Eng.* **2018**, *60*, 566–593. [CrossRef]
8. Akhundi, A.; Habibi-Yangjeh, A.; Abitorabi, M.; Rahim-Pouran, S. Review on photocatalytic conversion of carbon dioxide to value-added compounds and renewable fuels by graphitic carbon nitride-based photocatalysts. *Catal. Rev. Sci. Eng.* **2019**, *61*, 595–628. [CrossRef]
9. Jiang, X.; Nie, X.; Guo, X.; Song, C.; Chen, J.G. Recent advances in carbon dioxide hydrogenation to methanol via heterogenous catalysis. *Chem. Rev.* **2020**, *120*, 7984–8034. [CrossRef]
10. Saad, D.M.; Bilbeisi, R.A.; Alnouri, S.Y. Optimizing network pathways of CO_2 conversion processes. *J. CO2 Util.* **2021**, *45*, 101433. [CrossRef]
11. Saini, S.; Prajapati, P.K.; Jain, S.L. Transition metal-catalyzed carboxylation of olefins with Carbon dioxide: A comprehensive review. *Catal. Rev. Sci. Eng.* **2022**, *64*, 631–677. [CrossRef]
12. Latsiou, A.I.; Charisiou, N.D.; Frontistis, Z.; Bansode, A.; Goula, M.A. CO_2 hydrogenation for the production of higher alcohols: Trends in catalyst developments, Challenges and opportunities. *Catal. Today* **2023**, *420*, 114179. [CrossRef]
13. Sancho-Sanz, I.; Korili, S.A.; Gil, A. Catalytic valorization of CO_2 by hydrogenation: Current status and future trends. *Catal. Rev. Sci. Eng.* **2023**, *65*, 698–772. [CrossRef]
14. Sibi, M.G.; Verma, D.; Kim, J. Direct conversion of CO_2 into aromatics over multifunctional heterogeneous catalysts. *Catal. Rev. Sci. Eng.* **2022**; in press.
15. Xie, S.; Li, Z.; Li, H.; Fang, Y. Integration of carbon capture with heterogeneous catalysis toward methanol production: Chemistry, challenges, and opportunities. *Catal. Rev. Sci. Eng.* **2023**; in press. [CrossRef]
16. Torrez-Herrera, J.J.; Korili, S.A.; Gil, A. Recent progress in the application of Ni-based catalysts for the dry reforming of methane. *Catal. Rev. Sci. Eng.* **2021**; in press.
17. Sahu, A.K.; Zhao, X.S.; Upadhyayula, S. Ceria-based photocatalysts in water-splitting for hydrogen production and carbon dioxide reduction. *Catal. Rev. Sci. Eng.* **2023**; in press.
18. Komiyama, M. Design and preparation of impregnated catalysts. *Catal. Rev. Sci. Eng.* **2006**, *27*, 341–372. [CrossRef]
19. Bourikas, K.; Kordulis, C.; Lycourghiotis, A. The role of the liquid-solid interface in the preparation of supported catalysts. *Catal. Rev. Sci. Eng.* **2007**, *48*, 363–444. [CrossRef]
20. Jeffry, L.; Ong, M.Y.; Nomanbhay, S.; Mofijur, M.; Mubashir, M.; Show, P.L. Greenhouse gases utilization: A review. *Fuel* **2021**, *301*, 121017. [CrossRef]
21. Santamaría, L.; Korili, S.A.; Gil, A. Layered double hydroxides for CO_2 adsorption at moderate temperatures: Synthesis and amelioration strategies. *Chem. Eng. J.* **2023**, *455*, 140551. [CrossRef]
22. Mustafa, A.; Lougou, B.G.; Shuai, Y.; Wang, Z.; Tan, H. Current technology development for CO_2 utilization into solar fuels and chemicals. A review. *J. Energy Chem.* **2020**, *49*, 96–123. [CrossRef]
23. Rahman, F.A.; Aziz, M.M.A.; Saidur, R.; Abu Bakar, W.A.W.; Hainin, M.R.; Putrajaya, R.; Hassan, N.A. Pollution to solution: Capture and sequestration of carbon dioxide (CO_2) and its utilization as a renewable energy source for a sustainable future. *Renew. Sustain. Energy Rev.* **2017**, *71*, 112–126. [CrossRef]
24. Yentekakis, I.V.; Dong, F. Grand challenges for catalytic remediation in environmental and energy applications toward a cleaner and sustainable future. *Front. Environ. Chem.* **2020**, *1*, 5. [CrossRef]
25. Ye, R.-P.; Ding, J.; Gong, W.; Argyle, M.D.; Zhong, Q.; Wang, Y.; Russell, C.K.; Xu, Z.; Russell, A.G.; Li, Q.; et al. CO_2 hydrogenation to high-value products via heterogeneous catalysis. *Nat. Commun.* **2019**, *10*, 5698. [CrossRef]
26. Ashok, J.; Pati, S.; Hongmanorom, P.; Tianxi, Z.; Junmei, C.; Kawi, S. A review of recent catalyst advances in CO_2 methanation processes. *Catal. Today* **2020**, *356*, 471–489. [CrossRef]
27. Saeidi, S.; Najari, S.; Fazlollahi, F.; Nikoo, M.K.; Sefidkon, F.; Klemeš, J.J.; Baxter, L.L. Mechanisms and kinetics of CO_2 hydrogenation to value-added products: A detailed review on current status and future trends. *Renew. Sustain. Energy Rev.* **2017**, *80*, 1292–1311. [CrossRef]
28. Mucsi, Z.; Chass, G.A.; Viskolcz, B.; Csizmadia, I.G. Quantitative scale for the extent of conjugation of carbonyl groups: "Carbonylicity" percentage as a chemical driving force. *J. Phys. Chem. A* **2008**, *112*, 9153–9165. [CrossRef] [PubMed]
29. De, S.; Zhang, J.; Luque, R.; Yan, N. Ni-based bimetallic heterogeneous catalysts for energy and environmental applications. *Energy Environ. Sci.* **2016**, *9*, 3314–3347. [CrossRef]

30. Shen, L.; Xu, J.; Zhu, M.; Han, Y.-F. Essential role of the support for nickel-based CO_2 methanation catalysts. *ACS Catal.* **2020**, *10*, 14581–14591. [CrossRef]
31. Le, T.A.; Kang, J.K.; Park, E.D. CO and CO_2 methanation over Ni/SiC and Ni/SiO_2 catalysts. *Top. Catal.* **2018**, *61*, 1537–1544. [CrossRef]
32. Fan, W.K.; Tahir, M. Recent trends in developments of active metals and heterogenous materials for catalytic CO_2 hydrogenation to renewable methane: A review. *J. Environ. Chem. Eng.* **2021**, *9*, 105460. [CrossRef]
33. Mihet, M.; Dan, M.; Barbu-Tudoran, L.; Lazar, M.D. CO_2 Methanation using multimodal Ni/SiO_2 catalysts: Effect of support modification by MgO, CeO_2, and La_2O_3. *Catalysts* **2021**, *11*, 443. [CrossRef]
34. Rahat, J.; Tetsuya, N. Effect of texture and physical properties of catalysts on ammonia synthesis. *Catal. Today* **2022**, *397*, 592–597.
35. Rahat, J.; Tetsuya, N. Effect of preparation method and reaction parameters on catalytic activity for ammonia synthesis. *Int. J. Hydrogen Energy* **2021**, *46*, 35209–35218.
36. Cauqui, M.A.; Rodriguez-Izquierdo, J.M. Application of the sol-gel methods to catalyst preparation. *J. Non-Cryst. Solids* **1992**, *147–148*, 724–738. [CrossRef]
37. Pinna, F. Supported metal catalysts preparation. *Catal. Today* **1998**, *41*, 129–137. [CrossRef]
38. Campanati, M.; Fornasari, G.; Vaccari, A. Fundamentals in the preparation of heterogeneous catalysts. *Catal. Today* **2003**, *77*, 299–314. [CrossRef]
39. Serp, P.; Kalck, P.; Feurer, R. Chemical vapor deposition methods for the controlled preparation of supported catalytic materials. *Chem. Rev.* **2002**, *192*, 3085–3128. [CrossRef]
40. Kumar, V.A.; Nagaraja, B.M.; Shashikala, V.; Padmasri, A.H.; Madhavendra, S.S.; Raju, B.D.; Rao, K.S.R. Highly efficient Ag/C catalyst prepared by electro-chemical deposition method in controlling microorganisms in water. *J. Mol. Catal. A Chem.* **2004**, *223*, 313–319. [CrossRef]
41. Gil, A.; García, A.M.; Fernández, M.; Vicente, M.A.; González-Rodríguez, B.; Rives, V.; Korili, S.A. Effect of dopants on the structure of titanium oxide used as a photocatalyst for the removal of emergent contaminants. *J. Ind. Eng. Chem.* **2017**, *53*, 183–191. [CrossRef]
42. Siegbahn, K. Electron spectroscopy for atoms, molecules and condensed matter—An overview. *J. Electron Spectrosc. Relat. Phenom.* **1985**, *36*, 113–129. [CrossRef]
43. Amiri, M.J.; Roohi, R.; Gil, A. Numerical simulation of Cd(II) removal by ostrich bone ash supported nanoscale zero-valent iron in a fixed-bed column system: Utilization of unsteady advection-dispersion-adsorption equation. *J. Water Process. Eng.* **2018**, *25*, 1–14. [CrossRef]
44. Niehus, H.; Bauer, E. Quantitative aspects of ion scattering spectroscopy (ISS). *Surf. Sci.* **1975**, *47*, 222–233. [CrossRef]
45. Aono, M.; Souda, R.; Oshima, C.; Ishizawa, Y. Structure analysis of the Si(111)7 × 7 surface by low-energy ion scattering. *Phys. Rev. Lett.* **1983**, *50*, 1293. [CrossRef]
46. Soares, G.V.; Bastos, K.P.; Pezzi, R.P.; Miotti, L.; Driemeier, C.; Baumvol, I.J.R.; Hinkle, C.; Lucovsky, G. Nitrogen bonding, stability, and transport in AlON films on Si. *Appl. Phys. Lett.* **2004**, *84*, 4992–4994. [CrossRef]
47. Egerton, R.F. An Introduction to TEM, SEM, and AEM. In *Physical Principles of Electron Microscopy*; Springer: Berlin/Heidelberg, Germany, 2016.
48. Gil, A.; Fernández, M.; Mendizábal, I.; Korili, S.A.; Soto-Armañanzas, J.; Crespo-Durante, A.; Gómez-Polo, C. Fabrication of TiO_2 coated metallic wires by the sol–gel technique as a humidity sensor. *Ceram. Int.* **2016**, *42*, 9292–9298. [CrossRef]
49. Cardona, Y.; Korili, S.A.; Gil, A. A nonconventional aluminum source in the production of alumina-pillared clays for the removal of organic pollutants by adsorption. *Chem. Eng. J.* **2021**, *425*, 130708. [CrossRef]
50. Yanyan, C.; Caineng, Z.; Mastalerz, M.; Suyun, H.; Gasaway, C.; Xiaowan, T. Applications of micro-fourier transform infrared spectroscopy (FTIR) in the geological sciences-A review. *Int. J. Mol. Sci.* **2015**, *16*, 30223–30250.
51. De la Puente, G.; Centeno, A.; Gil, A.; Grange, P. Interactions between molybdenum and activated carbons on the preparation of activated carbon-supported molybdenum catalysts. *J. Colloid Interface Sci.* **1998**, *202*, 155–166. [CrossRef]
52. Greenwood, N.N.; Gibb, T.C. *Mössbauer Spectroscopy*; University of Leeds: Leeds, UK, 1971.
53. Gil, A.; Korili, S.A.; Trujillano, R.; Vicente, M.A. A review on characterization of pillared clays by specific techniques. *Appl. Clay Sci.* **2011**, *53*, 97–105. [CrossRef]
54. Moura, J.A.; Araujo, A.S.; Coutinho, A.C.S.L.S.; Aquino, J.M.F.B.; Silva, A.O.S.; Souza, M.J.B. Thermal analysis applied to characterization of copper and nickel catalysts. *J. Therm. Anal. Calorim.* **2005**, *79*, 435–438. [CrossRef]
55. Tursunov, O.; Zubek, K.; Czerski, G.; Dobrowolski, J. Studies of CO_2 gasification of the Miscanthus giganteus biomass over Ni/Al_2O_3-SiO_2 and Ni/Al_2O_3-SiO_2 with K_2O promoter as catalysts. *J. Therm. Anal. Calorim.* **2020**, *139*, 3481–3492. [CrossRef]
56. Jiménez, A.; Misol, A.; Morato, A.; Rives, V.; Vicente, M.A.; Gil, A. Optimization of hydrocalumite preparation under microwave irradiation for recovering aluminium from a saline slag. *Appl. Clay Sci.* **2021**, *212*, 106217. [CrossRef]
57. Vizcaíno, A.J.; Carrero, A.; Calles, J.A. Hydrogen production by ethanol steam reforming over Cu-Ni supported catalysts. *Int. J. Hydrogen Energy* **2007**, *32*, 1450–1461. [CrossRef]
58. Jiménez, A.; Misol, A.; Morato, A.; Rives, V.; Vicente, M.A.; Gil, A. Synthesis of pollucite and analcime zeolites by recovering aluminum from a saline slag. *J. Clean. Prod.* **2021**, *297*, 126667. [CrossRef]
59. Webb, P.A.; Orr, C. *Analytical Method in Fine Particle Technology*; Micromeritics Instrument Corporation: Norcross, GA, USA, 1995.
60. Gregg, S.J.; Sing, K.S.W. *Adsorption, Surface Area and Porosity*, 2nd ed.; Academic Press: Cambridge, MA, USA, 1982.

61. Gil, A. *Análisis Textural de Solidos Porosos Mediante Adsorción Física de Gases*; Universidad Pública de Navarra: Pamplona, Spain, 2020.
62. Bond, G.C. *Heterogeneous Catalysts: Principles and Applications*; Clarendon Press: Oxford, UK, 1987.
63. Baiker, A. Experimental methods for the characterization of catalysts. I. Gas adsorption methods, pycnometry and porosimetry. *Int. Chem. Eng.* **1985**, *25*, 16.
64. Delannay, F. *Characterization of Heterogeneous Catalysts*; Marcel Dekker: New York, NY, USA, 1984.
65. Yunes, S. Characterization and Catalytic Tests for Laterite and Bauxite. Ph.D. Thesis, Université Catholique de Louvain, Louvain-la-Neuce, Belgium, 1984.
66. Somorjai, G.A. *Introduction to Surface Chemistry and Catalysis*; John Wiley & Sons, Inc.: Hoboken, NJ, USA, 1994.
67. Langmuir, I. The constitution and fundamental properties of solids and liquids. Part I. Solids and liquids. *J. Am. Chem. Soc.* **1916**, *38*, 2221–2295. [CrossRef]
68. Langmuir, I. The adsorption of gases on plane surfaces of glass, mica and platinum. *J. Am. Chem. Soc.* **1918**, *40*, 1361–1403. [CrossRef]
69. Brunauer, S.; Emmett, P.H.; Teller, E. Adsorption of gases in multimolecular layers. *J. Am. Chem. Soc.* **1938**, *60*, 309–319. [CrossRef]
70. Lippens, B.C.; de Boer, J.H. Studies on pore systems in catalysts: V. The t method. *J. Catal.* **1965**, *4*, 319–323. [CrossRef]
71. Lippens, B.C.; de Boer, J.H. Studies on pore systems in catalysts I. The adsorption of nitrogen; apparatus and calculation. *J. Catal.* **1964**, *3*, 32–37. [CrossRef]
72. Pierce, C.; Ewings, B. Areas of uniform graphite surfaces. *J. Phys. Chem.* **1964**, *68*, 2562–2568. [CrossRef]
73. Pierce, C. Localized adsorption on graphite and absolute surface areas. *J. Phys. Chem.* **1969**, *73*, 813–817. [CrossRef]
74. Davis, B.W.; Varsanik, R.G. A study of crystalline titanium carbide. *J. Colloid Interface Sci.* **1971**, *37*, 870–878. [CrossRef]
75. Gurvitsch, L. Physicochemical attractive force. *J. Phys. Chem. Soc. Russ.* **1915**, *47*, 805–827.
76. Kissinger, H.E. Reaction kinetics in differential thermal analysis. *Anal. Chem.* **1957**, *29*, 1702–1706. [CrossRef]
77. Hofmann, A.W. Researches into the molecular constitution of the organic bases. *Philos. Trans. R. Soc. Lond.* **1851**, *141*, 357–398.
78. Hofmann, A.W. Übergang der flüchtigen basen in eine reihe nichtflüchtige alkaloide. *Ann. Chem. Pharms.* **1851**, *78*, 253–286.
79. Hofmann, A.W. Continuation of Übergang der flüchtigen basen in eine reihe nichtflüchtige alkaloide. *Ann. Chem. Pharms.* **1851**, *79*, 11–39. [CrossRef]
80. Wachs, I.E. Number of surface sites and turnover frequencies for oxide catalysts. *J. Catal.* **2022**, *405*, 462–472. [CrossRef]
81. Washburn, E.W. Note on a method of determining the distribution of pore sizes in a porous material. *Proc. Natl. Acad. Sci. USA* **1921**, *7*, 115–116. [CrossRef]
82. Gil, A. Classical and new insights into the methodology for characterizing adsorbents and metal catalysts by chemical adsorption. *Catal. Today* **2023**, *423*, 114016. [CrossRef]
83. Sabatier, P.; Senderens, J. Direct hydrogenation of oxides of carbon in presence of various finely divided metals. *CR Acad Sci.* **1902**, *134*, 689–691.

Disclaimer/Publisher's Note: The statements, opinions and data contained in all publications are solely those of the individual author(s) and contributor(s) and not of MDPI and/or the editor(s). MDPI and/or the editor(s) disclaim responsibility for any injury to people or property resulting from any ideas, methods, instructions or products referred to in the content.

Article

Performance of Thermal-, Acid-, and Mechanochemical-Activated Montmorillonite for Environmental Protection from Radionuclides U(VI) and Sr(II)

Iryna Kovalchuk [1,2]

[1] Institute for Sorption and Problems of Endoecology, National Academy of Science of Ukraine, 03164 Kyiv, Ukraine; kowalchukiryna@gmail.com
[2] Department of Chemical Technology of Ceramics and Glass, National Technical University of Ukraine "Igor Sikorsky Kyiv Polytechnic Institute", 03056 Kyiv, Ukraine

Abstract: Low-cost sorption materials based on the clay mineral of the smectite group—montmorillonite—were used for the removal of radionuclides uranium (VI) and strontium (II) from contaminated water. A wide range of industrial methods such as thermal treatment, acid activation, and mechanochemical activation were applied. Complex methods, such as SEM microscopy analysis, X-ray powder diffraction (XRD), thermal analysis, and nitrogen adsorption–desorption at −196 °C, were used to assess the characteristics of the structure of the obtained materials. The thermal treatment, acid activation, and mechanochemical activation resulted in changes in the surface properties of the clay minerals: specific surface area, porosity, and distribution of active sites. It was established that the mechanochemical activation of montmorillonite significantly increases the sorption characteristics of the material for U(VI) and Sr(II) and the acid activation of montmorillonite increases it for U(VI). The appropriateness of the experimental adsorption values for U(VI) and Sr(II) on modified montmorillonite to Langmuir and Freundlich models was found. Independently of the changes induced by acid attack, calcinations, or milling, the sorption of U(VI) and Sr(II) ions on treated montmorillonite occurs on a homogeneous surface through monolayer adsorption in a similar fashion to natural montmorillonite. Water purification technologies and modern environmental protection technologies may successfully use the obtained clay-based sorbents.

Keywords: montmorillonite; clay treatment; radionuclides; adsorption; water removal

1. Introduction

The uranium mining and uranium processing industry leads to the pollution of environmental uranium compounds [1]. Radioactive waste disposal sites form from the blocked waters of the radioactive sites of nuclear power plants, and nuclear power plant accidents are a potential source of radionuclides in water areas [2]. The decommissioning of several industrial enterprises and the formation of a significant amount of dangerous pollutants has taken place as a result of the war in Ukraine. Military weapon strikes damage the sites of extraction, processing, storage, and disposal of radioactive substances.

In addition, damage caused by military weapons leads to the release of toxic and radioactive substances into the environment as a result of explosions. Radioactive elements enter drinking water sources and they are a threat to public health. According to the World Health Organization, the maximum allowable limit in water for uranium consists of 0.03 mg/L, and for strontium-90, it consists of 2.0 Bk/L. The total beta activity for drinking water consists of ≤1.0 Bk/L [3]. Therefore, the high level of inorganic pollutants in the water is a significant ecological problem and needs a practical solution for sphere environmental protection.

Taking into account the very large areas with contaminated water and the huge volumes of waste generated during the mining and processing of radioactive ores, it becomes especially important to search for the cheapest reagents and materials used in appropriate environmental protection technologies [4,5].

Among the different methods for the purification of contaminated waters, sorption has an advantage because the selective removal of small amounts of toxicants can occur. Today, a wide variety of natural and synthetic sorbents are offered for water purification [6,7]. However, the feasibility of using natural clay minerals is due to their availability and low cost.

Clay mineral montmorillonite is the main representative of the smectite group. The base structural unit of montmorillonite comprises two tetrahedrally coordinated sheets of silicon ions surrounding octahedrally coordinated sheets of aluminum ions (2:1 type). As a result of the isomorphous substitution of Si^{4+} for Al^{3+} in the tetrahedral layer and Al^{3+} for Mg^{2+} in the octahedral layer, a net negative surface charge on the clay occurs [8]. These structure characteristics conduct the excellent sorption properties of montmorillonite, which possesses available sorption sites at the basal surface, edge sites, and within its interlayer space [9–11]. For the removal of Co, Cu, Zn, Cd, and U, montmorillonite was used [12,13]. The large adsorption capacity of montmorillonite for these metals was proved. Uranium (VI) sorption onto bentonite colloids in carbonate-containing groundwater was investigated [14,15]. The removal of Sr^{2+} ions from contaminated water was investigated [16].

The physical and chemical modification of the surface of clay minerals makes it possible to significantly improve their sorption and technological characteristics. Various methods of physical and chemical surface modification to increase the sorption capacity of layered silicates with the aim of their further use in industries have been widely used [17,18]. The most common are acid [19], mechanochemical activation [20], hydrothermal treatment, thermal activation [21], surface modification with organic substances [22], obtaining pillared clay minerals [23], surface modification with nano-dispersed metal powders [24,25], and chemical grafting on the surface of complex compounds [26], among other methods [27,28].

In industrial applications using clays as sorbent materials, one of the simplest technological operations to increase their surface activity is acid activation. At the same time, the cheap mineral acids HCl and H_2SO_4 are used as reagents [19,29]. An important result of the acid treatment of clay minerals (from the point of view of their further use in catalysis and sorption technologies) is the development of their meso- and microporous structure [19]. As a result of treatment with acids, first of all, the octahedral networks of layered silicates are destroyed. The final product, under harsh processing conditions (high acid concentration, processing temperature) and the sufficient duration of the process, is always highly porous amorphous silica [19,30]. The effect of the acid treatment on the clay structure is that it opens up the edges of the platelets. Therefore, the diameter of the pores and the surface area increases. The acid-modified clays have been found particularly useful for the adsorption of As, Cd, Cr, Co, Cu, Fe, Pb, Mn, Ni, and Zn in their ionic forms from an aqueous medium [10,31,32].

Fine dispersion is one of the main technological operations in the silicate industry and in the processing of mineral raw materials. The development and application of energy-intensive units (disintegrators, centrifugal-planetary mills, etc.) in the process of grinding solids not only ensures the achievement of the high dispersion of the final products but also, to a large extent, influences the concentration and nature of surface defects [33–35]. This circumstance provides broad prospects for the application of mechanochemical activation in catalyst technology and sorption processes [36,37]. The mechanochemical treatment appeared to decrease the structure order and increase the specific surface area of the clays. These phenomena: fragmentation, abrasion, and the amorphization of the particles lead to a higher cation exchange capacity and thus to an improved sorption capability [38]. Mechanochemically treated montmorillonite and kaolinite proved to be effective sorbents for the removal of heavy metals [39].

Along with milling processes, thermal treatment at different temperatures (drying, dehydroxylation, annealing) is one of the key operations in silicate technology. Drying processes usually include the loss of mechanically trapped, capillary, and adsorption-bound water and occur in a range of up to 200 °C. The dehydroxylation of clay minerals, which is accompanied by a partial restructuring of the structure, takes place in the temperature range of 400–700 °C. During annealing, a deep restructuring of the mineral structure occurs with the formation of new amorphous and crystalline phases [21,40]. The thermal dehydration of montmorillonite is accompanied by a decrease in its interplanar distance, as well as a partial decrease in its cation-exchange capacity. Water, removed from the structure of montmorillonite when heated, frees the pore space, thereby increasing the textural characteristics of the mineral [41]. The calcination of montmorillonite leads to the loss of the ability of the mineral structure to interlayer swelling and, thus, a decrease in the size of the active surface that participates in ion exchange. As a result, the amount of sorbed cobalt is decreased [42].

The applications of thermally modified and acid-activated clay minerals for environmental protection were demonstrated [43]. Contaminants such as pesticides, heavy metals, dyes, benzene, toluene, ethyl-benzene, xylenes, and methyl tertiary butyl ether were removed to a highly efficient degree.

The objective of the present study is the removal of U(VI) and Sr(II) from contaminated water by the sorption on thermal-, acid-, and mechanochemical-treated montmorillonite. X-ray diffraction, scanning electron microscopy, and low-temperature N_2 adsorption were used to explain the structure and surface morphology of the sorbents after the treatment of the montmorillonite. The mechanism of the sorption removal of these radionuclides from water was found.

2. Materials and Methods

2.1. Materials and Chemicals

Montmorillonite (MMT) is the mineral of the smectite group (2:1 type). MMT was taken from the Cherkasy bentonite deposit (Dashukivka site), PJSC "Dashukiv Bentonites", Lysyanka city, Cherkasy region, Ukraine. Its structural formula is $(Ca_{0.12}Na_{0.03}K_{0.03})_{0.18}(Al_{1.39}Mg_{0.13}Fe_{0.44})_{1.96}(Si_{3.88}Al_{0.12})_{4.0}O_{10}(OH)_2 \cdot nH_2O$ and the cation exchange capacity is 1.0 mmol/g.

Sedimentation from aqueous clay suspension and subsequent centrifugation of montmorillonite was carried out to purify coarse mineral impurities (quartz, feldspars, carbonates, aluminum, iron oxides, etc.). The obtained material was dried at 105 °C, and ground. The fraction < 0.1 mm was used for the following modification.

Hydrochloric acid (HCl), nitric acid (HNO_3), sodium hydroxide (NaOH), Arsenazo III and salts of sodium chloride (NaCl), uranyl sulfate ($UO_2SO_4 \cdot 3H_2O$), and strontium chloride ($SrCl_2 \cdot 6H_2O$) were obtained from Sigma–Aldrich, Burlington, MA, USA. All the chemicals were of analytical reagent grade and used without further purification. Distilled water was used for all experiments for the preparation of various concentrations of solutions.

2.2. Preparation of the Samples of Treated Montmorillonite

The natural montmorillonite has been treated in thermal, acid, and mechanochemical ways demonstrated in Figure 1.

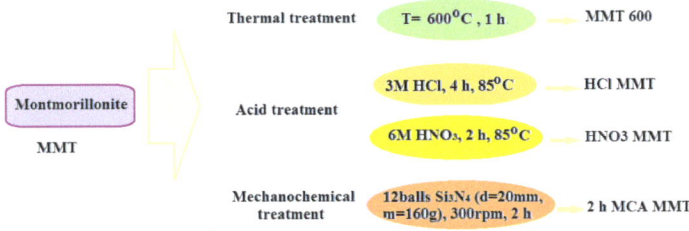

Figure 1. Diagram of the experiment of the thermal, acid, and mechanochemical montmorillonite treatment.

Thermal treatment of samples of natural MMT consisted of calcining the samples in the air in a muffle furnace SNOL 8.2/1100 (SnolTherm, UAB, Narkūnai, Lithuania) at 600 °C for 1 h.

Acid activation of the surface of MMT was carried out in solutions of 3 M and 6 M nitric (HNO_3)/hydrochloric (HCl) acids at 2 and 4 h without heating/at 85 °C with constant stirring using a magnetic stirrer. The clay mineral acid ratio was 1:4. After the synthesis, the samples were washed with distilled water from acid to negative reaction onto chlorides/nitrates.

Mechanochemical activation of MMT was carried out using planetary ball mill Pulverisette-6 (Fritsch, Idar-Oberstein, Germany) with a rotation speed of 300 rpm. The diameter of the Si_3N_4 balls was 20 mm, and the total weight was 160 g. The mass ratio of the balls/sample consists of about 10. Milling was performed in 10 min cycles; subsequently, a reverse was carried out after each cycle. The duration of milling was 2 h.

2.3. Characterization of Samples

Thermogravimetric analysis was performed on F. Paulik, J. Paulic, L. Erdey derivatograph (Hungary) in a temperature range of 20–1000 °C at a heating rate of 10 min^{-1}.

X-ray powder diffraction (XRD) of the natural clay mineral and thermal-, acid-, and mechanochemically activated MMT were recorded on DRON-4-07 diffractometer (SPE Burevestnik, Nizhniy Novgorod, Russia) under the following conditions: range of 2–60° (2θ) using CuKα radiation, λ = 0.154 nm, and were acquired at constant pass energy of U = 30 kV, I = 30 mA [44].

The surface morphology of montmorillonite samples was investigated using SEM method on Jeol JSM-6060 (Tokyo, Japan) scanning electron microscope in secondary electron mode with an accelerating voltage (AV) of 30 kV. The samples were sputter coated with gold for 3 min.

Low-temperature nitrogen adsorption isotherms were determined on a volumetric automatic apparatus (Quantachrome, Nova 2200e Surface Area and Pore Size Analyzer, Boynton Beach, FL, USA) at −196 °C. Parameters of the porous structure of montmorillonite samples such as specific surface area (S_{BET}, m^2/g), total pore volume (V, cm^3/g), average pore radius (r, nm), and distribution of pore sizes of the natural and treated montmorillonite were determined using BJH and DFT methods [45].

2.4. Adsorption Experiments

The solutions prepared from salts of uranyl sulfate ($UO_2SO_4 \cdot 3H_2O$) and strontium chloride ($SrCl_2 \cdot 6H_2O$) were used in the sorption experiments for the removing of uranium (VI) and strontium (II) ions from water. The 0.01 M ionic strength was created with a solution of NaCl. Solutions of 0.1 M NaOH and HCl were used for pH adjustment on the Ionomer -160M. Sorption isotherms were obtained at pH 6.

The batch sorption experiments (the mass of sorbents 0.1 g; the volume of the aqueous phase 50 mL) were carried out in a thermostated cell (Environmental shaker ES-20, SIA Biosan, Riga, Latvia) during 1 h continuous shaking at 25 °C.

After establishing the adsorption equilibrium, the separation of the liquid phase from the solid one occurred following centrifugation at 6000 rpm for 30 min. The equilibrium uranium (VI) concentration was determined using spectrophotometrical method on instrument UNICO 2100UV (United Products and Instruments, Suite E Dayton, NJ, USA) with Arsenazo III as reagent at a wavelength of 665 nm. The atomic adsorption spectroscopy method on the AA-6300 Shimadzu, Shimadzu Corporation, Tokyo, Japan, instrument was used for Sr(II) determination of the solution. The sorption U(VI) and Sr(II)) q, µmol/g, was calculated by Equation (1):

$$q = (C_{in} - C_{eq}) \times V/m, \quad (1)$$

where C_{in}, C_{eq}—the initial and equilibrium concentration of the metal, µmol/L; V—a volume of solution, L; m—a mass of the sample of sorbent, g. All experiments were undertaken in triplicate. The processing of the experimental isotherms of radionuclide adsorption was carried out using the Langmuir mathematical model for a homogeneous surface [46] and the Freundlich model for a heterogeneous surface [47].

3. Results

The sample of thermal-treated MMT (MMT 600) was used for comparison with the ones obtained by another method of treatment [42]. The acid-activated samples (HNO_3 MMT) (6 M HNO_3, 2 h, 85 °C) and (HCl MMT) (3 M HCl, 4 h, 85 °C) demonstrated better surface and sorption characteristics [48]. The sample with better surface and sorption characteristics was obtained after mechanochemical activation of 2 h [48].

The number of water molecules in the interlayer of MMT decreases at temperatures from 25 °C to 200 °C, the interlayer spacing decreases, the interlayers become highly ordered, and the crystal structure becomes more ordered [40]. The interlayer structure of the MMT collapses at 500–700 °C. The decomposition of the MMT unstable components (silica-alumina oxide crystals) leads to the collapse of the crystal layer (Figure 2).

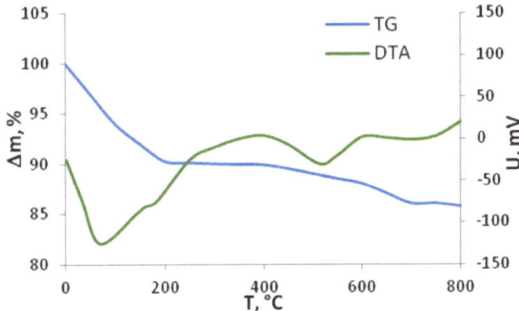

Figure 2. Thermogravimetry (TG) of natural montmorillonite.

Figure 3 shows the comparison of the X-ray diffraction patterns of the MMT, thermal-treated montmorillonite (MMT 600), acid-treated montmorillonite (HNO_3 MMT and HCl MMT), and mechanochemical-treated montmorillonite (2 h MCA MMT).

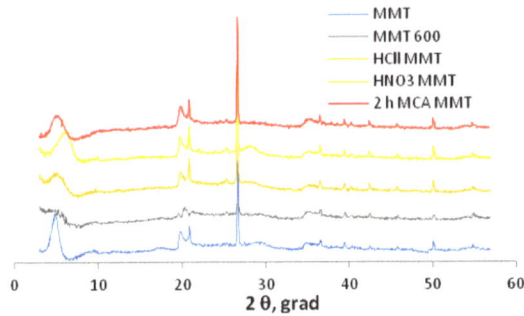

Figure 3. X-ray diffraction of natural and thermal-, acid-, and mechanochemical-treated montmorillonite.

During the thermal, acidic, and milling processes, structural changes occurred. The degree of crystallinity in the samples decreased. This was evidenced by a progressive decrease in the XRD reflection intensities: d_{001} were 1.824 nm for MMT and 1.724, 1.501, and 1.766 nm for HNO_3 MMT, HCl MMT, and 2 h MCA MMT, respectively. The broadening of the XRD reflections (Figure 3) indicated that, along with structural changes, a reduction in particle size occurred. In particular, the amorphization of structure MMT 600 occurred (Figure 3, curve 2). The mechanochemical treatment (2 h MCA MMT), similar to acid one (HNO_3 MMT, HCl MMT), led to the decrease in the intensity of the 001 peak and the shift of the 001 peak, with concomitant broadening in an asymmetrical fashion. This indicates that the stacking of the layers was disrupted and lost, and the initial structural destabilization of the basal plane began [29,34,36].

The morphology of the natural (MMT) and different treatment methods of the montmorillonite samples (MMT 600, HNO$_3$ MMT, 2 h MCA MMT) was observed by using high-resolution SEM. The morphology of the natural MMT consisted of the platy clay particles (Figure 4a). For the series of montmorillonite samples, the microstructure changed with the treatment performed (Figure 4b–d). The thermal treatment changed the structure of the MMT (Figure 4b); there was a more significant decrease in particles' size in the MMT 600 sample. The acid activation of the MMT led to the loosening of the edges of the mineral structure (HNO$_3$ MMT) and the formation of flaky edges (Figure 4c). Contrary to natural MMT, the morphology of the mechanochemically activated sample (MCA MMT 2 h) was more uniform and was dominated by spherulitic and (quasi-)regular particles. In the results of the mechanochemical treatment, flaky edges appeared (Figure 4d).

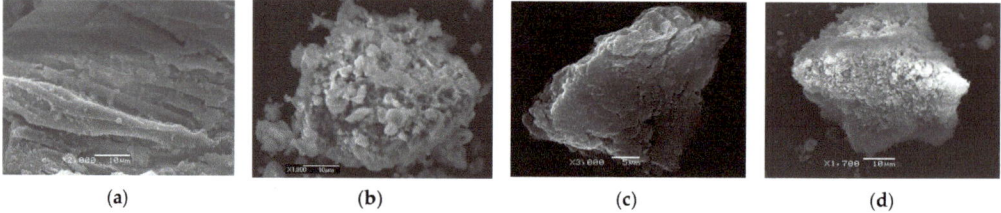

Figure 4. SEM microphotographs of the particles of natural (**a**) and thermal- (**b**), acid- (**c**), and mechanochemical- (**d**) treated montmorillonite.

The nitrogen sorption–desorption isotherm (Figure 5a) on the natural montmorillonite, according to the IUPAC classification, is attributed to type II [45]. The nitrogen adsorption curves of the MMT are similar to those on the samples of MMT 600, HNO$_3$ MMT, HCl MMT, and 2 h MCA MMT. This is typical for nonporous sorbents with a small macroporous component. The isotherms MMT, MMT 600, HNO$_3$ MMT, HCl MMT, and 2 h MCA MMT in the range of values $p/p0 > 0.4$ have the hysteresis loops of the H3 type, indicating a well-developed structure with slit-like pores [49,50].

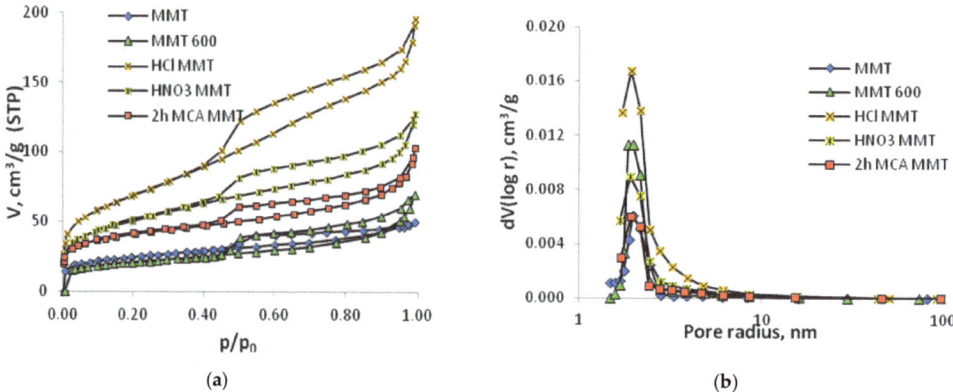

Figure 5. Nitrogen adsorption–desorption isotherms (**a**) and differential size distribution curves of the pores (BJH method) (**b**) of natural and thermal-, acid-, and mechanochemical-treated montmorillonite.

The mesopore-size distributions (Figure 5b) from the low-temperature nitrogen adsorption data were calculated using the currently used Barrett–Joyner–Halenda (BJH) method [45].

The maximum in the pore-size distribution of the natural montmorillonite was centered at 1.98 nm (Figure 5). This value slightly decreased after treatment to 1.87 ÷ 1.93 nm. The narrow peaks (r = 1.5 ÷ 3.0 nm) in the montmorillonite samples indicated narrow pore volume distribution by size. The calculations of the nitrogen isotherm-based characteristics of the porous structure presented in Table 1 indicated that the specific surface area of the samples, determined using the BET method, increased after the surface modification. The specific surface area (S_{BET}) of the MMT (89.1 m^2/g) increased after the surface modification and consisted of 179.4 and 244.0 m^2/g for HNO$_3$ MMT and HCl MMT, respectively, and 146.8 m^2/g for 2 h MCA MMT. The insignificant decreasing surface area of the MMT for the thermally treated mineral MMT 600 occurred to 75.4 m^2/g (Table 1).

Table 1. Surface characteristics of natural and thermal-, acid-, and mechanochemical-treated montmorillonite.

Sample	S, m^2/g	V, cm^3/g	r, nm	Distribution of Pore Sizes, nm				
				BJH dV(r)		DFT dV(r)		
				r_1	r_2	r_1	r_2	r_3
MMT	89.1	0.078	1.753	1.98	-	1.41	2.75	-
MMT 600	75.4	0.108	2.867	1.87	-	0.83	2.59	-
HNO$_3$ MMT	179.4	0.198	2.213	1.91	-	1.25	2.54	0.69
HCl MMT	244.0	0.304	2.500	1.93	-	0.72	2.64	-
2 h MCA MMT	146.8	0.161	2.187	1.92	-	1.13	2.64	0.67

The sorption isotherms of the uranium ions indicate a significant influence of the treatment using different methods on the sorption magnitude (Figure 6a). The obtained sorption characteristics for strontium ions in the modified samples of montmorillonite (Figure 6b) were changed to high values for the mechanochemical-activated MMT only. Langmuir and Freundlich models were used to estimate the adsorption capacity of the natural and treated montmorillonite. The obtained data are presented in Table 2.

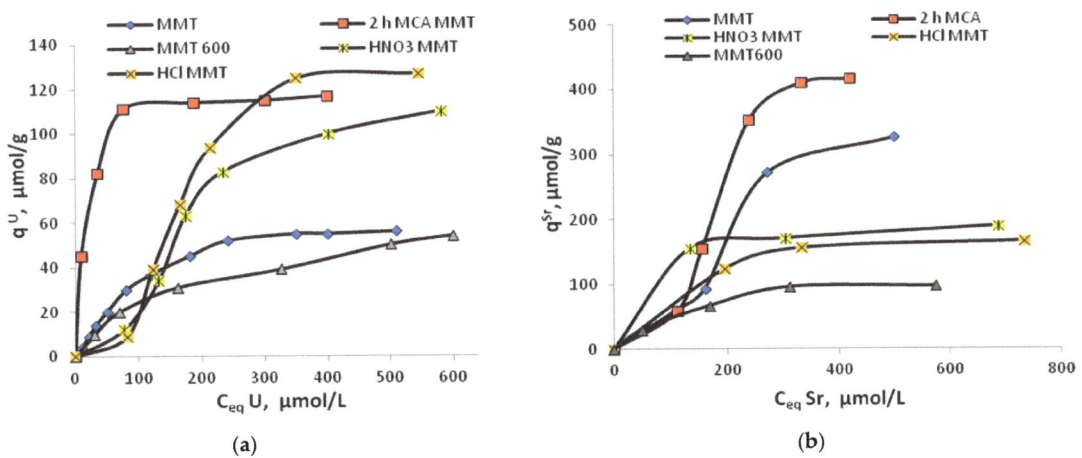

Figure 6. Isotherms of sorption of U(VI) (a) and Sr(II) (b) by the natural and thermal-, acid-, and mechanochemical-treated montmorillonite.

Table 2. Langmuir and Freundlich parameters for the adsorption of uranyl and strontium ions onto natural and thermal-, acid-, and mechanochemical-treated montmorillonite.

Me	Sample	Langmuir			Freundlich		
		q_m, μmol g^{-1}	K_L, L μmol^{-1}	R^2	n^{-1}	K_F, L μmol^{-1}	R^2
U(VI)	MMT	67.6	0.048	0.994	0.90	1.189	0.988
	MMT 600	55.0	0.032	0.997	0.57	3.911	0.992
	HCl MMT	161.3	0.033	0.973	0.75	1.040	0.829
	HNO$_3$ MMT	140.9	0.025	0.999	0.70	0.776	0.818
	2 h MCA MMT	120.5	0.349	0.999	0.25	0.692	0.844
Sr(II)	MMT	434.8	0.024	0.999	1.19	3.786	0.914
	MMT 600	163.9	0.018	0.990	0.51	0.630	0.931
	HCl MMT	185.2	0.045	0.996	0.21	0.767	0.828
	HNO$_3$ MMT	200.0	0.105	0.999	0.12	0.830	0.999
	2 h MCA MMT	555.6	0.038	0.982	1.47	8.335	0.878

q_m—the amount of adsorbate corresponding to complete monolayer coverage; K_L—Langmuir bonding energy coefficient; K_F—Frendlich coefficient of adsorption capacity; n—coefficient of adsorption intensity; R—correlation coefficient.

The Langmuir model better agrees with the experimental data, as evidenced by the higher correlation coefficient values (R^2 = 0.973 ÷ 0.999) compared with the Freundlich model (R^2 = 0.818 ÷ 0.999). The equation of monomolecular Langmuir sorption assumes the energy homogeneity of the active centers and, accordingly, the ion sorption energies close as the surface fills. The empirical Freundlich equation is suitable mainly for describing the beginning areas of isotherms. The value of K_F is associated with the interaction energy between the adsorbent and the adsorbate. The higher values of K_F indicate stronger interactions and mean a greater affinity or possible selectivity.

4. Discussion

There are two main types of active sites situated on the surface of layer silicates where the sorption of metal ions takes place. The first one is the exchange cations localized on the basal surfaces of the particles, the appearance of which is conditioned by the nonstoichiometric isomorphous substitutions in the tetrahedral and octahedral sheets of the structural layers. The second one is the Si–OH and Al–OH groups that are situated on the side edges of the particles at the place of disrupted Si–O–Si- and Al–O–Al-structured sheets.

The uranium uptake of the studied samples of montmorillonite is quite a complicated phenomenon associated with the aqueous chemistry of the elements and the nature of the materials. The chemical form of uranium in water is one of the main conditions that define its adsorption on natural minerals. In aqueous solutions, the uranium (VI) ions characteristically form a different complex at indicated pH values. In natural water, at pH near neutral values, uranium (VI) can exist in the form of a series of mono- and polynuclear hydroxo-complexes, such as UO_2^{2+}, UO_2OH^+, $(UO_2)_2(OH)_2^{2+}$, $(UO_2)_3(OH)_4^{2+}$, $(UO_2)_3(OH)_5^+$, $(UO_2)_4(OH)_7^+$, and others. In groundwater, the contents of neutral and negatively charged forms $UO_2(OH)_2$, UO_2CO_3, $(UO_2)_2CO_3(OH)_3^-$, and others are enhanced because the concentrations of dissolved CO_2 increase. The main amounts of strontium in waters are transported in the ionic form [51–53].

The interaction of uranium cations with the surface of layered silicates takes place primarily with the formation of strong surface complexes due to the Si–OH, Al–OH, and Mg–OH groups localized on the side faces of minerals. Sorption isotherms were obtained at neutral pH (pH 6), where the dissociation of the surface groups of the Si(Al)OH groups on the side faces of the mineral particles takes place. The sorption of strontium ions on layered silicates includes two different mechanism: the cation exchange in the interlayers resulting from the interactions between ions and negative permanent charge and the formation of inner-sphere complexes through Si–O– and Al–O– groups at the clay particle edges [54].

The aforementioned results confirmed that thermal, acid, and mechanochemical treatment significantly changed the structure and morphology of the particles of montmorillonite, which are demonstrated in the general schema (Figure 7).

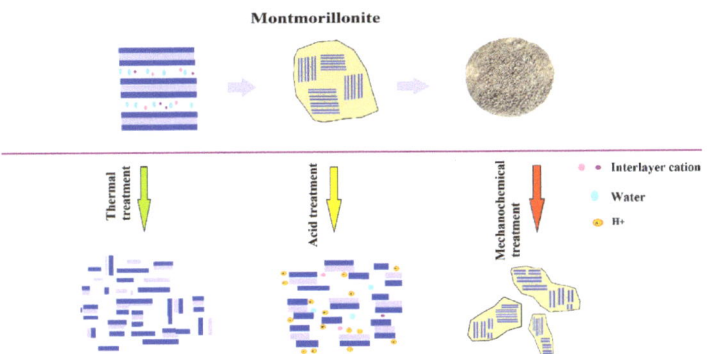

Figure 7. Schematic illustration of the effect of the different methods of montmorillonite treatment: thermal, acid, and mechanochemical.

The values of uranium sorption (Figure 6a, Table 2) correlate with the values of the specific surface area of montmorillonite for all the obtained samples (Table 1). For Sr^{2+} ions (Figure 6b, Table 2), the increase in sorption correlates with the 2 h MCA MMT sample.

During acid activation, the destruction of the crystalline structure of the silicates occurred due to the ion exchange of surface cations for acid protons. At further transformations, H^+ protons penetrated the hexagonal holes of the tetrahedral networks and attacked the structural cations of the octahedral networks of the mineral. The magnesium octahedra are the most vulnerable to proton attack, whereas the Al^{3+} and Fe^{3+} octahedra are somewhat more resistant to acids. However, they also undergo structural degradation and leaching if the process continues long enough [19]. Acid-activated montmorillonites HNO_3 MMT and HCl MMT were gradually destroyed and replaced by a crystalline structure by the amorphous phase of the silica gel, which was confirmed by the data of the X-ray phase analysis. At the same time, the porosity of the acid-activated montmorillonite changed significantly due to the partial dissolution of its oxide structure. There was an increase in the specific surface from 89.1 to 179.4 and 244.0 m^2/g and the volume of pores from 0.078 cm^3/g to 0.198 and 0.304 cm^3/g. The new mesopores are gaps between the particles of the amorphous silica gel and non-destructed aluminosilicate layers of the mineral [19,29]. The increase in the uranium sorption values on the acid-activated montmorillonite samples (161.3 and 140.9 µmol/g for HCl MMT and HNO_3 MMT samples, respectively) is due to the appearance of new sorption centers formed by proton attack. However, some negative effects for strontium were observed on acid-treated montmorillonite. The positively charged strontium ions underwent a more difficult reaction with the surface of the sorbent and so decreasing sorption was observed. The first reason for this effect is that the negative charge on the surface of the montmorillonite was partly neutralized by the acid treatment. And the next reason was as a result of the protonation of the Si–OH groups or the acceptance of protons by octahedrally coordinated Al^{3+} or that Fe^{3+} positively charged sites were generated [55].

In the case of layered silicates, due to the low rigidity of their structural frame, even with relatively small exposure energy and exposure time, significant changes occurred. The latter affected not only the specific surface of the material being crushed (increasing from 89.1 m^2/g to 146.8 m^2/g for 2 h MCA MMT) but also the character of its porous structure, which is also manifested in deeper deformations of the tetrahedral and octahedral meshes of the elementary aluminosilicate packets. In the process of mechanical activation, the porous structure of the layered silicates undergoes significant changes, with the total pore volume increasing from 0.078 to 0.161 cm^3/g. At the same time, the changes concern both the primary porosity, determined by the structure of the minerals themselves, and the secondary porosity in aggregates of clay particles [34–36]. With fine dispersion, the formation of "fluffy" side faces occurs. At the same time, the surface areas easily accessible to the radionuclides in the interlayer space of minerals near the side faces increases, which leads to an increase in their sorption values (from 67.6 to 120.5 for U(VI) and from 434.8 to 555.6 µmol/g and Sr(II) on 2 h MCA MMT) [36,56]. These improvements can be attributed to the increased surface available for adsorption, because of the decrease in the clay particle size, as well as the exfoliation of the clay mineral particles.

For the thermal-treated montmorillonite, the decrease in the sorption values of the uranyl ions from 67.6 to 55.0 µmol/g and strontium ions from 434.0 to 163.9 µmol/g correlated with the decrease in the specific surface (75.4 m^2/g). The thermal treatment of the natural MMT leads to a decrease in the mass that is connected with the removal of different forms of bound water. During thermal modification, there is a loss in the ability to interlayer the swelling of the montmorillonite structure, and, thus, a decrease in the size of the active surface, which can participate in ion exchange. The heat treatment of montmorillonite also affects the acidic nature of its surface [40]. The partial thermal removal of interlayer water increases the proton-donating capacity of its molecules. A gradual increase in temperature is accompanied by the transition of silanol groups to siloxane groups with their subsequent irreversible destruction at temperatures >600 °C [41,57]. The main part of the strontium ions, thus, binds on the thermally treated montmorillonite on the side faces of the mineral particles with the formation of strong surface complexes.

5. Conclusions

Thermal, acid, and mechanochemical activation significantly change the structural properties and sorption characteristics of montmorillonite. Various changes in these properties, such as the destruction of the crystalline phase and its transformation into amorphous, the particle size reduction, the domination of spherulitic and (quasi-)regular particles, and the increase in specific surface area and total pore volume, were observed following the treatment of the temperature, acids, and milling.

It was established that the thermal treatment of montmorillonite at 600 °C leads to a collapse followed by the dehydroxylation of the layered sheets. Chemical treatments with HNO_3 or HCl also destroy the mineral structure and the cause the dissolution of the tetrahedrally coordinated Al and Si ions. Mechanochemical activation leads to the partial amorphization of the structure too. The samples of montmorillonite with the obtained physical–chemical characteristics demonstrate high sorption properties to uranium ions. The effective removal of strontium ions can be produced by mechanoactivated montmorillonite.

Thus, sorption materials on a base of thermal-, acid-, and mechanochemical-modified montmorillonite can be used for the removal of radionuclides uranium (VI) and strontium (II) from contaminated water as effective cheap sorbents.

Funding: This research received no external funding.

Institutional Review Board Statement: Not applicable.

Informed Consent Statement: Not applicable.

Data Availability Statement: Not applicable.

Acknowledgments: The author is grateful to the staff of the Institute for Sorption and Problems of Endoecology and the Institute of General and Inorganic Chemistry of the National Academy of Sciences of Ukraine for their help and support in the experiment.

Conflicts of Interest: The author declares no conflict of interest.

References

1. Landa, E.R. Uranium mine tailings: Nuclear waste and natural laboratory for geochemical and radioecological investigations. *J. Environ. Radioact.* **2004**, *77*, 1–27. [CrossRef] [PubMed]
2. Chen, Z.; Wang, S.; Hou, H.; Chen, K.; Gao, P.; Zhang, Z.; Jin, Q.; Pan, D.; Guo, Z.; Wu, W. China's progress in radionuclide migration study over the past decade (2010–2021): Sorption, transport and radioactive colloid. *Chin. Chem. Lett.* **2022**, *33*, 3405–3412. [CrossRef]
3. World Health Organization. *Guidelines for Drinking-Water Quality: Fourth Edition Incorporating the First and Second Addenda*; World Health Organization: Geneva, Switzerland, 2022.
4. Kornilovych, B.Y.; Sorokin, O.G.; Pavlenko, V.M.; Koshyk, Y.I. *Environmental Protection Technologies in the Uranium Mining and Processing Industry*; Norma: Kyiv, Ukraine, 2011; p. 154. (In Ukrainian)
5. Chakraborty, R.; Asthana, A.; Singh, A.K.; Jain, B.; Susan, A.B.H. Adsorption of heavy metal ions by various low-cost adsorbents: A review. *Int. J. Environ. Anal. Chem.* **2020**, *102*, 342–379. [CrossRef]
6. Jun, B.-M.; Lee, H.-K.; Park, S.; Kim, T.-J. Purification of uranium-contaminated radioactive water by adsorption: A review on adsorbent materials. *Sep. Purif. Technol.* **2022**, *278*, 119675. [CrossRef]
7. Kravchenko, M.V.; Khodakovska, T.A.; Kovtun, M.F.; Romanova, I.V. Inorganic sorbents based on magnesium silicates obtained by two synthetic routes. *Environ. Earth Sci.* **2022**, *81*, 549. [CrossRef]
8. Bergaya, F.; Theng, B.K.; Lagaly, G. (Eds.) *Handbook of Clay Science*; Elsevier Ltd.: Amsterdam, The Netherlands, 2006; Volume 1.
9. Komadel, P.; Madejova, J.; Bujdak, J. Preparation and properties of reduced–charge smectites—A review. *Clays Clay Miner.* **2005**, *53*, 313–334. [CrossRef]
10. Bhattacharyya, K.G.; Gupta, S.S. Adsorption of a few heavy metals on natural and modified kaolinite and montmorillonite: A review. *Adv. Colloid Interface Sci.* **2008**, *140*, 114–131. [CrossRef]
11. Novikau, R.; Lujaniene, G. Adsorption behaviour of pollutants: Heavy metals, radionuclides, organic pollutants, on clays and their minerals (raw, modified and treated): A review. *J. Environ. Manag.* **2022**, *309*, 114685. [CrossRef]
12. Yuan, G.D.; Theng, B.K.G.; Churchman, G.J.; Gates, W.P. Clays and Clay Minerals for Pollution Control. In *Handbook of Clay Sciences*; Elsevier: Amsterdam, The Netherlands, 2013; Volume 5, pp. 587–644.
13. Misaelides, P. Clay minerals and zeolites for radioactive waste immobilization and containment. In *Modified Clay and Zeolite Nanocomposite Materials*; Elsevier: Amsterdam, The Netherlands, 2019; pp. 243–274.
14. Youssef, W.M. Uranium Adsorption from Aqueous Solution Using Sodium Bentonite Activated Clay. *J. Chem. Eng. Process Technol.* **2017**, *8*, 157–170.
15. Tran, E.L.; Teutsch, N.; Klein-BenDavid, O.; Weisbrod, N. Uranium and Cesium sorption to bentonite colloids under carbonate-rich environments: Implications for radionuclide transport. *Sci. Total Environ.* **2018**, *643*, 260–269. [CrossRef]
16. El-Kamash, A.M. Evaluation of zeolite A for the sorptive removal of Cs^+ and Sr^{2+} ions from aqueous solutions using batch and fixed bed column operations. *J. Hazard. Mater.* **2008**, *151*, 432–445. [CrossRef]
17. Tamayo, A.; Kyziol-Komosinska, J.; Sánchez, M.; Calejas, P.; Rubio, J.; Barba, M. Characterization and properties of treated smectites. *J. Eur. Ceram. Soc.* **2012**, *322*, 831–2841. [CrossRef]
18. Kovalchuk, I. Clay-Based Sorbents for Environmental Protection from Inorganic Pollutants. *Environ. Sci. Proc.* **2023**, *25*, 34. [CrossRef]
19. Komadel, P.; Madejova, J. Acid activation of clay minerals. In *Handbook of Clay Science*; Elsevier: Amsterdam, The Netherlands, 2013; Volume 5, pp. 385–409.
20. Mucsi, G. A review on mechanical activation and mechanical alloying in stirred media mill. *Chem. Eng. Res. Des.* **2019**, *148*, 460–474. [CrossRef]
21. Heller-Kallai, L. Thermally modified clay minerals. In *Handbook of Clay Sciences*; Elsevier: Amsterdam, The Netherlands, 2013; Volume 5, pp. 411–433.
22. Kovalchuk, I.A.; Laguta, A.N.; Kornilovych, B.Y.; Tobilko, V.Y. Organophilic layered silicates for sorption removal of uranium(VI) from mine water. *Chem. Phys. Surf. Technol.* **2020**, *2*, 215–227. [CrossRef]
23. Mnasri-Ghnimi, S.; Frini-Srasra, N. Removal of heavy metals from aqueous solutions by adsorption using single and mixed pillared clays. *Appl. Clay Sci.* **2019**, *179*, 105–151. [CrossRef]
24. Zhao, X.; Liu, W.; Cai, Z.; Han, B.; Qian, T.; Zhao, D. An overview of preparation and applications of stabilized zero-valent iron nanoparticles for soil and groundwater remediation. *Water Res.* **2016**, *100*, 245–266. [CrossRef]
25. Kornilovych, B.; Kovalchuk, I.; Tobilko, V.; Ubaldini, S. Uranium Removal from Groundwater and Wastewater Using Clay-Supported Nanoscale Zero-Valent Iron. *Metals* **2020**, *10*, 1421. [CrossRef]
26. Tobilko, V.Y.; Spasonova, L.M.; Kovalchuk, I.A.; Kornilovych, B.Y.; Kholodko, Y.M. Adsorption of Uranium (VI) from Aqueous Solutions by Amino-functionalized Clay Minerals. *Colloids Interfaces* **2019**, *3*, 41. [CrossRef]
27. Takahashi, C.; Shirai, T.; Fuji, M. Study on intercalation of ionic liquid into montmorillonite and its property evaluation. *Mater. Chem. Phys.* **2012**, *135*, 681–686. [CrossRef]
28. Kovalchuk, I.; Tobilko, V.; Kholodko, Y.; Zahorodniuk, N.; Kornilovych, B. Purification of mineralized waters from U(VI) compounds using bentonite/iron oxide composites. *Technol. Audit. Prod. Reserves* **2020**, *3*, 12–18. [CrossRef]
29. Önal, M.; Sarikaya, Y. Preparation and characterization of acid-activated bentonite powders. *Powder Technol.* **2007**, *172*, 14–18. [CrossRef]

30. Yassin, J.M.; Shiferaw, Y.; Tedla, A. Application of acid activated natural clays for improving quality of Niger (*Guizotia abyssinica* Cass) oil. *Heliyon* **2022**, *8*, e09241. [CrossRef] [PubMed]
31. Bhattacharyya, K.G.; Gupta, S.S. Adsorptive accumulation of Cd(II), Co(II), Cu(II), Pb(II), and Ni(II) from water on montmorillonite: Influence of acid activation. *J. Colloid Interface Sci.* **2007**, *310*, 411–424. [CrossRef]
32. Gupta, S.S.; Bhattacharyya, K.G. Adsorption of heavy metals on kaolinite and montmorillonite: A review. *Phys. Chem. Chem. Phys.* **2012**, *19*, 6698–6723. [CrossRef] [PubMed]
33. Yariv, S.; Lapides, I. The Effect of Mechanochemical Treatments on Clay Minerals and the Mechanochemical Adsorption of Organic Materials onto Clay Minerals. *J. Mater. Synth. Process.* **2000**, *8*, 223–233. [CrossRef]
34. Xia, M.; Jiang, Y.; Zhao, L.; Li, F.; Xue, B.; Sun, M.; Liu, D.; Zhang, X. Wet grinding of montmorillonite and its effect on the properties of mesoporous montmorillonite. *Colloids Surf. A Physicochem. Eng. Asp.* **2010**, *356*, 1–9. [CrossRef]
35. Vdović, N.; Jurina, I.; Škapin, S.D.; Sondi, I. The surface properties of clay minerals modified by intensive dry milling—Revisited. *Appl. Clay Sci.* **2010**, *48*, 575–580. [CrossRef]
36. Kornilovych, B.Y. *Structure and Surface Properties of Mechanochemically Activated Silicates and Carbonates*; Naukova Dumka: Kyiv, Ukraine, 1994; p. 127. (In Ukrainian)
37. Godet-Morand, L.; Chamayou, A.; Dodds, J. Talc grinding in an opposed air jet mill: Start-up, product quality and production rate optimization. *Powder Technol.* **2002**, *128*, 306–313. [CrossRef]
38. Djukić, A.; Jovanović, U.; Tuvić, T.; Andrić, V.; Novaković, J.G.; Ivanović, N.; Matović, L. The potential of ball-milled Serbian natural clay for removal of heavy metal contaminants from wastewaters: Simultaneous sorption of Ni, Cr, Cd and Pb ions. *Ceram. Int.* **2013**, *39*, 7173–7178. [CrossRef]
39. Kumrić, K.R.; Đukić, A.B.; Trtić-Petrović, T.M.; Vukelić, N.S.; Stojanović, Z.; Grbović Novaković, J.D.; Matović, L.L. Simultaneous removal of divalent heavy metals from aqueous solutions using raw and mechanochemically treated interstratified montmorillonite/kaolinite clay. *Ind. Eng. Chem. Res.* **2013**, *52*, 7930–7939. [CrossRef]
40. Hong, W.; Meng, J.; Li, C.; Yan, S.; He, X.; Fu, G. Effects of Temperature on Structural Properties of Hydrated Montmorillonite: Experimental Study and Molecular Dynamics Simulation. *Adv. Civil Eng.* **2020**, *2020*, 8885215. [CrossRef]
41. Aytas, S.; Yurtlu, M.; Donat, R. Adsorption characteristic of U(VI) ion onto thermally activated bentonite. *J. Hazard. Mater.* **2009**, *172*, 667–674. [CrossRef]
42. Tobilko, V.Y.; Kovalchuk, I.A.; Kornilovych, B.Y.; Denisova, T.I.; Kravchenko, O.V. Sorption of uranium and cobalt ions by thermally modified layered silicates. *Rep. Natl. Acad. Sci. Ukr.* **2010**, *5*, 150–155.
43. Espaca, V.A.A.; Sarkar, B.; Biswas, B.; Rusmin, R.; Naidu, R. Environmental applications of thermally modified and acid activated clay minerals: Current status of the art. *Environ. Technol. Innov.* **2019**, *13*, 383–397. [CrossRef]
44. Brindley, G. *Crystal Structures of Clay Minerals and Their X-ray Identification*; Brown, G., Ed.; The Mineralogical Society of Great Britain and Ireland: London, UK, 1980; p. 496.
45. Rouquerol, F.; Rouquerol, J.; Sing, K.S.W.; Llewellyn, P.; Maurin, G. *Adsorption by Powders and Porous Solids*; Elsevier: Amsterdam, The Netherlands, 2014.
46. Langmuir, I. The adsorption of gases on plane surfaces of glass, mica and platinum. *J. Am. Chem. Soc.* **1918**, *40*, 1361–1403. [CrossRef]
47. Freundlich, H.; Heller, W. The Adsorption of cis- and trans-Azobenzene. *J. Am. Chem. Soc.* **1939**, *61*, 2228–2230. [CrossRef]
48. Kovalchuk, I.A.; Spasyonova, L.M.; Zakutevskyi, O.I. Sorptive purification of radioactively contaminated waters by acid and mechanical activation montmorillonite. In Proceedings of the Materials of the Seventh International Conference on Nuclear Decommissioning and Environment Recovery INUDECO 22, Slavutych, Ukraine, 27–28 April 2022.
49. Sun, L.; Liang, X.; Liu, H.; Cao, H.; Liu, X.; Jin, Y.; Li, X.; Chen, S.; Wu, X. Activation of Co-O bond in (110) facet exposed Co_3O_4 by Cu doping for the boost of propane catalytic oxidation. *J. Hazard. Mater.* **2023**, *452*, 131319. [CrossRef]
50. Feng, X.; Hu, H.; Chen, D. Key nanomaterials for industrial chemical process. *Nano Res.* **2023**, *16*, 6212–6219. [CrossRef]
51. Merkel, B.J.; Hasche-Berger, A. *Uranium, Mining and Hydrogeology*; Springer: Berlin/Heidelberg, Germany, 2008; p. 955.
52. Liu, B.; Peng, T.; Hong-Juan, S.; Yue, H. Release behavior of uranium in uranium mill tailings under environmental conditions. *J. Environ. Radioact.* **2017**, *171*, 160–168. [CrossRef]
53. Atwood, D.A. (Ed.) *Radionuclides in the Environment*; Wiley: Hoboken, NJ, USA, 2010; p. 522.
54. Kraepiel, A.M.L.; Keller, K.; Morel, F.M.M. A model for metal adsorption on montmorillonite. *J. Colloid Interface Sci.* **1999**, *210*, 43–54. [CrossRef] [PubMed]
55. Pradas, E.G.; Sánchez, M.V.; Cruz, F.C.; Viciana, M.S.; Pérez, M.F. Adsorption of cadmium and zinc from aqueous solution on natural and activated bentonite. *J. Chem. Technol. Biotechnol.* **1994**, *59*, 289–295. [CrossRef]
56. Bekri-Abbes, I.; Srasra, E. Effect of mechanochemical treatment on structure and electrical properties of montmorillonite. *J. Alloys Compd.* **2016**, *671*, 34–42. [CrossRef]
57. Gao, Y.; Wang, Y.; Chen, C.; Zhou, J.; Cheng, Y.; Shi, L. Preparation of Montmorillonite Nanosheets with a High Aspect Ratio through Heating/Rehydrating and Gas-Pushing Exfoliation. *Langmuir* **2022**, *38*, 10520–10529. [CrossRef] [PubMed]

Disclaimer/Publisher's Note: The statements, opinions and data contained in all publications are solely those of the individual author(s) and contributor(s) and not of MDPI and/or the editor(s). MDPI and/or the editor(s) disclaim responsibility for any injury to people or property resulting from any ideas, methods, instructions or products referred to in the content.

Article

Implementation of Cloud Point Extraction Using Surfactants in the Recovery of Polyphenols from Apricot Cannery Waste

Ioannis Giovanoudis [1,2], Vassilis Athanasiadis [1], Theodoros Chatzimitakos [1], Dimitrios Kalompatsios [1], Eleni Bozinou [1], Olga Gortzi [2], George D. Nanos [2] and Stavros I. Lalas [1,*]

[1] Department of Food Science and Nutrition, University of Thessaly, Terma N. Temponera Street, 43100 Karditsa, Greece; gio@uth.gr (I.G.); vaathanasiadis@uth.gr (V.A.); tchatzimitakos@uth.gr (T.C.); dkalompsios@uth.gr (D.K.); empozinou@uth.gr (E.B.)

[2] Department of Agriculture Crop Production and Rural Environment, School of Agricultural Sciences, University of Thessaly, 38446 Volos, Greece; olgagortzi@uth.gr (O.G.); gnanos@uth.gr (G.D.N.)

* Correspondence: slalas@uth.gr; Tel.: +30-24-4106-4783

Abstract: The objective of this study was to investigate the feasibility of using Cloud Point Extraction (CPE) to isolate natural antioxidants (polyphenols) from apricot cannery waste (ACW). Four different food-grade surfactants (Genapol X-080, PEG 8000, Tween 80, and Lecithin) were tested at varying concentrations to evaluate the effectiveness of the technique. It was observed that low concentrations of surfactants in one-step CPE resulted in less than 65% polyphenol recovery, which necessitated further extraction steps. However, high concentrations of surfactants were found to significantly improve polyphenol extraction from ACW for all surfactants tested. Among the four surfactants, PEG 8000 was found to be the most effective in most circumstances; specifically, adding only 2% of the surfactant per step in a two-step CPE was enough to effectively extract polyphenols with recovery rates better than 99%. When 10% w/v of PEG 8000 was used, recoveries greater than 92% were obtained. Since PEG 8000 is a reagent with low toxicity and the CPE method is simple, rapid, cheap, sensitive, and selective, the extracted organic compounds from ACW can be used as natural antioxidants in food technology. This has important implications for the development of natural and sustainable food additives.

Keywords: cloud point extraction; polyphenols; surfactants; apricot cannery waste; food industry

1. Introduction

Free radical chemistry has been a subject that has attracted a lot of attention recently owing to the potential negative impact of reactive oxygen species (ROS) on both food systems and human health. Antioxidants are crucial in reducing oxidative processes and mitigating the consequences of ROS [1]. In food and pharmaceutical processing and storage, lipid peroxidation is a major cause of product degradation, but antioxidants can scavenge free radicals and extend shelf life by slowing this process [2]. In the human body, antioxidants are also beneficial, as they can help slow the progression of many chronic illnesses [3]. As a safer alternative to synthetic antioxidant compounds such as butylated hydroxytoluene (BHT) and butylated hydroxyanisole (BHA), natural antioxidants have gained increasing attention [4]. There has been an extensive awareness in studying natural additives, particularly fruit and vegetable residues, which are abundant in beneficial chemicals. The valorization of these byproducts can not only provide a sustainable solution to waste management but also produce natural antioxidants that may be included into food technology to improve the health and well-being of consumers [5].

Apricot (*Prunus armeniaca* L.) is a highly sought-after fruit in the market due to its vibrant color, unique flavor, and impressive nutritional profile. This fruit is rich in carotenoids, which are responsible for its distinctive yellow and orange peel color, making it visually appealing to consumers [6]. While the apricot originated in China, it found its way to

Europe through Armenia, leading to its scientific name [7]. Additionally, apricots are an abundant source of phenolic compounds, which are mainly found in the fruit's skin and pulp [8]. Rutin, catechin, epicatechin, and chlorogenic acid are among the dominant phenolic substances present in apricot cultivars, and their levels vary among different varieties [9]. Polyphenols derived from fruit waste are widely used as natural additives to food and preservatives due to their unique biological properties [10]. While solid/liquid (S/L) extraction is commonly used in the industry to recover polyphenols, this method has several drawbacks, including the extensive use of organic solvents, high manufacturing costs, and time-consuming processes [11,12]. Similarly, other methods such as microwave-assisted extraction, membrane processes, or supercritical fluid extraction are not suitable for bulk operations due to their high energy needs or expensive equipment [13,14]. Furthermore, conventional solvents are unable to extract both polar and non-polar bioactive compounds at the same time, making it challenging to extract all chemical constituents of the plant. Therefore, a major demand arises for a low-priced, high-throughput method to analyze apricot-derived bioactive compounds.

Novel liquid–liquid extraction (LLE) methods, such as two-phase (or multi-phase) separation, are gaining popularity for the extraction and concentration of active chemicals from natural materials. Aqueous two-phase extraction, dispersive liquid–liquid extraction, micellar extraction, and cloud-point extraction (CPE) are some of the novel approaches that have emerged [15,16]. Among them, CPE is regarded to be both an ecologically benign and biocompatible approach for extracting and concentrating active chemicals from plant-based resources, with possible applications in the food and pharmaceutical industries. Nonetheless, several drawbacks include the formation of emulsions, the requirement of hazardous organic solvents, and hence the production of vast amounts of pollutants make liquid–liquid extraction techniques laborious, costly, and ecologically unfriendly. In addition, aqueous two-phase extraction uses are restricted to laboratory environments and remain in pilot-size operations, as seen by the abundance of the literature reviews [17,18]. On the other hand, CPE is a simple and inexpensive technique for extracting bioactive chemicals from liquid matrices that employ surfactants [19]. It also gives the opportunity to use food-grade surfactants so that food industries can directly insert the extracted compounds into their products [20]. Above a certain micellar concentration, these molecules can form spontaneous aggregates (micelles) in aqueous solutions [21]. These formed structures can bind with either hydrophobic or hydrophilic compounds via dipole–dipole interactions and hydrogen bonding to be deployed for separation [22]. The micellar system characteristics and CPE factors that influence the extraction of high nutritional value compounds were investigated by Carabias-Martinez et al. [23]. The essential characteristics of a sample handled by CPE are the pH level and ionic strength, along with its temperature and the amount of surfactant used [24]. Most ionic surfactants are non-volatile and considered to be either relatively non-toxic or harmless chemicals, with the least toxicological or dermatological concerns [25]. Several surfactants, including Triton X-114, Triton X-100, and Brij 30, have been successfully employed to isolate bioactive plant components [26]. Several surfactants that might be employed in the CPE approach are found in nature. Lecithin is a natural surfactant that is widely used in the food sector. It is an inexpensive and low-toxicity compound [27]. El-Abbassi et al. [28] established an efficient and quick CPE process for extracting polyphenols from olive mill effluent utilizing Triton X-100 as the solvent for the extraction process. After one step of CPE, the recovery was 66.5%.

To the best of our knowledge, there is a scarcity of studies concerning the extraction of polyphenols from apricot cannery waste using CPE. To address the current gap in knowledge, this study aims to explore the potential of CPE using low biological hazard surfactants for polyphenol extraction from apricot cannery waste. This study investigated the efficiency of different food-grade surfactants, including Genapol X-080, PEG 8000, Tween 80, and natural surfactant Lecithin, at varying concentrations for polyphenol extraction from apricot waste. The effectiveness of CPE extraction cycles, as well as the antiradical activity of the extracted polyphenols, were evaluated for each surfactant. By conducting these experiments, the study sought to identify the optimal surfactant concentration for the efficient extraction of polyphenols from apricot waste using CPE and to determine the most effective surfactant for this purpose.

2. Materials and Methods

2.1. Chemicals, Reagents, and Materials

Methanol, 1,1-diphenyl-2-picrylhydrazyl (DPPH), and Genapol X-080 were all obtained from Sigma-Aldrich (Steinheim, Germany). Gallic acid, anhydrous sodium carbonate, and Folin–Ciocalteu reagent were purchased from Penta (Prague, Czech Republic). PEG 8000 was obtained from Alfa Aesar (Karlsruhe, Germany). Citric acid anhydrous was purchased from Merck (Darmstadt, Germany). Tween 80 was purchased from Panreac (Barcelona, Spain). Sodium chloride and soya lecithin (>97%) were purchased from Carlo Erba (Milano, Italy). To produce the deionized water used in the experiments, a deionizing column was employed.

Apricot (*Prunus armeniaca* 'Bebeco' variety) cannery wastewater (ACW) was obtained from the stream resulting from the peeling step (for the reason of being the main contributor to the total waste stream) from ELBAK S.A. (Falani, Larissa, Greece).

2.2. CPE Procedure

The CPE method was carried out with slight modification from Chatzilazarou et al. [29]. Selection of the experimental parameters (i.e., pH, temperature, etc.) was carried out based on preliminary experiments. A Remi Neya 16R (Remi Elektrotechnik Ltd., Palghar, India) was used to centrifuge 70 g ACW for 20 min at 4500 rpm, so as to remove the solids. Prior to CPE, solid-free ACW samples were adjusted to a pH value of 3.5 with 2 N citric acid [30]. To accelerate the phase separation process by increasing the bulk density of the aqueous phase, 3% w/v sodium chloride was added to the sample. Sodium chloride also decreases the cloud point temperature [31]. The concentration of surfactants tested were 2, 5, and 10% w/w. A magnetic stirrer Heidolph MR Hei-Standard was used to equilibrate the temperature and stir the samples during CPE. The samples were stirred at 800 rpm and equilibrated at 65 °C for 20 min. After centrifuging the mixture for 5 min at 3500 rpm at 30 °C (first extraction stage), the phases were separated by decanting. The surfactant-rich phase had high viscosity. The volumes of both surfactant and aqueous phases were measured after centrifugation. The unextracted polyphenols in the aqueous phase were then decanted and either extracted once (second CPE step) or twice using the same method (third CPE step). Since each CPE experiment was repeated three times under identical conditions, the recovery findings represent the means of three extraction trials.

2.3. Polyphenol Recovery by CPE

The % polyphenol recovery was measured using a polyphenol mass balance. The surfactant recovery was estimated in accordance with previous descriptions [29,32].

$$\text{Recovery } (\%) = \frac{Cs \cdot Vs}{Co \cdot Vo} \times 100 = Co \cdot Vo - \frac{Cw \cdot Vw}{Co \cdot Vo} \times 100 \qquad (1)$$

where Co is the concentration of polyphenols in the initial sample volume Vo (10 mL), Cw is the concentration of polyphenols in the water phase volume Vw, and Cs is the concentration of polyphenols in the surfactant phase volume Vs.

2.4. Total Polyphenol Content

Total polyphenols were measured photometrically using a modified Folin–Ciocalteu method by Katsoyannos et al. [33]. An amount of 100 μL of the sample was mixed with 100 μL of the Folin–Ciocalteu reagent, and after 2 min, 800 μL (5% w/v) of sodium carbonate solution was added. Finally, a Shimadzu spectrophotometer (UV-1700, Shimadzu Europa GmbH, Duisburg, Germany) was utilized to measure the absorbance of the solution at 750 nm after 20 min incubating at 40 °C in the absence of light.

2.5. Determination of Antioxidant Activity

The DPPH technique established by Tsaknis and Lalas [34] was employed to calculate the antioxidant activity of both extracted polyphenols in the surfactant phase and polyphenols that remained in the sample after CPE treatment. Briefly, 4 mL of sample was combined with 1 mL of 0.1 mM DPPH solution in methanol. The mixture was thoroughly mixed and allowed to remain in the dark for 30 min at room temperature. The absorbance was determined at 517 nm. The following equation was used to calculate the % scavenging:

$$\% \text{ Scavenging} = A_{\text{control}} - \frac{A_{\text{sample}}}{A_{\text{control}}} \times 100 \qquad (2)$$

where A_{control} and A_{sample} represent the corresponding absorbances.

2.6. Statistical Analysis

All analyses were carried out in triplicate. The results were reported as the standard deviation of the three replicate mean values. After assessing the data with the Kolmogorov–Smirnov test, statistically significant differences were investigated using the Kruskal–Wallis test. Statistically significant differences were assessed for $p < 0.05$.

3. Results and Discussion

The objective of this work was to evaluate the potential of using CPE with low biological hazard surfactants to extract polyphenols from ACW. Four different food-grade surfactants (Genapol X-080, PEG 8000, Tween 80, and Lecithin) were tested, and the recovery of polyphenol extraction was measured. Considering the findings from other studies (vide infra), we opted to increase polyphenol extraction. To ensure consistency, the CPE method was conducted at 65 °C, with a sodium chloride concentration of 3% w/v, and at a pH level of 3.5. Previous research by Kiai et al. [35] suggested that the optimum temperature for CPE is between 50 and 70 °C when using surfactants such as Genapol X-080, Tween 80, and Triton-X. Additionally, Shi et al. [36] suggested that concentrations below 20% w/v of sodium chloride would not result in effective phase separation. However, in our case, a much lower sodium chloride concentration (3% w/v) proved to be adequate for the completion of separation, possibly due to the different material used (apricot peeling waste in our case) and procedure. Furthermore, the optimum pH level was set to 3.5 because polyphenols are protonated at low pH values, and thus, they interact extensively with micellar clusters of non-ionic surfactants, making them quite soluble in the micelle [28]. In contrast, polyphenols are deprotonated at high pH values, which reduces their solubility in hydrophobic micelles [37]. The recovery of polyphenols was tested in three different concentrations (2, 5, and 10% w/w) and with three extraction steps. According to Santana et al. [30], higher polyphenol extraction yields require high concentrations of surfactants. In the first step of extraction, each surfactant at a 10% concentration had a statistically significant ($p < 0.05$) higher extraction recovery than any other concentration used. However, this pattern did not apply to the other extraction cycles. Therefore, it is important to compare the extraction yields in different steps between the surfactants. Finally, the method employed herein was a modified (based on preliminary experiments) version of the method discussed by Chatzilazarou et al. [29], who investigated the usage of surfactant Genapol X-080 in order to isolate lycopene and total carotenoids from red-fleshed orange. Authors used a 0.5–15% v/v concentration of surfactant. The optimum CPE conditions were 30 min extraction at 55 °C, 35% sodium chloride, and pH level at 2.53. The recorded recoveries in carotenoids and lycopene were both slightly above 90% after three CPE steps.

3.1. Effect of Surfactant Concentration via CPE with Genapol X-080

Genapol X-080 belongs to the class of alkylpolyethylene glycol ethers, and is a non-ionic surfactant. The chemical structure of Genapol X-080 consists of a hydrophilic polyethylene glycol (PEG) chain attached to a lipophilic alkyl chain. It is commonly used in various sectors, including the food, cosmetic, and pharmaceutical industries, as an emulsifier, solubilizer, and wetting agent [38]. Figure 1 depicts the results of Genapol X-080, which demonstrate a clear association between the amount of surfactant employed and the recovery of polyphenols. It was observed that the use of higher concentrations of Genapol X-080 led to higher polyphenol recoveries, as expected. However, the recovery rate remained below 60% in samples with low surfactant content, indicating the importance of additional CPE steps.

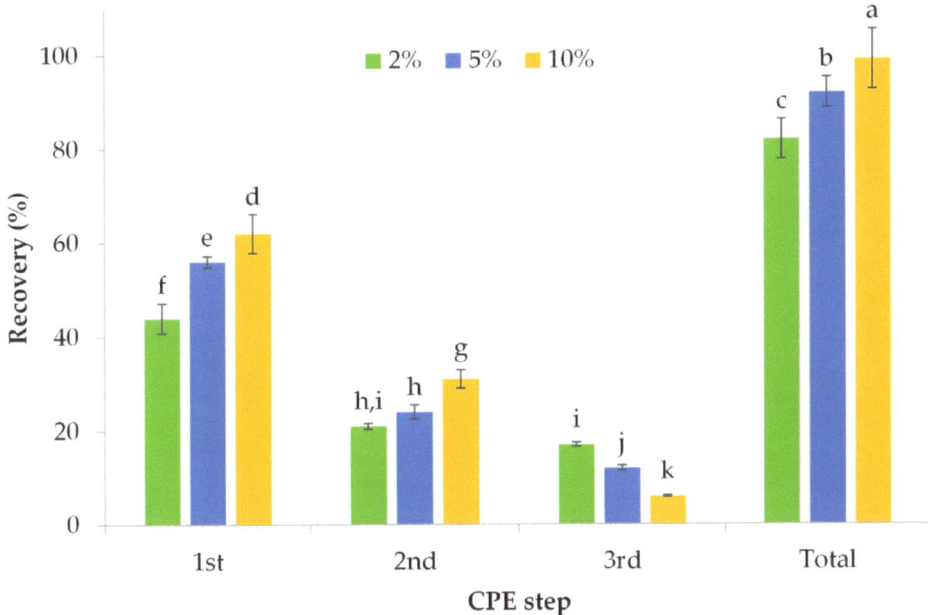

Figure 1. Percentage of polyphenol recovery with Genapol X-080 in different concentrations and extraction steps; standard deviation is shown with error bars; different letters (i.e., a–k) indicate samples that differ significantly (statistical difference for $p < 0.05$) using Student's t-test.

The first step of extraction using 2, 5, and 10% w/v Genapol X-080 resulted in extraction yields of 44, 56, and 62%, respectively. Statistically significant ($p < 0.05$) differences were found in every extraction step. The most efficient way to recover 82% of the total polyphenols was by using 2% Genapol X-080 thrice, whereas the most cost-effective approach was to use 5% Genapol X-080 twice, resulting in 92% polyphenol recovery. Our findings are similar to Kiai et al. [35], who investigated the extraction of table olive processing wastewater polyphenols with the use of the surfactant Genapol X-080 and CPE. At 70 °C equilibrium time and 4.0 pH level for 30 min CPE, they reached an approximate 60% recovery while using a 10% concentration of the surfactant. These results highlight the importance of carefully selecting the concentration of the surfactant and the extraction steps needed to optimize the recovery of polyphenols.

3.2. Effect of Surfactant Concentration via CPE with PEG 8000

PEG 8000 is a polyether, water-soluble, waxy solid that is commonly used as a thickening agent, lubricant, solvent, and surfactant in a wide range of usages, including pharmaceuticals, cosmetics, and food products. PEG 8000 has low toxicity and is considered safe for human use [39]. The results of polyphenols extracted via CPE from ACW samples using three different concentrations of PEG 8000 are presented in Figure 2. As seen in the figure, the recovery of polyphenols is highly dependent on the concentration of the surfactant used. Specifically, increasing the concentration of PEG 8000 led to a higher recovery of polyphenols. However, when using low concentrations of PEG 8000, the recovery rate remained below 60%, indicating the importance of additional CPE steps. The extraction yields of 2, 5, and 10% w/v PEG 8000 in the first step of extraction were 55, 70, and 92%, respectively. Statistically significant differences ($p < 0.05$) were noted in most cases. It is worth noting that the second step of extraction using 2% PEG 8000 provided a high recovery rate of 44%. Interestingly, the recovery of polyphenols in the first two steps of extraction using 2 and 10% PEG 8000 was almost the same (~99%). Thus, a satisfactory recovery rate of polyphenols could be obtained by conducting a two-step extraction by using 2% of the surfactant in each step, which was as effective as the extraction using 10% PEG 8000. Our results are consistent with those of Chatzilazarou et al. [32], who also used the CPE method at 65 °C and obtained a recovery rate of 55.2% using 5% w/v PEG 8000 for one step of extraction from wine sludge. These findings suggest that using 2% PEG 8000 in ACW can be a cost-effective method for extracting a significant number of polyphenols. A wide variety of food industries could use the CPE technique because it is compatible with most food matrices. For instance, an apricot cannery industry could use this method with PEG 8000 in a two-step extraction with 2% w/v surfactant in each step. It would need around 20 Kg of PEG 8000 for each CPE step with 2% w/v concentration. Given that PEG 8000 costs around EUR 100/Kg, it would cost approximately EUR 4000 for a two-step CPE per ton of ACW. Other methodology reported by De Marco et al. [40] requires LLE with mixing the same volumes of olive mill wastewater with solvents such as hexane or ethyl acetate. Despite the encouraging results, it has a considerably higher cost and it could not be used directly in food.

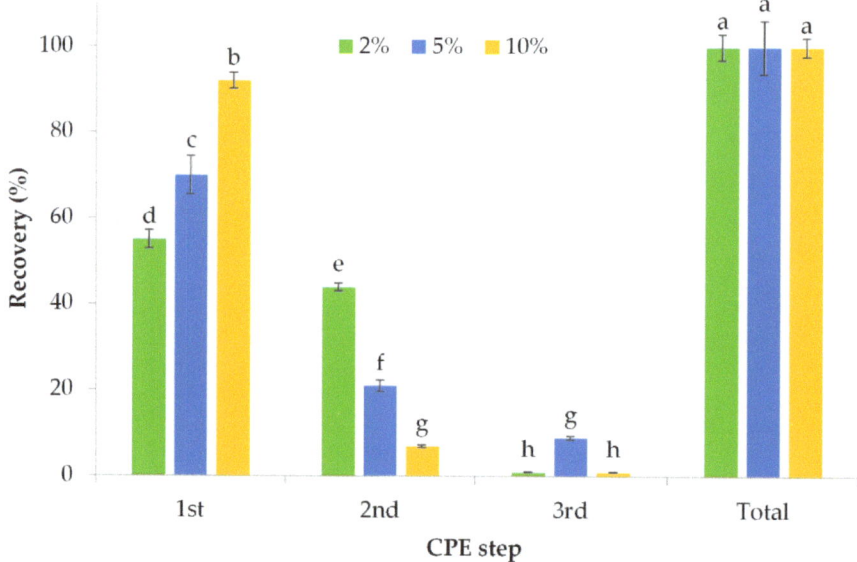

Figure 2. Percentage of polyphenol recovery with PEG 8000 in different concentrations and extraction steps; standard deviation is shown with error bars; different letters (i.e., a–h) indicate samples that differ significantly (statistical difference for $p < 0.05$) using Student's t-test.

3.3. Effect of Surfactant Concentration via CPE with Tween 80

Tween 80, frequently referred as Polysorbate 80, acts as a non-ionic surfactant of the polysorbate family. It is a water-soluble liquid that is widely utilized as a stabilizer, an emulsifier, and solubilizer in various sectors, including food, pharmaceuticals, and personal care. It stems from natural sources such as sorbitol, ethylene oxide, and oleic acid. Tween 80 is known for its ability to increase both the bioavailability and the solubility of insoluble medicines, as well as for its emulsifying properties in food products. It is generally regarded as safe by regulatory agencies such as the FDA and has a wide range of applications due to its versatile properties [41]. Figure 3 shows the findings of the polyphenol CPE in ACW samples using three distinct concentrations of Tween 80. The results revealed that polyphenol recovery was related to the concentration of Tween 80 utilized. Statistically significant differences ($p < 0.05$) were recorded among the various tested concentration for the two first steps of the extraction. The first extraction step yielded 66, 75, and 88% of polyphenols with 2, 5, and 10% w/v Tween 80, respectively. Among the 5 and 10% w/v Tween 80 concentrations, the first two extraction steps yielded high recovery rates of approximately 97 and 100%, respectively. An economically feasible method for extracting polyphenols from ACW is to use 5% Tween 80 twice. Stamatopoulos et al. [42] conducted a study on the isolation and CPE of polyphenols from olive leaf extract with some modifications. They employed 35% w/v sodium sulfate as salt, pH 2.6, and 4% w/v Tween 80, achieving excellent extraction yields in a single extraction step (>90%).

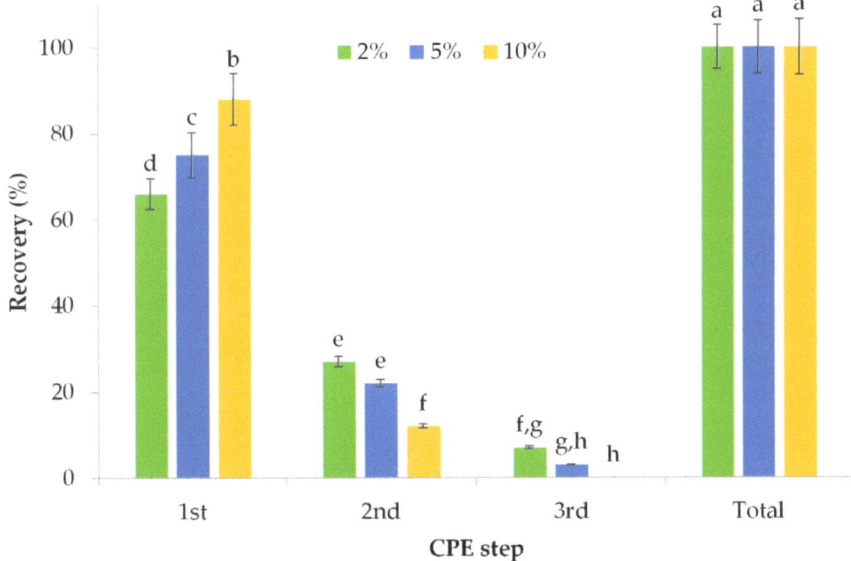

Figure 3. Percentage of polyphenol recovery with Tween 80 in different concentrations and extraction steps; standard deviation is shown with error bars; different letters (i.e., a–h) indicate samples that differ significantly (statistical difference for $p < 0.05$) using Student's t-test.

3.4. Effect of the Surfactant Concentration via CPE with Lecithin

Lecithins are naturally extracted amphiphilic molecules composed of phosphatidylcholine (PC), phosphatidylethanolamine (PE), phosphatidylinositol (PI), and phosphatidic acid (PA). They are often used as common emulsifiers in the food industry and have no maximum level restrictions. Lecithins are quite a preferable alternative to synthetic components, which is attributed to both regulatory demands and also to the health benefits of phospholipids [43]. The results of polyphenol CPE in ACW samples using three different concentrations of lecithin are illustrated in Figure 4. The recoveries of polyphenols with 2, 5, and 10% w/v lecithin were relatively low, at 40, 56, and 73%, respectively. Statistically significant differences ($p < 0.05$) were recorded in most cases. Although the extraction yield in each step was not as high as with other surfactants, lecithin is a cheap, natural, edible, and non-toxic surfactant compared to other options [27]. Our results were similar to those of Alibade et al. [44], who investigated the usage of lecithin in the CPE technique for extracting antioxidant compounds from winery sludges. They conducted experiments to optimize the CPE parameters and found that after three steps of extraction with 5% w/v lecithin, the recovery ranged from approximately 65 to 87%. Additionally, Karadag et al. [45] used lecithin to extract polyphenols from olive mill effluent and achieved a yield of approximately 50% by employing 12.5% w/v lecithin.

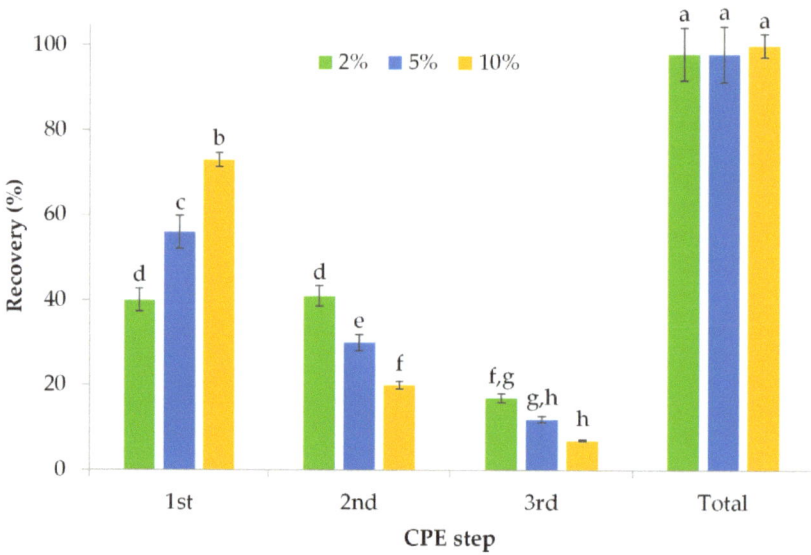

Figure 4. Percentage of polyphenols recovery with Lecithin in different concentrations and extraction steps; standard deviation is shown with error bars; different letters (i.e., a–h) indicate samples that differ significantly (statistical difference for $p < 0.05$) using Student's t-test.

3.5. Antioxidant Activity of the Recovered Polyphenols

Compound extraction from a sample might be desirable provided that the isolated compounds preserve their characteristics. Consequently, it was essential to determine if the extracted polyphenols maintain their antioxidant activity or if it was impacted by CPE extraction. Total polyphenol content (TPC) was measured as mg of gallic acid equivalents per liter (mg GAE/L) using the Folin–Ciocalteu technique, and the DPPH test was used to evaluate the antiradical activity. No statistically significant differences ($p > 0.05$) were observed in both tests for the polyphenols prior to and after the CPE step. We identified 2% PEG 8000 as the most efficient and cost-effective surfactant to examine polyphenol recovery and scavenging activity. As such, a two-step CPE was applied, and Table 1 illustrates the results of these two methods. TPC of initial ACW was measured at 55.2 mg GAE/L and 54.6 mg GAE/L after the CPE method, where high recovery yields were achieved. Our findings are similar to Hong et al. [46], who measured 0.65 mg GAE/g from apricot waste but defined as waste whole fruit samples that are typically discarded by the clients owing to their poor quality. As for the DPPH test, the percentage of scavenging activity of the initial ACW was fairly low. However, the CPE method caused the percentage of scavenging activity to be close to the initial ACW sample. Due to the scarcity of other studies on recovering polyphenols in ACW via CPE, we could only compare our results with Cheaib et al. [47], who investigated the isolations of polyphenols from apricot pomace waste via solid–liquid extraction. While exploring the optimal conditions for solid–liquid extraction (temperature, extraction solvent), they achieved an approximate 4% scavenging activity in 90 min extraction with water as the solvent at 25 °C.

Table 1. Total polyphenolic content (TPC) and antioxidant activity in ACW.

Phase	TPC (mg GAE/L)	Percentage of Scavenging
Initial ACW sample	55.2 ± 1.7 [a,*]	6.3 ± 0.5 [a]
CPE extract from ACW	54.6 ± 1.5 [a]	5.9 ± 0.3 [a]

* Data represent mean values ± standard deviation of three replicates; no statistically significant differences ($p > 0.05$) were found between the samples (e.g., [a]).

4. Conclusions

The feasibility of using the CPE procedure to separate natural antioxidants (polyphenols) from ACW was studied. The efficiency of the method was tested using a range of food-grade surfactants at various concentrations. Owing to its efficiency, minimum cost and duration, and the utilization of harmless extraction solvent, the CPE method is an appealing alternative to the liquid–liquid or liquid–solid solvent extraction of polyphenols. Further CPE steps might be employed to improve the acquired polyphenol recoveries. However, the optimization of CPE conditions (equilibration time, salt addition, pH value) may result in an even simpler polyphenol isolation technique, even avoiding the need for the second extraction step. PEG 8000 was found to be the most cost-effective surfactant to attain a high polyphenol yield. Just 2% w/v PEG 8000 when added twice in a two-step extraction was enough to acquire ~99% recovery. The natural surfactant lecithin did not provide statistically significant differences in polyphenol recovery when compared to the other surfactants. Furthermore, many low-toxicity and specifically greener surfactants should be investigated to see whether they are more effective than the previous examples.

Author Contributions: Conceptualization, I.G. and S.I.L.; methodology, V.A. and T.C.; software, V.A. and E.B.; validation, I.G. and V.A.; formal analysis, I.G., V.A., and T.C.; investigation, I.G. and V.A.; resources, S.I.L.; data curation, I.G. and V.A.; writing—original draft preparation, I.G., V.A., T.C., and D.K.; writing—review and editing, I.G., V.A., T.C., D.K., E.B., O.G., G.D.N., and S.I.L.; visualization, V.A.; supervision, O.G., G.D.N., and S.I.L.; project administration, S.I.L. All authors have read and agreed to the published version of the manuscript.

Funding: This research received no external funding.

Institutional Review Board Statement: Not applicable.

Informed Consent Statement: Not applicable.

Data Availability Statement: Not applicable.

Acknowledgments: The authors would like to thank ELBAK S.A. fruit canning factory, Falani, Larissa, Greece, for donating the apricot cannery wastewater (ACW).

Conflicts of Interest: The authors declare no conflict of interest.

References

1. Gulcin, İ. Antioxidants and Antioxidant Methods: An Updated Overview. *Arch. Toxicol.* **2020**, *94*, 651–715. [CrossRef] [PubMed]
2. Halliwell, B. Antioxidants in Human Health and Disease. *Annu. Rev. Nutr.* **1996**, *16*, 33–50. [CrossRef]
3. Gülçin, I.; Mshvildadze, V.; Gepdiremen, A.; Elias, R. Screening of Antiradical and Antioxidant Activity of Monodesmosides and Crude Extract from Leontice Smirnowii Tuber. *Phytomed. Int. J. Phytother. Phytopharm.* **2006**, *13*, 343–351. [CrossRef] [PubMed]
4. Flieger, J.; Flieger, W.; Baj, J.; Maciejewski, R. Antioxidants: Classification, Natural Sources, Activity/Capacity Measurements, and Usefulness for the Synthesis of Nanoparticles. *Materials* **2021**, *14*, 4135. [CrossRef] [PubMed]
5. Cheaib, D.; El Darra, N.; Rajha, H.N.; Ghazzawi, I.E.; Maroun, R.G.; Louka, N. Biological Activity of Apricot Byproducts Polyphenols Using Solid–Liquid and Infrared-Assisted Technology. *J. Food Biochem.* **2018**, *42*, e12552. [CrossRef]
6. Zhou, W.; Niu, Y.; Ding, X.; Zhao, S.; Li, Y.; Fan, G.; Zhang, S.; Liao, K. Analysis of Carotenoid Content and Diversity in Apricots (*Prunus armeniaca* L.) Grown in China. *Food Chem.* **2020**, *330*, 127223. [CrossRef] [PubMed]
7. Wani, S.M.; Jan, N.; Wani, T.A.; Ahmad, M.; Masoodi, F.A.; Gani, A. Optimization of Antioxidant Activity and Total Polyphenols of Dried Apricot Fruit Extracts (*Prunus armeniaca* L.) Using Response Surface Methodology. *J. Saudi Soc. Agric. Sci.* **2017**, *16*, 119–126. [CrossRef]
8. Veberic, R.; Stampar, F. Selected Polyphenols in Fruits of Different Cultivars of Genus Prunus. *Phyton-Ann. Rei. Bot. A* **2005**, *45*, 375–383.

9. Madrau, M.A.; Piscopo, A.; Sanguinetti, A.M.; Del Caro, A.; Poiana, M.; Romeo, F.V.; Piga, A. Effect of Drying Temperature on Polyphenolic Content and Antioxidant Activity of Apricots. *Eur. Food Res. Technol.* **2009**, *228*, 441–448. [CrossRef]
10. Sun, J.; Chu, Y.-F.; Wu, X.; Liu, R.H. Antioxidant and Antiproliferative Activities of Common Fruits. *J. Agric. Food Chem.* **2002**, *50*, 7449–7454. [CrossRef]
11. Cheaib, D.; El Darra, N.; Rajha, H.; El-Ghazzawi, I.; Mouneimne, Y.; Jammoul, A.; Maroun, R.; Louka, N. Study of the Selectivity and Bioactivity of Polyphenols Using Infrared Assisted Extraction from Apricot Pomace Compared to Conventional Methods. *Antioxidants* **2018**, *7*, 174. [CrossRef] [PubMed]
12. Wang, L.; Weller, C.L. Recent Advances in Extraction of Nutraceuticals from Plants. *Trends Food Sci. Technol.* **2006**, *17*, 300–312. [CrossRef]
13. Conidi, C.; Drioli, E.; Cassano, A. Membrane-Based Agro-Food Production Processes for Polyphenol Separation, Purification and Concentration. *Curr. Opin. Food Sci.* **2018**, *23*, 149–164. [CrossRef]
14. Herrero, M.; Sánchez-Camargo, A.d.P.; Cifuentes, A.; Ibáñez, E. Plants, Seaweeds, Microalgae and Food by-Products as Natural Sources of Functional Ingredients Obtained Using Pressurized Liquid Extraction and Supercritical Fluid Extraction. *TrAC Trends Anal. Chem.* **2015**, *71*, 26–38. [CrossRef]
15. Peng, X.; Xu, H.; Yuan, X.; Leng, L.; Meng, Y. Mixed Reverse Micellar Extraction and Effect of Surfactant Chain Length on Extraction Efficiency. *Sep. Purif. Technol.* **2016**, *160*, 117–122. [CrossRef]
16. Zhang, W.; Liu, X.; Fan, H.; Zhu, D.; Wu, X.; Huang, X.; Tang, J. Separation and Purification of Alkaloids from Sophora Flavescens Ait. by Focused Microwave-Assisted Aqueous Two-Phase Extraction Coupled with Reversed Micellar Extraction. *Ind. Crops Prod.* **2016**, *86*, 231–238. [CrossRef]
17. Pereira, J.F.B.; Freire, M.G.; Coutinho, J.A.P. Aqueous Two-Phase Systems: Towards Novel and More Disruptive Applications. *Fluid Phase Equilibria* **2020**, *505*, 112341. [CrossRef]
18. Grilo, A.L.; Raquel Aires-Barros, M.; Azevedo, A.M. Partitioning in Aqueous Two-Phase Systems: Fundamentals, Applications and Trends. *Sep. Purif. Rev.* **2016**, *45*, 68–80. [CrossRef]
19. Al_Saadi, M.R.; Al-Garawi, Z.S.; Thani, M.Z. Promising Technique, Cloud Point Extraction: Technology & Applications. *J. Phys. Conf. Ser.* **2021**, *1853*, 012064. [CrossRef]
20. Racheva, R.; Rahlf, A.F.; Wenzel, D.; Müller, C.; Kerner, M.; Luinstra, G.A.; Smirnova, I. Aqueous Food-Grade and Cosmetic-Grade Surfactant Systems for the Continuous Countercurrent Cloud Point Extraction. *Sep. Purif. Technol.* **2018**, *202*, 76–85. [CrossRef]
21. Yazdi, A.S. Surfactant-Based Extraction Methods. *TrAC Trends Anal. Chem.* **2011**, *30*, 918–929. [CrossRef]
22. Sharma, S.; Kori, S.; Parmar, A. Surfactant Mediated Extraction of Total Phenolic Contents (TPC) and Antioxidants from Fruits Juices. *Food Chem.* **2015**, *185*, 284–288. [CrossRef] [PubMed]
23. Carabias-Martínez, R.; Rodríguez-Gonzalo, E.; Moreno-Cordero, B.; Pérez-Pavón, J.L.; García-Pinto, C.; Fernández Laespada, E. Surfactant Cloud Point Extraction and Preconcentration of Organic Compounds Prior to Chromatography and Capillary Electrophoresis. *J. Chromatogr. A* **2000**, *902*, 251–265. [CrossRef] [PubMed]
24. Katsoyannos, E.; Gortzi, O.; Chatzilazarou, A.; Athanasiadis, V.; Tsaknis, J.; Lalas, S. Evaluation of the Suitability of Low Hazard Surfactants for the Separation of Phenols and Carotenoids from Red-Flesh Orange Juice and Olive Mill Wastewater Using Cloud Point Extraction. *J. Sep. Sci.* **2012**, *35*, 2665–2670. [CrossRef]
25. Martin, G. Surfactant Systems: Their Chemistry, Pharmacy and Biology. *Biochem. Soc. Trans.* **1984**, *12*, 719–720. [CrossRef]
26. Tang, X.; Zhu, D.; Huai, W.; Zhang, W.; Fu, C.; Xie, X.; Quan, S.; Fan, H. Simultaneous Extraction and Separation of Flavonoids and Alkaloids from *Crotalaria Sessiliflora* L. by Microwave-Assisted Cloud-Point Extraction. *Sep. Purif. Technol.* **2017**, *175*, 266–273. [CrossRef]
27. van Nieuwenhuyzen, W. Lecithin and Other Phospholipids. In *Surfactants from Renewable Resources*; Kjellin, M., Johansson, I., Eds.; John Wiley & Sons: Chichester, UK, 2010; pp. 191–212, ISBN 978-0-470-68660-7.
28. El-Abbassi, A.; Kiai, H.; Raiti, J.; Hafidi, A. Cloud Point Extraction of Phenolic Compounds from Pretreated Olive Mill Wastewater. *J. Environ. Chem. Eng.* **2014**, *2*, 1480–1486. [CrossRef]
29. Chatzilazarou, A.; Katsoyannos, E.; Lagopoulou, M.; Tsaknis, J. Application of Cloud Point Extraction with the Aid of Genapol X-080 in the Pre-Concentration of Lycopene and Total Carotenoids from Red Fleshed Orange. *Ernährung/Nutrition* **2011**, *35*, 10.
30. Santana, C.M.; Ferrera, Z.S.; Rodríguez, J.J.S. Use of Non-Ionic Surfactant Solutions for the Extraction and Preconcentration of Phenolic Compounds in Water Prior to Their HPLC-UV Detection. *Analyst* **2002**, *127*, 1031–1037. [CrossRef]
31. Sosa-Ferrera, Z. The Use of Micellar Systems in the Extraction and Pre-Concentration of Organic Pollutants in Environmental Samples. *TrAC Trends Anal. Chem.* **2004**, *23*, 469–479. [CrossRef]
32. Chatzilazarou, A.; Katsoyannos, E.; Gortzi, O.; Lalas, S.; Paraskevopoulos, Y.; Dourtoglou, E.; Tsaknis, J. Removal of Polyphenols from Wine Sludge Using Cloud Point Extraction. *J. Air Waste Manag. Assoc.* **2010**, *60*, 454–459. [CrossRef] [PubMed]
33. Katsoyannos, E.; Chatzilazarou, A.; Gortzi, O.; Lalas, S.; Konteles, S.; Tataridis, P. Application of cloud point extraction using surfactants in the isolation of physical antioxidants (phenols) from olive mill wastewater. *Fresenius Environ. Bull.* **2006**, *15*, 4.
34. Tsaknis, J.; Lalas, S. Extraction and Identification of Natural Antioxidant from *Sideritis euboea* (Mountain Tea). *J. Agric. Food Chem.* **2005**, *53*, 6375–6381. [CrossRef] [PubMed]
35. Kiai, H.; Raiti, J.; El-Abbassi, A.; Hafidi, A. Recovery of Phenolic Compounds from Table Olive Processing Wastewaters Using Cloud Point Extraction Method. *J. Environ. Chem. Eng.* **2018**, *6*, 1569–1575. [CrossRef]

36. Shi, Z.; Zhu, X.; Zhang, H. Micelle-Mediated Extraction and Cloud Point Preconcentration for the Analysis of Aesculin and Aesculetin in Cortex Fraxini by HPLC. *J. Pharm. Biomed. Anal.* **2007**, *44*, 867–873. [CrossRef]
37. Gortzi, O.; Chatzilazarou, A.; Katsoyannos, E.; Papaconstandinou, S.; Dourtoglou, E. Recovery of Natural Antioxidants from Olive Mill Wastewater Using Genapol-X080. *J. Am. Oil Chem. Soc.* **2008**, *85*, 133–140. [CrossRef]
38. Kori, S. Cloud Point Extraction Coupled with Back Extraction: A Green Methodology in Analytical Chemistry. *Forensic Sci. Res.* **2021**, *6*, 19–33. [CrossRef]
39. Ibrahim, M.; Ramadan, E.; Elsadek, N.E.; Emam, S.E.; Shimizu, T.; Ando, H.; Ishima, Y.; Elgarhy, O.H.; Sarhan, H.A.; Hussein, A.K.; et al. Polyethylene Glycol (PEG): The Nature, Immunogenicity, and Role in the Hypersensitivity of PEGylated Products. *J. Control. Release* **2022**, *351*, 215–230. [CrossRef]
40. De Marco, E.; Savarese, M.; Paduano, A.; Sacchi, R. Characterization and Fractionation of Phenolic Compounds Extracted from Olive Oil Mill Wastewaters. *Food Chem.* **2007**, *104*, 858–867. [CrossRef]
41. Kaur, G.; Mehta, S.K. Developments of Polysorbate (Tween) Based Microemulsions: Preclinical Drug Delivery, Toxicity and Antimicrobial Applications. *Int. J. Pharm.* **2017**, *529*, 134–160. [CrossRef]
42. Stamatopoulos, K.; Katsoyannos, E.; Chatzilazarou, A. Antioxidant Activity and Thermal Stability of Oleuropein and Related Phenolic Compounds of Olive Leaf Extract after Separation and Concentration by Salting-Out-Assisted Cloud Point Extraction. *Antioxidants* **2014**, *3*, 229–244. [CrossRef] [PubMed]
43. Wang, M.; Yan, W.; Zhou, Y.; Fan, L.; Liu, Y.; Li, J. Progress in the Application of Lecithins in Water-in-Oil Emulsions. *Trends Food Sci. Technol.* **2021**, *118*, 388–398. [CrossRef]
44. Alibade, A.; Batra, G.; Bozinou, E.; Salakidou, C.; Lalas, S. Optimization of the Extraction of Antioxidants from Winery Wastes Using Cloud Point Extraction and a Surfactant of Natural Origin (Lecithin). *Chem. Pap.* **2020**, *74*, 4517–4524. [CrossRef]
45. Karadag, A.; Kayacan Cakmakoglu, S.; Metin Yildirim, R.; Karasu, S.; Avci, E.; Ozer, H.; Sagdic, O. Enrichment of Lecithin with Phenolics from Olive Mill Wastewater by Cloud Point Extraction and Its Application in Vegan Salad Dressing. *J. Food Process. Preserv.* **2022**, *46*, e16645. [CrossRef]
46. Hong, Y.; Wang, Z.; Barrow, C.J.; Dunshea, F.R.; Suleria, H.A.R. High-Throughput Screening and Characterization of Phenolic Compounds in Stone Fruits Waste by LC-ESI-QTOF-MS/MS and Their Potential Antioxidant Activities. *Antioxidants* **2021**, *10*, 234. [CrossRef] [PubMed]
47. Cheaib, D.; El Darra, N.; Rajha, H.N.; Maroun, R.G.; Louka, N. Systematic and Empirical Study of the Dependence of Polyphenol Recovery from Apricot Pomace on Temperature and Solvent Concentration Levels. *Sci. World J.* **2018**, *2018*, 8249184. [CrossRef]

Disclaimer/Publisher's Note: The statements, opinions and data contained in all publications are solely those of the individual author(s) and contributor(s) and not of MDPI and/or the editor(s). MDPI and/or the editor(s) disclaim responsibility for any injury to people or property resulting from any ideas, methods, instructions or products referred to in the content.

Article

Prediction Model for Optimal Efficiency of the Green Corrosion Inhibitor Oleoylsarcosine: Optimization by Statistical Testing of the Relevant Influencing Factors

Saad E. Kaskah [1,2,*], Gitta Ehrenhaft [3], Jörg Gollnick [3] and Christian B. Fischer [1,4,*]

1. Department of Physics, University of Koblenz, 56070 Koblenz, Germany
2. General Directorate of Industrial Development, Ministry of Industry and Minerals, Baghdad 22017, Iraq
3. Institute of Mechanics and Material Science, TH Mittelhessen University of Applied Sciences, 35390 Giessen, Germany
4. Materials Science, Energy and Nano-Engineering Department, Mohammed VI Polytechnic University, Ben Guerir 43150, Morocco
* Correspondence: saadelias82@gmail.com (S.E.K.); chrbfischer@uni-koblenz.de (C.B.F.); Tel.: +964-7726095804 (S.E.K.); +49-2612872345 (C.B.F.)

Citation: Kaskah, S.E.; Ehrenhaft, G.; Gollnick, J.; Fischer, C.B. Prediction Model for Optimal Efficiency of the Green Corrosion Inhibitor Oleoylsarcosine: Optimization by Statistical Testing of the Relevant Influencing Factors. *Eng* 2023, *4*, 635–649. https://doi.org/10.3390/eng4010038

Academic Editor: Tomasz Lipiński

Received: 6 January 2023
Revised: 9 February 2023
Accepted: 13 February 2023
Published: 15 February 2023

Copyright: © 2023 by the authors. Licensee MDPI, Basel, Switzerland. This article is an open access article distributed under the terms and conditions of the Creative Commons Attribution (CC BY) license (https://creativecommons.org/licenses/by/4.0/).

Abstract: Optimization and statistical methods are used to minimize the number of experiments required to complete a study, especially in corrosion testing. Here, a statistical Box–Behnken design (BBD) was implemented to investigate the effects of four independent variables (inhibitor concentration [I], immersion time t, temperature ϑ, and NaCl content [NaCl]) based on the variation of three levels (lower, middle, and upper) on the corrosion protection efficiency of the green inhibitor oleoylsarcosine for low-carbon steel type CR4 in salt water. The effects of the selected variables were optimized using the response surface methodology (RSM) supported by the Minitab17 program. Depending on the BBD analytical tools, the largest effects were found for ϑ, followed by [I]. The effect of interactions between these variables was in the following order: [I] and $\vartheta > t$ and $\vartheta > $ [I] and [NaCl]. The second-order model used here for optimization showed that the upper level (+1) with 75 mmol/L for [I], 30 min for t, and 0.2 mol/L [NaCl] provided an optimal protective effect for each of these factors, while the lower level (−1) was 25 °C for ϑ. The theoretical efficiency predicted by the RSM model was 99.4%, while the efficiency during the experimental test procedure with the best-evaluated variables was 97.2%.

Keywords: Box–Behnken design; response surface methodology; potentiodynamic polarization; prediction of corrosion protection efficiency; normal probability distribution

1. Introduction

Oleoylsarcosine (O, Scheme 1) is classified as an *N*-acylsarcosine derivative that has a high efficiency to inhibit corrosion of low-carbon steel CR4 in 0.1 M NaCl [1]. Organic compounds, such as O, are of increasing interest because of their low cost and potential properties for adsorption to the metal surface [2–4]. Furthermore, these derivatives are known to effectively inhibit corrosion and are aerobically and anaerobically biodegradable [5–7].

Scheme 1. Chemical structure of the tested inhibitor oleoylsarcosine (O).

Several parameters affect the corrosion inhibition process, such as the concentration of the used inhibitor [I], the immersion time t of the metal in an anticorrosive solution, the degree of acidity of the surrounding environment, the temperature ϑ, etc. [8,9]. In our recent studies [1,10,11], the classical experimental principle was used to determine the trend of concentration and time dependency of dip coating (immersion time of metal samples). The usual procedure for experimental work, used primarily in research, is to study the effects of changing one factor over time while holding other factors constant. However, these classical methods require a large number of experiments, which consume a lot of materials, such as substrate samples and (inhibitor) solutions, as well as time and effort, thus resulting in high costs [12,13]. In addition, simple trial-and-error methods implemented in such experiments for electrochemical processes cannot easily overcome the complex analytical challenges [14]. To determine the effect of multiple variables on a process, even when there is a complex interaction between them, optimization can help to eliminate the classical methods of experimentation [15]. Before starting such experiments, it is worthwhile to conscientiously plan the experimental work and then follow the developed plan [16]. In this context, there are many recent studies that focus on the application of experimental design and its tools to better explain the effects of different independent variables.

Chemometric tools are used to analyze and optimize the effects of variables and their control on the process [17]. The statistical design of experiment (DOE) is a technique that can be utilized to create an experimental model to reduce the number of tests required compared to the usual experimental methods while taking into account selected or even all relevant parameters in the experiment [17]. In addition to reducing the number of experiments, DOE develops mathematical models that will allow statistical evaluation of the influencing factors and show whether there are significant interactions between the variables in the process. DOE receives the variables as input and gives the result of the experimentation process as output. For optimization strategies there are many mathematical techniques using univariate and multivariate methods [17–19]. To avoid errors that may occur in the univariate method when the effect of one variable depends on the level of the other variables involved in the optimization process, the levels of all variables are changed simultaneously, which is processed in multivariate optimization. In multivariate optimization, the first step is to fulfill the screening effect of the studied variables in the fractional factorial design, which depends on the statistical method used in the optimization, to know the main effect of the levels for each variable [20–22]. The response surface methodology (RSM) is one of the most important optimization strategies, and it has been shown in many studies to be a very useful tool for application in many areas of industry [23–28].

The step after determining the influencing factors that have a significant effect is to apply more complex experimental designs, such as the Box–Behnken design, at three levels [29–31]. These may be qualitative factors, such as concentration, temperature, immersion time, etc., or quantitative factors, such as steel type and grade, type of inhibitor, etc. The responses are the independent variables, and their values depend on the level of the factors. BBD is a type of RSM that can be used to determine the best conditions for an optimal result in the experiment. BBD considers an independent quadratic design, and the treatment is located at the midpoint of the edges of the process space and in the center [32]. This technique is designed so that each variable has three levels. There are a number of designs for BBD, and selection of the specific design depends on the number of variables and their levels. The main advantage of the BBD method is the smaller number of experiments compared to other response surface designs, e.g., the central composite design [18,33].

In the present work, the BBD method was used to optimize the influence of four independent variables affecting the protection of steel CR4 against corrosion in salt water. The aim was to determine the optimum protective effect that can be achieved using the examined levels of the selected factors. The selected independent variables and their denomination were the concentration of inhibitor [I], the immersion time t of the steel samples in the anticorrosive inhibitor stock solution, the temperature ϑ of the test solution, and the NaCl content [NaCl] in the testing solution. Each independent variable had three levels (low, mid, and upper) and was designed as coded value (−1, 0, and +1). The dependent variable (objective function) to be determined was the efficiency of the inhibitor to reduce the current density and to achieve the best protection.

2. Materials and Methods

2.1. Metal Samples, Inhibitor, and Experimental Setup

Low-carbon steel type CR4, which corrodes relatively quickly at ambient environment without protection, was tested against corrosion in salt water with three different NaCl contents (0.05, 0.1, and 0.2 M). The single steel sheets were laser-cut in the dimensions 25 × 25 × 1 mm and delivered without further treatment by the supplier (Janssen CNC-Blechbearbeitungs GmbH, 65604 Elz, Germany). The chemical composition was analyzed by optical emission spectroscopy (Vario Lab, Belec Spektrometrie Opto-Elektronik GmbH, 49124 Osnabrück, Germany). The result of the metal analysis was in line with the standard range of this steel type (CR4, also denominated as DC01- [1,34]).

Proper preparation, such as grinding, cleaning, drying of metal samples, and coating with the selected inhibitor, was carried out before testing. Details can be found in [1,11,12]. Briefly, the specifics of this investigation were as follows. After grinding according to DIN EN ISO 9227:2006 and DIN EN ISO 1514:2005-02 [35,36] with silicon carbide paper (120 and 220, WS FLEX 18c waterproof, HERMES, Hamburg, Germany), the samples were cleaned ultrasonically in isopropanol (10 min) and dried with pressurized air. Subsequently, they could be used immediately for immersion in the inhibitor solutions or stored at 60 °C for slightly later use. The final step prior to corrosion testing was coating with an inhibitor at three different concentrations [I] (25, 50, and 75 mmol/L in toluene) and at three different duration times t (1, 10, and 30 min). According to our previous studies of N-acyl-sarcosine derivatives [1,11], the most effective inhibitor oleoylsarcosine (O) was used.

Potentiodynamic polarization (Galvanostate Wenking, working electrode 1 cm^2, thermostated cell, LPG03 system, Bank Electronic—Intelligent Controls GmbH, 35415 Hessen, Germany) was carried out following previous studies [1,10,11] to test the inhibitor O and find the optimum corrosion protection for steel CR4 according to the experimental design. This included variation of the variables [I], t, ϑ, and [NaCl], as mentioned above. Prior to any measurement, the open circuit potential (OCP) was equilibrated at free potential. The pH value was controlled to be constant at 6.5 ± 0.5 (EL20/EL2, Toledo Group, Zurich, Switzerland). Polarizations were recorded in duplicate at ±150 mV with respect to the OCP and a scan speed of 0.5 mV/s.

2.2. Experimental Design and Optimization

Design of experiments (DOE) involves subjecting certain experimental parameters, i.e., variables, to an organized treatment using a designed measurement matrix in order to obtain appropriate responses. Each experiment has one or more variables that represent an input to the DOE (A, B, C, etc.), which is then converted to a corresponding output (Y). After obtaining reasonable results from this experimental design and following proper analysis, the optimal experiment can be planned. All DOE techniques are nontraditional methods. Independent variables that represent effective combinations of each other are determined so that the tests can be carried out statistically in a limited number. The Box–Behnken design (BBD) is one such technique that reduces the required number of experiments while maintaining the accuracy of the results. The type of design depends on the levels used for the variables. A two-level matrix includes only a lower and upper limit (−1 and +1), while a three-level matrix includes an additional midpoint (−1, 0, and +1). For each classical experimental procedure, a certain number of experiments is required, which can be calculated from the number of variables and levels according to Equation (1):

$$S = l^k \tag{1}$$

where S is the number of experiments, l the number of levels, and k the number of variables.

In the present work, the number of experiments required in a conventional study with four variables and three levels each according to Equation (1) would have been as follows: 3^4 = 81. However, the required number was minimized with BBD. The independent variables with their levels were performed in coded values, as summarized in Table 1. The orthogonal three-level matrix for four variables with three levels each according to BBD therefore resulted in only 27 runs, as shown in Table 2.

Table 1. The four independent variables with their levels performed with coded values.

No.	Variable Code	Selected Variable	Coded Level		
			−1	0	+1
1	A	Inhibitor concentration [I] (mmol/L)	25	50	75
2	B	Immersion time t (min)	1	10	30
3	C	Temperature ϑ (°C)	25	40	55
4	D	NaCl content [NaCl] (mol/L)	0.05	0.1	0.2

Table 2. Three-level matrix for four variables (A, B, C, and D) and three levels (−1, 0, and +1) according to BBD with resulting 27 single experiments.

No.	Coded Value			
	A	B	C	D
1	−1	−1	0	0
2	−1	+1	0	0
3	+1	−1	0	0
4	+1	+1	0	0
5	0	0	−1	−1
6	0	0	−1	+1
7	0	0	+1	−1
8	0	0	+1	+1
9	−1	0	0	−1
10	−1	0	0	+1
11	+1	0	0	−1
12	+1	0	0	+1
13	0	−1	−1	0
14	0	−1	+1	0
15	0	+1	−1	0
16	0	+1	+1	0
17	−1	0	−1	0
18	−1	0	+1	0
19	+1	0	−1	0
20	+1	0	+1	0
21	0	−1	0	−1
22	0	−1	0	+1
23	0	+1	0	−1
24	0	+1	0	+1
25	0	0	0	0
26	0	0	0	0
27	0	0	0	0

The main objective of the DOE application is to optimize the experimental process and predict the maximum response Y depending on the selected variables and levels with a minimum number of experiments. There are many computer-aided optimization methods for DOE, but RSM is one of the most effective [37,38]. RSM combines statistical and mathematical techniques that are useful in developing, improving, and optimizing processes. The first aim in RSM is to find the optimal response Y. The second objective is to understand how Y changes in a particular direction by a graphical adaptation. Here, the efficiency of the inhibitor is the response variable expressed as a function of four independent variables, as shown in Equation (2) [37]:

$$Y = f(A, B, C, D) + \varepsilon \tag{2}$$

where A, B, C, and D are the variables, and ε is the experimental error.

RSM starts with a first-order model function if the response is a linear function of the independent variables. Here, the first-order model can be expressed as Equation (3):

$$Y = \beta_0 + \beta_1 A + \beta_2 B + \beta_3 C + \beta_4 D + \varepsilon \tag{3}$$

where β is the respective regression coefficient.

If the response surface has a curvature, then a second-order model should be used, given in Equation (4):

$$\begin{aligned} Y = &\beta_0 + \beta_1 A + \beta_2 B + \beta_3 C + \beta_4 D \\ &+ \beta_{11} A^2 + \beta_{22} B^2 + \beta_{33} C^2 + \beta_{44} D^2 \\ &+ \beta_{12} AB + \beta_{13} AC + \beta_{14} AD + \beta_{23} BC + \beta_{24} BD + \beta_{34} CD + \varepsilon \end{aligned} \tag{4}$$

In this work, the second-order model for RSM was used to identify all possible effects on the process. RSM was applied using Minitab 17 to perform 27 experiments and determine the responses based on the proposed predictive mode.

3. Results and Discussion

3.1. Experimental Corrosion Protection Efficiencies According to the BBD Matrix

Polarization measurements were carried out according to the aforementioned BBD matrix to analyze the effects of the four selected variables on the protection efficiency. The result of these 27 experimental runs is given in Table 3.

Table 3. Efficiency results of the 27 experiments according to the BBD matrix, including the coded and real values (compare Table 1).

No.	Coded Value				Real Value				Efficiency %
	A	B	C	D	A	B	C	D	
1	−1	−1	0	0	25	1	40	0.1	85.20 ± 4.1
2	−1	+1	0	0	25	30	40	0.1	77.99 ± 8.9
3	+1	−1	0	0	75	1	40	0.1	66.77 ± 2.1
4	+1	+1	0	0	75	30	40	0.1	69.87 ± 3.6
5	0	0	−1	−1	50	10	25	0.05	89.11 ± 0.6
6	0	0	−1	+1	50	10	25	0.2	92.06 ± 0.8
7	0	0	+1	−1	50	10	55	0.05	39.94 ± 0.2
8	0	0	+1	+1	50	10	55	0.2	37.86 ± 2.5
9	−1	0	0	−1	25	10	40	0.05	70.05 ± 8.0
10	−1	0	0	+1	25	10	40	0.2	52.40 ± 6.1
11	+1	0	0	−1	75	10	40	0.05	85.57 ± 1.0
12	+1	0	0	+1	75	10	40	0.2	74.59 ± 4.8
13	0	−1	−1	0	50	1	25	0.1	92.86 ± 0.1
14	0	−1	+1	0	50	1	55	0.1	50.15 ± 8.3
15	0	+1	−1	0	50	30	25	0.1	91.96 ± 1.0
16	0	+1	+1	0	50	30	55	0.1	41.80 ± 9.7
17	−1	0	−1	0	25	10	25	0.1	91.98 ± 1.6
18	−1	0	+1	0	25	10	55	0.1	54.26 ± 7.9
19	+1	0	−1	0	75	10	25	0.1	94.35 ± 0.5
20	+1	0	+1	0	75	10	55	0.1	46.03 ± 4.1

Table 3. Cont.

No.	Coded Value				Real Value				Efficiency %
	A	B	C	D	A	B	C	D	
21	0	−1	0	−1	50	1	40	0.05	86.46 ± 2.3
22	0	−1	0	+1	50	1	40	0.2	80.63 ± 0.2
23	0	+1	0	−1	50	30	40	0.05	76.51 ± 3.9
24	0	+1	0	+1	50	30	40	0.2	73.55 ± 4.6
25	0	0	0	0	50	10	40	0.1	74.23 ± 0.5
26	0	0	0	0	50	10	40	0.1	74.23 ± 0.5
27	0	0	0	0	50	10	40	0.1	74.23 ± 0.5

For the next step in optimization of the variables' influence, including their levels, RSM was applied on the data outlined in Table 3. The tools used to analyze and optimize the process are given in the following sections: general effects of variables (see Section 3.2.), response plot (see Section 3.3.), normal probability distribution (see Section 3.4.), and optimization by RSM (see Section 3.5.).

3.2. General Effects of the Variables and Their Levels

Before starting the optimization process, it is important to outline the main effects of each variable depending on its level (−1, 0, and +1)). Figure 1 displays the main effect diagrams (based on results of Table 4), which show the response of each variable as it changes according to the given level, for the current corrosion protection process with the selected inhibitor O. The following observations can be derived from Figure 1.

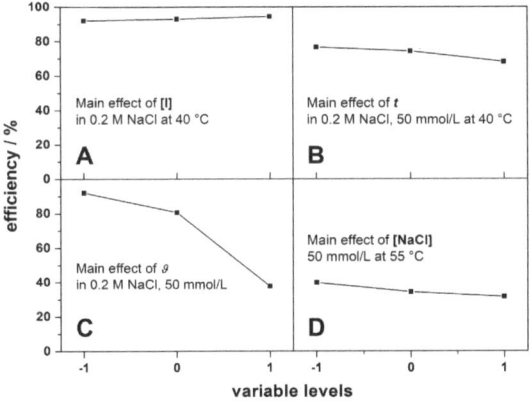

Figure 1. General effects of the four selected variables [I], t, ϑ, and [NaCl] with their three levels (−1, 0, and +1) on the corrosion protection of steel CR4: Main effect on the process (**A**) of [I], (**B**) of t, (**C**) of ϑ, and (**D**) of [NaCl], respectively, while the other tested variables remain unchanged.

The protection efficiency gradually increased as the inhibitor O concentration increased, reaching an optimum of 94% at the upper level (+1, Figure 1A). For the remaining variables t, ϑ, and [NaCl] (Figure 1B–D), the highest efficiency was achieved at their respective lower levels (−1). Considering all changes in the individual main effect diagrams, it can be noticed that the temperature (Figure 1C), had the greatest influence. The change in the efficiency was significantly more pronounced when switching between the levels than for any other variable (Figure 1C). The overall least effect was recorded for the salt content with efficiencies around 40% (Figure 1D).

This general overview only gives a quick indication of the behavioral effect of the tested independent variables. For a more in-depth analysis, BBD tools, such as response graphs, normal probability distribution, and interaction effect graphs, were used.

Table 4. Normal probability calculations.

Variable/Interaction	Estimated Effect E_f	Rank I	Probability (Pi) = 100(I − 0.5)/10
C	47.046	1	5
CD	9.805	2	15
BD	9.244	3	25
B	6.952	4	35
D	6.091	5	45
AB	6.021	6	55
AC	5.300	7	65
BC	5.186	8	75
AD	3.335	9	85
A	1.158	10	95
Mean	10.014	-	-

3.3. Response Plot

The result of the implantation runs in the BBD experimental design are shown in the response table (see Supporting Information S1). Each determined efficiency value represents the average of two independent experiments to ensure reproducibility. With the plot of the response graph in Figure 2, the individual and interactive effects of the variables on the process can be easily examined. The effect of the variables and their levels on efficiency is represented as E_f, which was obtained by the higher (l_H) and lower (l_L) values for each factor and interaction in the response table (see Supporting Information S1). The response graph (Figure 2) represents the estimated impact. Therefore, it is a valuable tool that uses the absolute impact to analyze the effect of factors on the efficiency of the process.

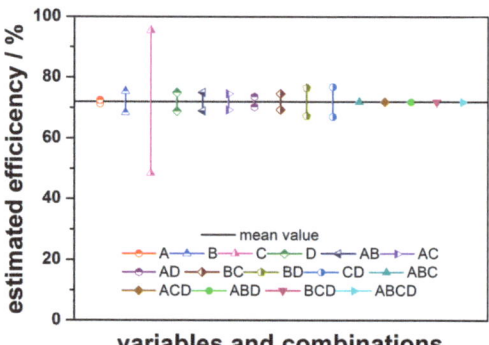

Figure 2. Response plot of the individual and interactive effects of the four selected variables with their three levels on the corrosion protection of steel CR4.

From the response graph (Figure 2), it is clear that the temperature had the greatest influence in reducing corrosion in the presence of inhibitor O, with E_f C = 47 (half-filled triangles up, magenta). This is consistent with the result of the main effect plot in Figure 1. The interactions between temperature and [NaCl], depicted as CD, ranked second for the influence on the process, along with the immersion time and [NaCl], depicted as BD (E_f = 9 for both, half-filled circles in blue and half-filled pentagons in dark yellow, respectively). The immersion time, depicted as B, followed on rank 3 (E_f = 7, blue, half-filled triangles up), while both the NaCl content and the interaction between [I] and *t*, depicted as D and

AB, respectively, were rank 4 ($E_f = 6$ for both, green half-filled diamonds and dark blue half-filled triangles left, respectively). Rank 5 was shared by the interactions between AC and BC, each with an effect of 5 (light purple half-filled triangles right and brown half-filled diamonds, respectively). The interaction between inhibitor concentration and NaCl content results came in at rank 6 with E_f AD = 3 (dark purple half-filled diamonds). In the last place of individual and combined two-factor interactions was the inhibitor concentration, depicted as A, with E_f of only = 1 (red half-filled circles), which meant that a change of inhibitor O to higher concentrations did not have a great influence on the response (and therefore the efficiency of corrosion protection). Interactions between three and four factors had the same value for the mean efficiency of the process, which in the end had no detectable effects on corrosion protection (Figure 2).

3.4. Normal Probability Plot

The results reviewed in the response graph (Figure 2) clearly indicated that some variables and interactions had a greater effect than others. To ensure that these effects were real and did not just occur by chance, a normal probability representation was applied. Calculations required to plot the normal probability are summarized in Table 4. The probability represents how far the variables and their respective interactions affected the corrosion protection.

The estimated effect of the variables and the interactive effects yielded the normal probability data given in Table 4 according to the order from maximum to minimum value. Figure 3 shows the normal probability diagram, which indicates which factors and interactions had a significant influence on the protection efficiency and which were rather insignificant. The effective value is based on the distance of the data point (colored diamond) from the mean line, with the effect being greater the further it is from it. The red diamonds refer to the variables and combinations that have the strongest effect, the green ones still have a good to moderate effect, while the blue ones indicate insignificant effects. From this, it can be clearly deduced that the temperature (C) and the inhibitor concentration (A) had the greatest influence on the efficiency (red diamonds) as they are most distant from the mean line. Accordingly, the interaction effects AB, AC, BC, and AD (green diamonds) also had good effects. The individual variables of the immersion time (B) and the NaCl content (D), shown as green diamonds, still indicated a good to moderate effect. The interactive effects between ϑ and [NaCl] (CD) as well as immersion time and [NaCl] (BD) (blue diamonds) had an insignificant effect as they are very close to the indicated mean line.

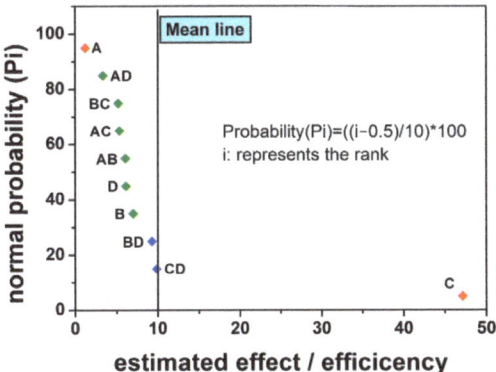

Figure 3. Normal probability plot for the distribution of individual and interaction effects of the four tested variables with their three levels on the CR4 steel protection in salt water using inhibitor O. The following color code applies: Red for high, green for good to medium, and of blue for negligible effects, respectively, of the tested variable on the process.

The normal probability distribution in Figure 3 clearly shows that the concentration of inhibitor O had a high effect on protection, while in the response graph in Figure 2, it showed a lower effect compared to the other variables and interactions. As mentioned before, the normal probability shows the actual impact of the factors affecting the process rather than the random impact. This also explains why some factors had a greater influence in the response graph than in the normal probability distribution and vice versa. For this reason, it was reasonable to use both methods for data analysis in the current study.

3.5. Optimization

The RSM (response surface methodology) was used here to determine the maximum protection efficiency for inhibitor O in accordance with the selected factors and their chosen levels. Optimization was implemented to statistically design the combinations, determine the coefficient by fitting the experimental results to the response function, predict the response according to the fitted model, and finally check the fit of the designed model.

3.5.1. Implementation of the Empirical Model

Here, the second-order model of Equation (4) was used, with only the double interaction effect included in the regression model. Regression coefficients were obtained using Minitab 17 software and are summarized in Table 5 (see Supporting Information S2).

Table 5. Obtained regression coefficients of the optimization model according to Equation (4).

No.	Coefficient	Coefficient Obtained	Symbol
1	Constant	74.23	β_0
2	A	0.44	β_1
3	B	−2.53	β_2
4	C	−23.52	β_3
5	D	−3.05	β_4
6	AA	−0.23	β_{11}
7	BB	2.86	β_{22}
8	CC	−6.07	β_{33}
9	DD	−1.52	β_{44}
10	AB	2.58	β_{12}
11	AC	−2.65	β_{13}
12	AD	1.67	β_{14}
13	BC	−1.86	β_{23}
14	BD	0.72	β_{24}
15	CD	−1.26	β_{34}

Using the regression coefficients listed in Table 6 and the regression equation generated by the program (see Supporting Information S3), the empirical model in Equation (4) can be rewritten as follows:

$$\text{Predicted efficiency \%} = 74.23 + 0.44 \times A - 2.53 \times B - 23.53 \times C - 3.05 \times D - 0.23 \times A^2 + 2.86 \times B^2 - 6.07 \times C^2 - 1.52 \times D^2 + 2.58 \times AB - 2.65 \times AC + 1.67 \times AD - 1.86 \times BC + 0.72 \times BD - 1.26 \times CD \quad (5)$$

The determination of multiple coefficients (R^2) for the proposed predictor model yielded 89.2%, which meant that only 10.8% was not considered by the proposed model, indicating a good acceptance of the model used in this study. The actual experimental efficiency, the predicted efficiency resulting from the application of the empirical model, and the error resulting from the difference between them are given in Table 6.

3.5.2. Statistical Simulation of the RSM

The simulation of RSM was carried out to understand the changes in response according to a particular direction by adjusting the selected variables. The response surface was plotted graphically in 3D and 2D (Figures 4 and 5, respectively). In the response surface plot, only two variables could be included in the diagram, so the other two factors were placed at the middle level (0 level). Figure 4 shows a 3D plot of efficiency for each of the two variables changing from a lower to a higher level.

Table 6. The actual experimental efficiency and the efficiency predicted by the model according to Equation (4).

Experiment No.	Experimental Efficiency [%]	Predicted Efficiency [%]	Error [%]
1	85.20	81.52	2.59
2	77.99	71.30	4.72
3	66.77	77.25	7.41
4	69.87	77.34	5.28
5	89.11	91.95	2.00
6	92.06	88.37	2.60
7	39.94	47.42	5.28
8	37.86	38.81	0.67
9	70.05	76.74	4.73
10	52.40	67.32	10.55
11	85.57	74.29	7.97
12	74.59	71.54	2.15
13	92.86	95.21	1.66
14	50.15	51.88	1.23
15	91.96	93.87	1.35
16	41.80	43.09	0.91
17	91.98	88.36	2.55
18	54.26	46.61	5.40
19	94.35	94.54	0.13
20	46.03	42.20	2.70
21	86.46	81.86	3.25
22	80.63	74.33	4.45
23	76.51	75.36	0.81
24	73.55	70.70	2.01
25	74.23	74.23	0
26	74.23	74.23	0
27	74.23	74.23	0

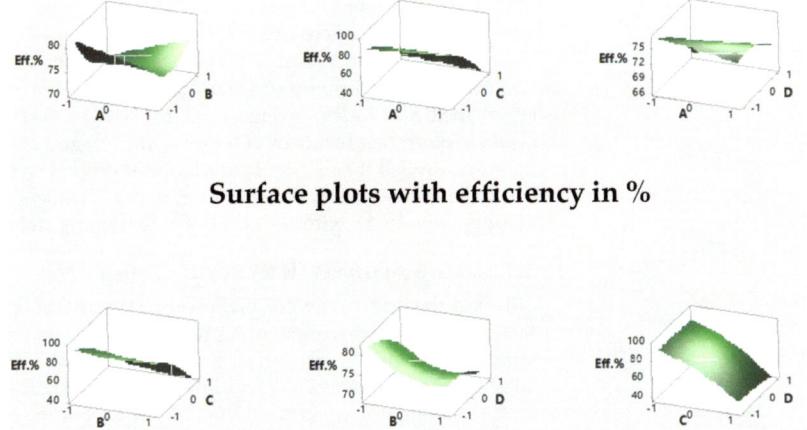

Figure 4. The 3D response surface plot for the interaction effect of two variables when crossed from the lower level (−1) to the upper level (+1) while keeping the other two factors at the middle level (0).

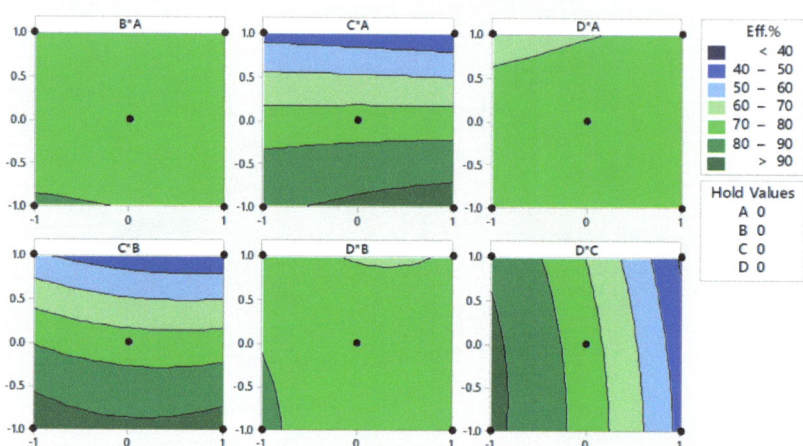

Figure 5. 2D contour plots related to the response surface graph (Figure 4). Each contour plot represents the interaction between two variables moving from the lower level (−1) to the upper level (+1) while keeping the other two factors at the middle level (0).

Contour plots were used to make the results of the response surface more visible and understandable, as shown in Figure 5. The contour plot was a 3D plot created for each two pairs of factors while keeping the others factors at the middle level. If the inhibitor concentration (A) versus the immersion time (B) changed from the lower to the upper level while C and D remained at the middle level, this essentially resulted in an efficiency of up to 80% and only partly up to 90% when A and B were at the lower level (Figure 5B*A). The efficiency was around 90% when the levels of C and A moved from lower (−1) to middle (0) and less than 50% at the (+1) level (Figure 5C*A). No significant effect could be detected for factors D and A from (−1) to (0) with 70–80% and only a decrease to 60% towards the upper level, corresponding to an increased NaCl content (Figure 5D*A). For the combination of temperature C and immersion time B, a continuous decrease from up to 90% at level −1 to as low as 40% at level +1 was observed (Figure 5C*B). For factors D and B, there was an efficiency of around 90% at the −1 levels, which decreased extensively to 80% to the 0 and +1 levels (Figure 5D*B). The last surface response plot (Figure 5D*C) showed an interesting behavior of temperature (C) and NaCl content (D) as the effect of C was more controlled than D. For factor C, the effect on D was stable at all levels as it moved from −1 to 0, with an efficiency of around 80–90%, while it dropped to 40% when C moved to the upper level (+1), with the effect of D remaining stable throughout (Figure 5D*C).

3.5.3. Find the Optimal Levels for Best Prediction

To find the best values for the levels that provide optimal efficiency in the current protection process with compound O, the proposed predicted model was used and applied by Minitab17. Using the software, it was determined that the upper level (+1) was optimal for inhibitor concentration (A) and immersion time (B), while the lower level (−1) was better for the temperature factor (C, see Figure 6). For the NaCl content (D), the value was 0.1919, which is approximately 0.2 (real value = 0.2 M) and thus corresponded to the upper level. Therefore, the most appropriate levels for the selected factors could be determined by the prediction model used for A (+1), B (+1), C (−1), and D (+1). Moreover, this series was not included in the BBD matrix, which is a successful indicator for the model proposed in this study. The overall predicted efficiency depending on the regression equation was 99.4% (see Figure 6).

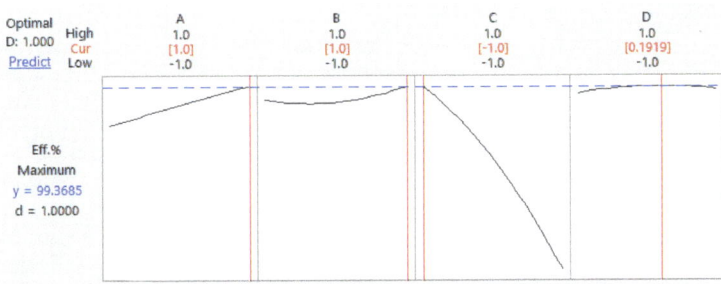

Figure 6. Prediction of the optimum efficiency (dotted blue line) of inhibitor O to protect steel CR4 against corrosion in salt water (course of the black line, best level indicated in red). Optimization and prediction were performed by RSM with support from Minitab17.

By substituting the obtained values and according levels from the RSM by Minitab17 for each tested variable into Equation (6), the optimal value of efficiency can be predicted as follows:

$$\text{Predicted efficiency \%} = 74.23 + 0.44 \times (1) - 2.53 \times (1) - 23.53 \times (-1) - 3.05 \times (0.1919) - 0.23 \times (1)^2 + 2.86 \times (1)^2 - 6.07 \times (-1)^2 - 1.52 \times (0.1919)^2 + 2.58 \times (1 \times 1) - 2.65 \times (1 \times -1) + 1.67 \times (1 \times 0.1919) - 1.86 \times (1 \times -1) + 0.72 \times (1 \times 0.1919) - 1.26 \times (-1 \times 0.1919) = 99.4 \quad (6)$$

The optimal protective effect predicted theoretically by RSM supported using Minitab17 was 99.4% according to the following combination of levels: A = 75 mmol/L (+1), B = 30 min (+1), C = 25 °C (−1), and D = 0.2 M NaCl (+1). Thus, the predicted efficiency was higher than the maximum experimental efficiency of 94.4% obtained in experiment No. 19 (for details, see Table S1 and Supporting Information S3). Even though the predicted combination was not included in the 27 single designed tests determined in the matrix, it was essential to confirm it experimentally. In Figure 7, the anodic and cathodic polarization for experiment No. 19 (red) is displayed, together with the predicted and experimentally confirmed one (optimum in blue) in comparison to the unprotected steel CR4 as reference (black).

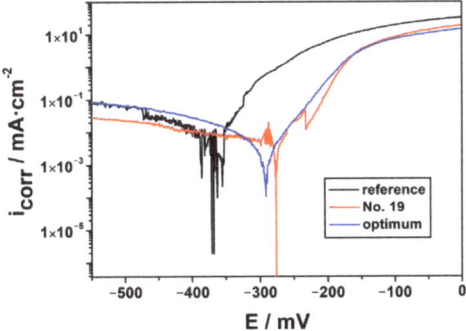

Figure 7. Polarization curves for experiment No. 19 and the predicted one (with levels +1, +1, −1, and +1) performed experimentally compared to unprotected steel CR4 (reference).

The experimentally confirmation, including the predicted combination of the levels (+1, +1, −1, and +1), was performed and resulted in a 97.2% efficiency. In comparison to the theoretical value of 99.4%, this results is an error of 2.2%, indicating high confirmation and acceptance of the proposed predictive model.

4. Conclusions

In this work, the efficiency of the compound oleoylsarcosine in inhibiting corrosion of steel type CR4 in aqueous solution with different NaCl contents was investigated. Furthermore, the variables that have the greatest influence on corrosion protection, such as inhibitor concentration, immersion time, temperature, and NaCl content, were evaluated accordingly. The individual tests were performed based on a design of experiments using the Box–Behnken design (BBD) technique. The optimization process was carried out using the response surface methodology (RSM) supported by the program Minitab17 to determine the optimal efficiency of the process according to selected variables and their respective levels. The obtained results showed that the BBD technique used here is a good systematic control tool to explore the process of corrosion protection by inhibitors. In this context, BBD is suitable as a very flexible means of representing changing environmental conditions, allowing the variables or their limits to be changed and controlled. The data obtained explain how the factors tested affect the process, with temperature having the greatest influence, followed by inhibitor concentration. The possible interactions between the selected variables was also examined in BBD, with the interaction of inhibitor concentration and temperature (AC), immersion time and temperature (BC), and inhibitor concentration and NaCl content (AD) having larger effects than all others.

In the subsequent optimization, the upper level (+1) proved to be the better limit for the inhibitor concentration (A), immersion time (B), and NaCl content (D), while the lower level (−1) was found to be optimal for the temperature (C). The RSM acceptance was up to 89.2% when using the program Minitab 17 to determine the predicted response in comparison to the experimental results.

The highest efficiency obtained on the basis of the experimental design was up to 94% in one test with the following combination of variables and levels: A +1, B 0, C −1, and D 0. The optimal efficiency theoretically predicted by RSM was 99.4% with the combination A +1, B +1, C −1, and D +1, which in turn was not included in the experimental design matrix. In summary, the error between the theoretical and experimental results of the optimum efficiency was only 2.2%, indicating good confirmation and acceptance of the proposed prediction model.

Supplementary Materials: The following supporting information can be downloaded at: https://www.mdpi.com/article/10.3390/eng4010038/s1, Supporting Information S1: The complete response table for three levels of the four independent variables. The efficiency is the output of 27 runs based on Box–Behnken design (BBD). Supporting Information S2: Details of optimization by response surface method (RSM) implementation in Minitab17 based on Box–Behnken Design (BBD). Supporting Information S3: Prediction setting and results.

Author Contributions: S.E.K.: conceptualization, data acquisition, visualization, writing—original draft, methodology, review, and editing. G.E.: data acquisition, OES analysis, review, and editing. J.G.: conceptualization, supervision, review, and editing. C.B.F.: conceptualization, visualization, writing, supervision, review, and editing. All authors have read and agreed to the published version of the manuscript.

Funding: This research was funded by the German Academic Exchange Service (DAAD), grant number 91549239-A13/95797.

Institutional Review Board Statement: Not applicable.

Informed Consent Statement: Not applicable.

Data Availability Statement: The data presented in this study are available on request from the corresponding authors. The data are not publicly available, except the supporting data, due to further studies.

Conflicts of Interest: The authors declare no conflict of interest.

References

1. Kaskah, S.E.; Pfeiffer, M.; Klock, H.; Bergen, H.; Ehrenhaft, G.; Ferreira, P.; Gollnick, J.; Fischer, C.B. Surface protection of low carbon steel with N-acyl sarcosine derivatives as green corrosion inhibitors. *Surf. Interfaces* **2017**, *9*, 70–78. [CrossRef]
2. Rahim, A.A.; Kassim, J. Recent development of vegetal tannins in corrosion protection of iron and steel. *Recent Pat. Mater. Sci.* **2008**, *1*, 223–231. [CrossRef]
3. Zhao, L.; Teng, H.K.; Yang, Y.S.; Tan, X. Corrosion inhibition approach of oil production systems in offshore oilfields. *Mater. Corros.* **2004**, *9*, 684–688. [CrossRef]
4. Moyo, F.; Tandlich, R.; Wilhelm, B.S.; Balaz, S. Sorption of hydrophobic organic compounds on natural sorbent and organoclays from aqueous and non-aqueous solutions: A mini-review. *Int. J. Environ. Res. Public Health* **2014**, *11*, 5020–5048. [CrossRef] [PubMed]
5. Lanigan, S. Final Report on the Safety Assessment of Cocoyl Sarcosine, Lauroyl Sarcosine, Myristoyl Sarcosine, Oleoyl Sarcosine, Stearoyl Sarcosine, Sodium Cocoyl Sarcosinate, Sodium Lauroyl Sarcosinate, Sodium Myristoyl Sarcosinate, Ammonium Cocoyl Sarcosinate, and Ammonium Lauroyl Sarcosinate. *Int. J. Toxicol.* **2001**, *20*, 1–14. [PubMed]
6. Frignani, A.; Trabanelli, G.; Wrubl, C.; Mollica, A. N-Lauroyl sarcosine sodium salt as a corrosion inhibitor for type 1518 carbon steel in neutral saline environments. *NACE Int. Corros.* **1996**, *52*, 177–182. [CrossRef]
7. Salensky, G.A.; Cobb, M.G.; Everhart, D.S. Corrosion-Inhibitor Orientation on Steel. *Ind. Eng. Chem. Prod. Res. Dev.* **1986**, *25*, 133–140. [CrossRef]
8. Popov, B.N. *Corrosion Engineering Principles and Solved Problems*; Elsevier, B.V.: Amsterdam, The Netherlands, 2015.
9. Sastri, V.S. *Challenges in Corrosion: Cost, Causes, Consequences, and Control*; John Wiley: Hoboken, NJ, USA, 2015.
10. Kaskah, S.E.; Ehrenhaft, G.; Gollnick, J.; Fischer, C.B. Concentration and coating time effects of N-acyl sarcosine derivatives for corrosion protection of low-carbon steel CR4 in salt water—Defining the window of application. *Corros. Eng. Sci. Technol.* **2019**, *3*, 216–224. [CrossRef]
11. Kaskah, S.E.; Ehrenhaft, G.; Gollnick, J.; Fischer, C.B. N-b-Hydroxyethyl Oleyl Imidazole as Synergist to Enhance the Corrosion Protection Effect of Natural Cocoyl Sarcosine on Steel. *Corros. Mater. Degrad.* **2022**, *3*, 536–552. [CrossRef]
12. Nist, N.I. Comparisons of response surface design. In *NIST/SEMATECH e-Handbook of Statistical Methods*; United States Department of Commerce: Washington, DC, USA, 2012. [CrossRef]
13. Penteado, R.B.; Hag Ui, J.C.; Faria, J.C.; Ribeiro, M.V. Application of Taguchi Method in process improvement of turning of a Superalloy NIMONIC 80A. *Int. J. Innov. Res. Eng. Manag.* **2015**, *2*, 81–88.
14. Rakić, T.; Kasagić-Vujanović, I.; Jovanović, M.; Jančić-Stojanović, B.; Ivanović, D. Comparison of Full Factorial Design, Central Composite Design, and Box-Behnken Design in Chromatographic Method Development for the Determination of Fluconazole and Its Impurities. *Anal. Lett.* **2014**, *47*, 1334–1347. [CrossRef]
15. Mourabet, M.; El Rhilassi, E.; El Boujaady, H.; Bennani-Ziatni, M.; El Hamri, R.; Taitai, A. Removal of fluorid from aqueous solution by adsorption on apatitic tricalcium phosphate using box-behnken design and desirability function. *Appl. Surf. Sci.* **2012**, *258*, 4402–4410. [CrossRef]
16. Montgomery, D.C. *Design and Analysis of Experiments*; SAS Insitute Inc.: Cary, NC, USA, 2013.
17. Ferreira, S.L.C.; Bruns, R.E.; Ferreira, H.S.; Matos, G.D.; David, J.M.; Brandao, G.C.; da Silva, E.G.P.; Portugal, L.A.; dos Reis, P.S.; Souza, A.S.; et al. Box-behnken design: An alternative for the optimization of analytical methods. *Anal. Chim. Acta* **2007**, *597*, 179–186. [CrossRef] [PubMed]
18. Rupi, E.; Mawonike, R. Response surface methodology for process monitoring of soft drink: A case of delta beverages in Zimbabwe. *J. Math. Stat. Sci.* **2015**, *2015*, 213–233.
19. Masoumi, H.R.F.; Kassim, A.; Basri, M.; Abdullah, D.K.K. Determining optimum condition for Lipase-Catalyzed synthesis of triethanolamine (TEA)-Based esterquat cationic surfactant by a Taguchi robust design method. *Molecules* **2011**, *16*, 4672–4680. [CrossRef] [PubMed]
20. Kumar, S.S.; Malyan, S.K.; Kumar, A.; Bishnol, N.R. Optimization of fenton's oxidation by box-behnken design of response surface methodology for landfill leachate. *J. Mater. Environ. Sci.* **2016**, *7*, 4456–4466.
21. Ahmadi, M.; Ghanbari, F. Optimizing COD removal from greywater by photoelectro-persulfate process using box-behnken design: Assessment of effluent quality and electrical energy consumption. *Environ. Sci. Pollut. Res.* **2016**, *23*, 19350–19361. [CrossRef]
22. Magdum, V.B.; Naik, V.R. Evaluation and optimization of machining parameter for turning of EN 8 steel. *Int. J. Eng. Trends Technol.* **2013**, *4*, 1564–1568.
23. Khuri, A.I. Response surface methodology and its application in agricultural and food sciences. *Biom. Biostat. Int. J.* **2017**, *5*, 1–11. [CrossRef]
24. Hill, W.J.; Huntar, W.G. *Response Surface Methodology: A Review*; Technical report; University of Wisconsin: Madison, WI, USA, 1966.
25. Olivi, L. *Response Surface Methodology, Handbook for Nuclear Reactor Safety*; Commission of the European Communities: Luxembourg, 1985.
26. Myers, R.H.; Montgomery, D.C.; Anderson-Cook, C.M. *Response Surface Methodology, Process and Product Optimization Using Designed Experiments*, 3rd ed.; John Wiley & Sons INC.: Hoboken, NJ, USA, 2009.

27. Nakhai, B.; Neves, J.S. The challenges of six sigma in improving service quality. *Int. J. Qual. Reliab. Manag.* **2009**, *26*, 663–684. [CrossRef]
28. Li, M.; Feng, C.; Zhang, Z.; Chen, R.; Xue, Q.; Gao, C.; Sugiura, N. Optimization of process parameters for electrochemical nitrate removal using Box-Behnken design. *Electrochim. Acta* **2010**, *56*, 265–270. [CrossRef]
29. Box, G.E.P.; Hunter, W.G.; Hunter, J.S. *Statistics for Experimenters, An Introduction to Design, Data Analysis and Model Building*; John Wiley & Sons: Hoboken, NJ, USA, 1978.
30. Bezerra, M.A.; Bruns, R.E.; Ferreira, S.L.C. Statistical design-principal component analysis optimization of a multiple response procedure using cloud point extraction and simultaneous determination of metals by ICP OES. *Anal. Chim. Acta* **2006**, *580*, 251–257. [CrossRef]
31. Vandeginste, B.G.M.; Massart, D.L.; Buydens, L.M.C.; De Jong, S.; Lewi, P.J.; Smeyers-Verbeke, J. Handbook of Chemometrics and Qualimetrics, Part B. In *Data Handling in Science and Technology–Volume 20*; Vandeginste, B.G.M., Rutan, S.C., Eds.; Elsevier: Amsterdam, The Netherlands, 1998.
32. Dwivedi, S.P.; Kumar, S.; Kumar, A. Effects of turning parameters on dimensional deviation of A356/5% SiC composite using Box-Behnken design and genetic algorithm. *Front. Manuf. Eng.* **2014**, *2*, 8–14.
33. Qureshi, M.J.; Phin, F.F.; Patro, S. Enhanced solubility and dissolution rate of clopidogrel by nanosuspension: Formulation via high pressure homogenization technique and optimization using Box-Behnken design response surface methodology. *J. Appl. Pharm. Sci.* **2017**, *7*, 106–113.
34. Wegst, C.; Wegst, M. *Stahlschlüssel*, 22nd ed.; Verlag Stahlschlüssel Wegst GmbH: Marbach am Neckar, Germany, 2010.
35. *DIN EN ISO 9227:2006*; Corrosion Tests in Artificial Atmospheres—Salt Spray Tests. European Committee for Standardization: Brussels, Belgium, 2006.
36. *DIN EN ISO 1514:2005-02*; Paints and Varnishes—Standard Panels for Testing. European Committee for Standardization: Brussels, Belgium, 2004.
37. Mondal, N.K.; Samanta, A.; Dutta, S.; Chattoraj, S. Optimization of Cr(VI) biosorption onto aspergillus niger using 3-level Box-Behnken design: Equilibrium, kinetic, thermodynamic and regeneration studies. *J. Genet. Eng. Biotechnol.* **2017**, *15*, 151–160. [CrossRef] [PubMed]
38. Chen, H.; Ma, D.; Li, Y.; Liu, Y.; Wang, Y. Optimization the process of microencapsulation of *Bifidobacterium bifidum* BB01 by Box-Behnken design. *Acta Univ. Cibiniensis Ser. E Food Technol.* **2016**, *20*, 17–28. [CrossRef]

Disclaimer/Publisher's Note: The statements, opinions and data contained in all publications are solely those of the individual author(s) and contributor(s) and not of MDPI and/or the editor(s). MDPI and/or the editor(s) disclaim responsibility for any injury to people or property resulting from any ideas, methods, instructions or products referred to in the content.

MDPI AG
Grosspeteranlage 5
4052 Basel
Switzerland
Tel.: +41 61 683 77 34

Eng Editorial Office
E-mail: eng@mdpi.com
www.mdpi.com/journal/eng

Disclaimer/Publisher's Note: The title and front matter of this reprint are at the discretion of the Guest Editor. The publisher is not responsible for their content or any associated concerns. The statements, opinions and data contained in all individual articles are solely those of the individual Editor and contributors and not of MDPI. MDPI disclaims responsibility for any injury to people or property resulting from any ideas, methods, instructions or products referred to in the content.

www.ingramcontent.com/pod-product-compliance
Lightning Source LLC
LaVergne TN
LVHW072350090526
838202LV00019B/2515